METHODS IN MOLECULAR BIOLOGY™

Series Editor
John M. Walker
School of Life Sciences
University of Hertfordshire
Hatfield, Hertfordshire, AL10 9AB, UK

For further volumes:
http://www.springer.com/series/7651

Therapeutic Peptides

Methods and Protocols

Edited by

Andrew E. Nixon

Discovery Research, Dyax Corp, Cambridge, MA, USA

Humana Press

Editor
Andrew E. Nixon
Discovery Research
Dyax Corp
Cambridge, MA, USA

ISSN 1064-3745 ISSN 1940-6029 (electronic)
ISBN 978-1-4939-5712-5 ISBN 978-1-62703-673-3 (eBook)
DOI 10.1007/978-1-62703-673-3
Springer New York Heidelberg Dordrecht London

Softcover reprint of the hardcover 1st edition 2014

Printed on acid-free paper

Humana Press is a brand of Springer
Springer is part of Springer Science+Business Media (www.springer.com)

Preface

There has been a tremendous amount of interest in peptide therapeutics in the last decade, which has resulted in six FDA-approved peptides in 2012 alone, with a strong pipeline of peptides in clinical development. Methods for the identification of therapeutic candidates are therefore of interest. This volume is concerned with biological methods for the preparation of peptide phage display libraries using both filamentous and lytic phage, and selection and screening of the prepared peptide libraries for peptides with the desired function and the subsequent characterization of the identified peptides. The relatively small size of peptides, <100 amino acids, makes them amenable to synthetic methods, and other volumes in this series deal with the synthetic chemistry of preparing peptides.

The ability to identify specific and potent peptide drugs requires the creation of high-quality libraries. The process of library design, construction, and the subsequent mining of these libraries through effective selection and screening techniques is described in the chapters by Chang and Yang, Fagerlund et al., Ryu et al., Krumpe and Mori, and Kulseth et al. Xiong and Scott describe an alternate approach to library generation, which uses incremental truncation of open reading frames to generate libraries of peptides that can then be mined for therapeutic candidates.

Shrivastava et al. discuss identification of peptides that are suitable for dimerizing to form heterovalent ligands. This is an interesting idea, in that the heterovalent peptide would be anticipated to have an increase in apparent affinity for the target protein and also to cover a larger area of the target protein and increase the potential for antagonistic activity.

One application of peptide libraries is the characterization of protease specificity. This information can be used to identify potential natural substrates and develop diagnostic probes for protease activity, and may also serve as the jumping off point for protease inhibitor development. The use of such "substrate phage" libraries is described by Hsu and Yang.

One common concern with the use of peptides as drugs in chronic diseases is half-life. Because of the small size of peptides, they are rapidly cleared by the renal system. Approaches to increase the apparent size of peptides are therefore of interest in contemplating the use of peptides where sustained levels are desired. The MIMETIBODY™ approach to solving this problem is described in Picha et al.

Evaluation of the biological properties of peptides identified from the library selection and screening can be achieved through a variety of in vitro and in vivo assays. Here Koistinen et al. describe assays that can be used to evaluate the role of peptides as protease inhibitors or activators, while Douglas et al. provide methods to assess the cytotoxicity of peptides towards cancer cells and untransformed cells.

Peptides have been shown to be very effective targeting agents; labeled with a drug, they can be effective therapeutics, while labeled with a suitable imaging agent, they can be useful diagnostic tools. Huhtala et al. and Kolodziej et al. describe strategies for labeling peptides for in vivo imaging and Ochocki et al. describe the use of prenylated peptides to study cell penetration and the enzymology of protein prenylation.

Finally, in a series of papers Bakashi et al. describe biophysical methods that can be used to characterize the identified peptides, tools that are an essential part of the characterization of peptides as they move along the clinical development path.

Cambridge, MA, USA *Andrew E. Nixon*

Contents

Contributors

EDGAR A. ARRIAGA • *Department of Chemistry, University of Minnesota, Minneapolis, MN, USA*

KUNAL BAKSHI • *Department of Pharmaceutical Chemistry, University of Kansas, Lawrence, KS, USA*

PETER BUGELSKI • *Centocor Research and Development, Inc., A Division of Johnson & Johnson Company, Radnor, PA, USA*

PETER CARAVAN • *Department of Radiology, Athinoula A. Martinos Center for Biomedical Imaging, Massachusetts General Hospital and Harvard Medical School, Charlestown, MA, USA*

HUNG-JU CHANG • *Chemical Biology and Molecular Biophysics Program, Taiwan International Graduate Program, Institute of Biological Chemistry, Genomics Research Center, Academia Sinica, Taipei, Taiwan; Institute of Biochemical Science, National Taiwan University, Taipei, Taiwan*

A. YEON CHO • *Bio Lab, Emerging Technology Center, SAIT, Samsung Electronics, Giheung-Gu, Yongin-Si, Gyeonggi-Do, Republic of Korea*

MARK D. DISTEFANO • *Department of Chemistry, University of Minnesota, Minneapolis, MN, USA*

SUSAN DOUGLAS • *Department of Microbiology & Immunology, Dalhousie University, Halifax, NS, Canada; Institute for Marine Biosciences, National Research Council, Halifax, NS, Canada*

ANNETTE FAGERLUND • *Department of Pharmaceutical Biosciences, University of Oslo, Oslo, Norway*

CAN HEKIM • *Department of Clinical Chemistry, Biomedicum, University of Helsinki and Helsinki University Central Hospital, Helsinki, Finland*

ASHLEY L. HILCHIE • *Department of Microbiology & Immunology, Dalhousie University, Halifax, NS, Canada*

DAVID W. HOSKIN • *Department of Microbiology & Immunology, Pathology, and Surgery, Dalhousie University, Halifax, NS, Canada*

HUNG-JU HSU • *Genomics Research Center, Academia Sinica, Taipei, Taiwan; Graduate Institute of Life Sciences, National Defense University, Taipei, Taiwan*

CHICHI HUANG • *Centocor Research and Development, Inc., A Division of Johnson & Johnson Company, Radnor, PA, USA*

TUULIA HUHTALA • *Department of Biotechnology and Molecular Medicine, University of Eastern Finland, Kuopio, Finland*

URULE IGBAVBOA • *Department of Pharmacology, Geriatric Research, Education and Clinical Center, VAMC, School of Medicine, University of Minnesota, Minneapolis, MN, USA*

VINCENT JACQUES • *EPIX Pharmaceuticals, Lexington, MA, USA*

HANNU KOISTINEN • *Department of Clinical Chemistry, Biomedicum, University of Helsinki and Helsinki University Central Hospital, Helsinki, Finland*

Andrew F. Kolodziej • *EPIX Pharmaceuticals, Lexington, MA, USA*
Lauren R.H. Krumpe • *Molecular Targets Laboratory, SAIC-Frederick, Inc., Frederick National Laboratory, Frederick, MD, USA*
Mari Ann Kulseth • *Department of Medical Genetics, Rikshospitalet, Oslo University Hospital, Oslo, Norway*
Mangala R. Liyanage • *Department of Pharmaceutical Chemistry, University of Kansas, Lawrence, KS, USA*
C. Russell Middaugh • *Department of Pharmaceutical Chemistry, University of Kansas, Lawrence, KS, USA*
Hosung Min • *Bio Lab, Emerging Technology Center, SAIT, Samsung Electronics, Giheung-Gu, Yongin-Si, Gyeonggi-Do, Republic of Korea*
Toshiyuki Mori • *Research Alliance Office, Pharmaceutical Research Division, Takeda Pharmaceutical Company, Kanagawa, Japan*
Astrid Hilde Myrset • *Omegatri AS, Oslo Innovation Centre, Oslo, Norway*
Ale Närvänen • *School of Pharmacy, University of Eastern Finland, Kuopio, Finland*
Adrian D. Nunn • *Ernst Felder Laboratories, Bracco Research USA, Princeton, NJ, USA*
Karyn O'Neil • *Centocor Research and Development, Inc., A Division of Johnson & Johnson Company, Radnor, PA, USA*
Joshua D. Ochocki • *Department of Chemistry, University of Minnesota, Minneapolis, MN, USA*
Kirsten Overoye-Chan • *EPIX Pharmaceuticals, Lexington, MA, USA*
Kristen Picha • *Centocor Research and Development, Inc., A Division of Johnson & Johnson Company, Radnor, PA, USA*
Jussi Rytkönen • *School of Pharmacy, University of Eastern Finland, Kuopio, Finland*
Jong Sang Ryu • *Bio Lab, Emerging Technology Center, SAIT, Samsung Electronics, Giheung-Gu, Yongin-Si, Gyeonggi-Do, Republic of Korea*
Charles Scott • *Department of Biochemistry & Molecular Biology, Thomas Jefferson University, Philadelphia, PA, USA*
Sang Won Seo • *Bio Lab, Emerging Technology Center, SAIT, Samsung Electronics, Giheung-Gu, Yongin-Si, Gyeonggi-Do, Republic of Korea*
Ajay Shrivastava • *Ernst Felder Laboratories, Bracco Research USA, Princeton, NJ, USA*
Ulf-Håkan Stenman • *Department of Clinical Chemistry, Biomedicum, University of Helsinki and Helsinki University Central Hospital, Helsinki, Finland*
Michael F. Tweedle • *Ernst Felder Laboratories, Bracco Research USA, Princeton, NJ, USA; Department of Radiology, The Ohio State University Medical School, Columbus, OH, USA*
David B. Volkin • *Department of Pharmaceutical Chemistry, University of Kansas, Lawrence, KS, USA*
Matthew von Wronski • *Ernst Felder Laboratories, Bracco Research USA, Princeton, NJ, USA*
Elizabeth V. Wattenberg • *Division of Environmental Health Sciences, University of Minnesota, Minneapolis, MN, USA*
Janne Weisell • *School of Pharmacy, University of Eastern Finland, Kuopio, Finland*
W. Gibson Wood • *Department of Pharmacology, Geriatric Research, Education and Clinical Center, VAMC, School of Medicine, University of Minnesota, Minneapolis, MN, USA*

PING WU • *Department of Urology, The University of Texas Health Science Center at San Antonio, San Antonio, TX, USA*
LING XIONG • *Department of Biochemistry & Molecular Biology, Thomas Jefferson University, Philadelphia, PA, USA*
AN-SUEI YANG • *Genomics Research Center, Academia Sinica, Taipei, Taiwan*
ZHAODA ZHANG • *EPIX Pharmaceuticals, Lexington, MA, USA*

Chapter 1

Design of Phage-Displayed Cystine-Stabilized Mini-Protein Libraries for Proteinaceous Binder Engineering

Hung-Ju Chang and An-Suei Yang

Abstract

Cystine-stabilized mini-proteins are important scaffolds in the combinatorial search of binders for molecular recognition. The structural determinants of a cystine-stabilized scaffold are the critical residues determining the formation of the native disulfide-bonding configuration, and thus should remain unchanged in the combinatorial libraries so as to allow a large portion of the library sequences to be compatible with the scaffold structure. A high-throughput molecular evolution procedure has been developed to select and screen for the polypeptide sequences folding into a specific cystine-stabilized structure. Patterns of sequence preference that emerge from the resultant sequence profiles provide structural determinant information, which facilitates the designs of combinatorial libraries for combinatorial approaches as in phage display. This methodology enables artificial cystine-stabilized proteins to be engineered with enhanced folding and binding properties.

Key words Phage display, Protein engineering, Protein design, Cystine-stabilized beta-sheet motif, Protein–protein interaction, Proteinaceous binder, Knottin, Directed evolution

1 Introduction

The knottin family is composed of a large number (~1,200 known sequences) of cystine-stabilized small (<~50 residues) proteins [1], which are a rich source of potentially useful scaffolds for constructing combinatorial protein libraries to be used for engineering molecular recognition. The cystine-stabilized beta-sheet (CSB) motif is a major structural motif common to many knottins, making the CSB motif appealing as an ideal model scaffold for protein engineering. The structural information dictating the 1–3/2–4 disulfide bonding configuration in CSB is known to be encoded in a subset of the residues in the polypeptide sequence [2]. The information is critical not only for understanding the structural determinants in the folding of the CSB but also for designing combinatorial libraries with variants that retain the CSB folding topology.

Andrew E. Nixon (ed.), *Therapeutic Peptides: Methods and Protocols*, Methods in Molecular Biology, vol. 1088,
DOI 10.1007/978-1-62703-673-3_1,

The 23-residue autonomous folding unit (Min-23) from EETI-II is the smallest CSB motif known so far [3], and is used as a model system herein to demonstrate a high-throughput methodology for protein folding engineering and design, leading to phage-displayed proteinaceous binders against human VEGF-A [2, 4]. The significance of the methodology is that the binders are selected and screened not only for binding to the target protein but also for retention of the structure of the parent scaffold, which confers extreme stability under harsh conditions. The methodology is particularly useful in designing small cystine-stabilized proteins for which consensus-based protein engineering approaches [5] are ineffective due to the lack of evolutionarily related sequences.

The key technique in the molecular evolution methodology is to incorporate a bovine fXa substrate sequence (for example, -IEGR-; other possible substrate sequences for bovine fXa are elucidated with a substrate phage display methodology described in the chapter in this volume; *see* also refs. 6, 7) to a strategically selected position in the mini-protein sequence. As such, the fXa cleavable site does not hinder the folding of the designated structure and is protected from fXa digestion in the cystine-stabilized structure; the site needs to become accessible for fXa cleavage upon protein unfolding due to reduction of the disulfide bonds. In addition to the incorporated cleavage site, no other fXa cleavable sites should exist in the phage-displayed sequences. An artificial intelligence computational program [7] (*see* also Chap. 1 on substrate phage display in this volume) can be used to scan protein sequences for potential fXa cutting sites.

With the strategically embedded fXa cutting site, the library sequences folded into cystine-stabilized structures can be selected and screened for stability by using varying levels of denaturant followed by fXa digestion [2]. The selection cycle is outlined in Fig. 1. The stability of the identified phage-displayed protein structures is measured by quantitating the intact fraction shown in Fig. 2 using a high-throughput platform; the intact fraction is the percentage of the phage-displayed proteins remaining folded (intact) after treatment with the protein denaturants followed by fXa digestion.

2 Materials

2.1 Phage Display Library Construction

2.1.1 Single-Strand Template DNA Preparation

1. Synthetic double-strand DNA inserts for pC-Mx2_TAA1 and pC-Mx2_TAA3 phagemid construction (synthesized by Integrated DNA technologies (IDT), USA) (*see* **Note 1**):

   Mx2-TAA1_F: 5′ GGC GGC TCT AGA CTA ATG CGA TGT AAG CAG TAA TA.A TAA TGT ATC GAA GGT AGG GTT TGT GGA CCT AAT GGT TTT TGT GGT GCG GCC GCA GGT GCG 3′.

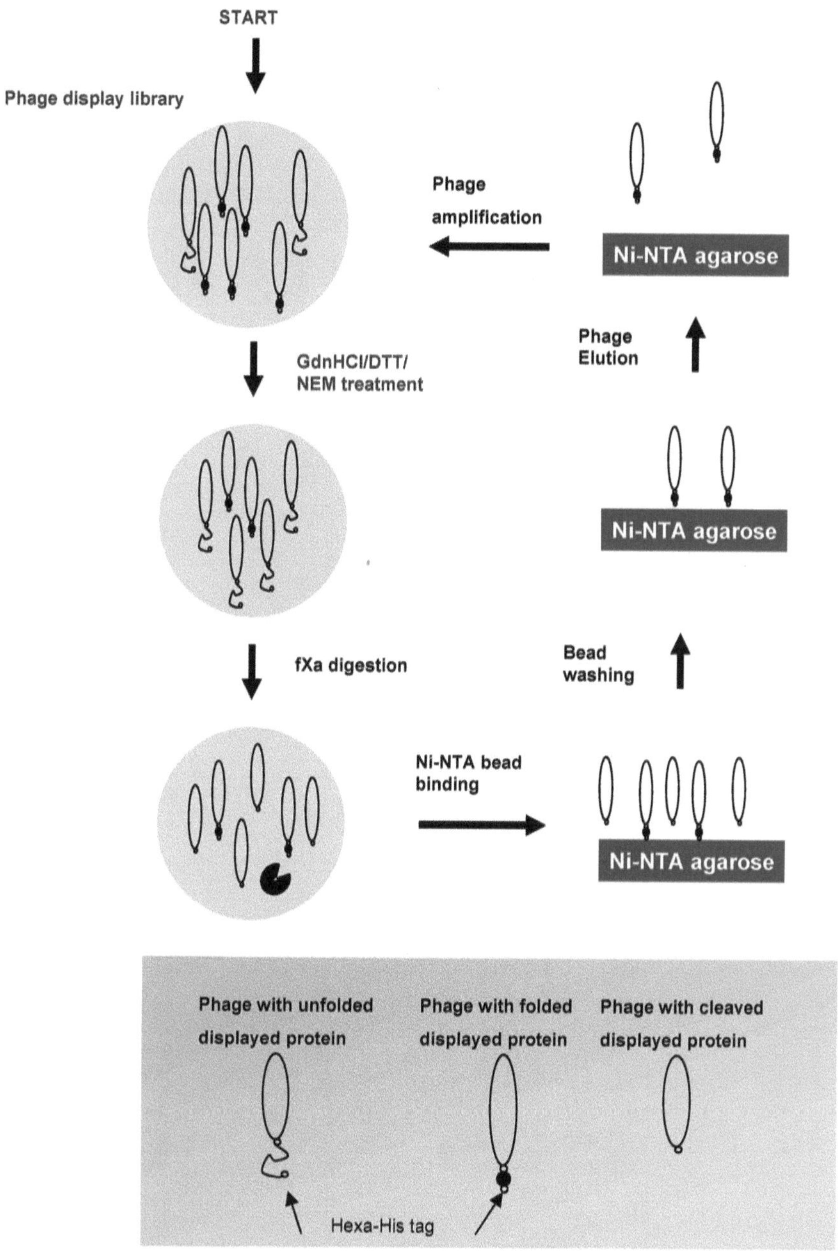

Fig. 1 The selection cycle for engineering cystine-stabilized phage-displayed mini-proteins. The phage-displayed proteins are unfolded by guanidine hydrochloride (GdnHCl) in the presence of dithiothreitol (DTT) (or tris(2-carboxyethyl)phosphine (TCEP)) to reduce disulfide bonds, followed by the *N*-ethylmaleimide (NEM) treatment to block cysteines from reforming the cystine-stabilized structure. Only the GdnHCl/DTT-unfolded and NEM-blocked polypeptides are cleaved at the exposed cutting site (-IEGR-) by fXa, and consequently the N-terminal hexa-His tag is removed from the phage particle. Phage particles with intact hexa-HIS tag are enriched with Ni-NTA beads. These phage-displayed protein sequences are stable species under the denaturant condition. The selected phage particles are amplified and used as input phage display library for the next selection cycle

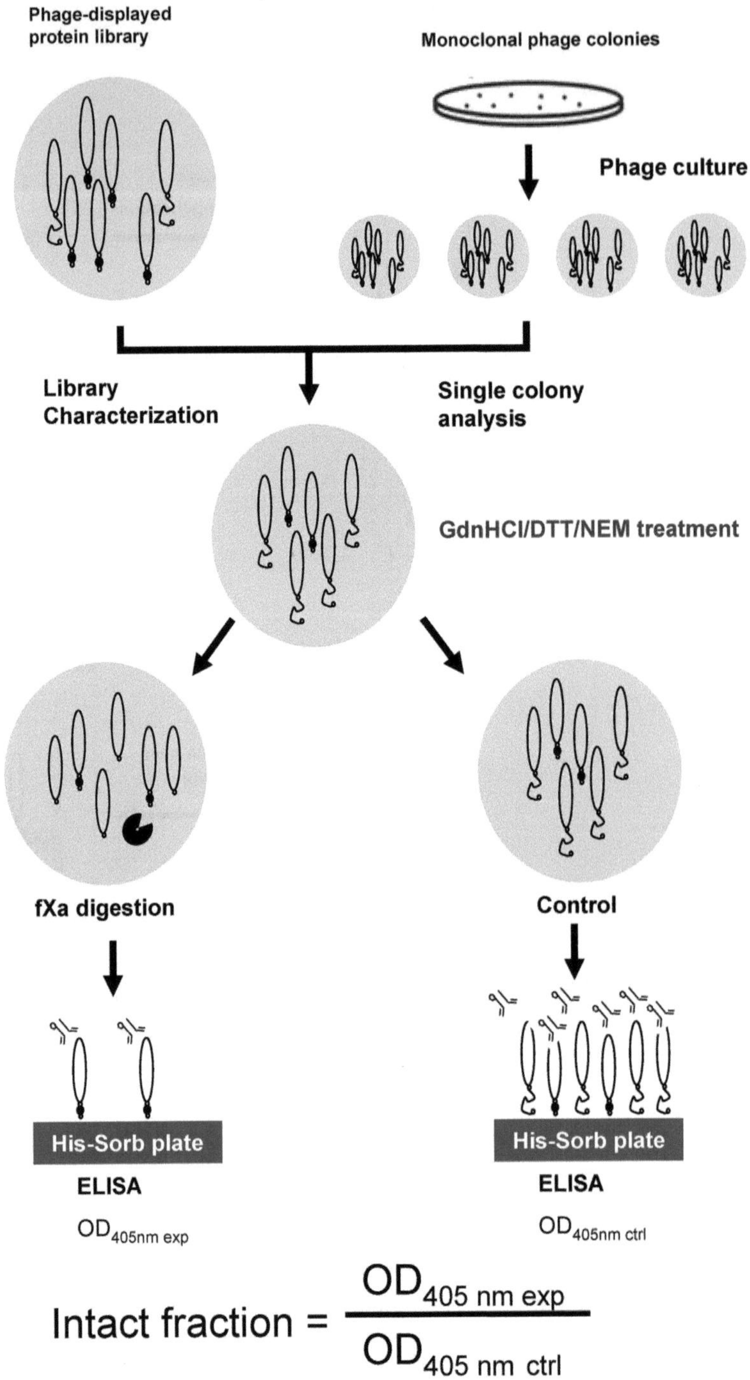

Fig. 2 Stability characterization of library and single-colony phage-displayed cystine-stabilized mini-proteins. Phage particles bearing displayed proteins are treated with denaturants. The treated phage solution is parted into two equal portions: one treated with fXa and the other one used as control. The intact fraction, which is the ratio of phages capable of binding to Ni^{2+}-coated surface from phage solutions before and after the fXa digestion, is determined by a pair of enzyme-linked immunosorbent assays (ELISA)

Mx2-TAA1_R: 5′ CGC ACC TGC GGC CGC ACC ACA AAA ACC ATT AGG TCC ACA AAC CCT ACC TTC GAT ACA TTA TTA TTA CTG CTT ACA TCG CAT TAG TCT AGA GCC GCC 3′.

Mx2-TAA3_F: 5′ GGC GGC TCT AGA CTA ATG CGA TGT AAG CAG GAC TCA GAT TGT ATC GAA GGT AGG GTT TGT GGA TAA TAA TAA TTT TGT GGT GCG GCC GCA GGT GCG 3′.

Mx2-TAA3_R: 5′ CGC ACC TGC GGC CGC ACC ACA AAA TTA TTA TTA TCC ACA AAC CCT ACC TTC GAT ACA ATC TGA GTC CTG CTT ACA TCG CAT TAG TCT AGA GCC GCC 3′.

2. *E. coli dut*$^-$/*ung*$^-$ strain CJ236 (NEB).
3. *E. coli* strain ER2738 (NEB).
4. pCANTAB5E phagemid (GE Healthcare).
5. Qiaprep spin M13 kit (Qiagen).
6. Restriction enzymes *Eag*I, *Xba*I (NEB): 10 U/μL.

2.1.2 Library Construction

1. DNA primers for pC-Lib1, pC-Lib3 libraries (synthesized by IDT, USA) (*see* **Note 2**):

   Lib1: 5′ CCA CAA ACC CTA CCT TCG ATA CAM NNM NNM NNM NNM NNA CAT CGC ATT AGT CTA GAG CC 3′.

   Lib3: 5′ CTG CGG CCG CAC CAC AMN NMN NMN NMN NMN NAC AAA CCC TAC CTT CGA TAC AAT C 3′.
2. Deoxyribonucleotide triphosphate mixture (dNTP) (Promega): 10 mM of dATP, dCTP, dTTP, and dGTP.
3. T4 polynucleotide kinase and 10× T4 polynucleotide kinase reaction buffer (NEB).
4. Adenosine 5′-triphosphate (ATP) (Sigma): 10 mM in ddH_2O, filter-sterilized. Store at −20 °C.
5. 10× TM buffer for primer annealing: 0.5 M Tris–HCl and 0.1 M $MgCl_2$, pH 7.5.
6. Dithiothreitol (DTT) (Sigma): 100 mM in ddH_2O, filter-sterilized. Store at −20 °C.
7. T4 DNA ligase (NEB): 400 U/μL.
8. T7 DNA polymerase (NEB): 10 U/μL.
9. QIAquick gel extraction kit (Qiagen).

2.1.3 Electroporation

1. Tetracycline (Sigma): 20 mg/mL at 70 % EtOH, filter-sterilized. Store at 20 °C.
2. SOC medium: 20 g of Bacto-tryptone (Becton and Dickinson, France), 5 g of Bacto-yeast extract (AMRESCO), 0.5 g of NaCl,

and 186 mg of KCl are dissolved in ddH_2O, the pH adjusted to 7.0 with NaOH to a final volume of 1 L, and sterilized with autoclave. After the medium cools down, add 1 mL 2 M $MgCl_2$ (filter-sterilized) and 20 mL 1 M glucose (filter-sterilized). Store at room temperature.

3. 2YT medium (Fluka): 16 g/L tryptone, 10 g/L yeast extract, and 10 g/L NaCl. Sterilize with autoclave. Store at room temperature.
4. Ampicillin (Sigma): 100 mg/mL in ddH_2O is filter-sterilized. Store at −20 °C.
5. Kanamycin (Sigma): 50 mg/mL in ddH_2O is filter-sterilized. Store at −20 °C.
6. LB medium (Sigma): 10 g/L tryptone, 5 g/L yeast extract, and 5 g/L sodium chloride in ddH_2O, sterilized with autoclave. Store at room temperature.
7. 10 % (v/v) glycerol (Merck): In ddH_2O, sterilized with autoclave.
8. pUC18 DNA (Stratagene): 10 ng/μL in ddH_2O. Store at −80 °C.
9. Glucose (Merck): 2 M glucose in sterilized ddH_2O, filter-sterilized. Store at room temperature.
10. 2YT plate: 2YT medium with 15 g/L bacto-agar. Sterilize with autoclave. Cool to 55–65 °C. Add proper antibiotics to the sterilized medium and mix thoroughly. Pour to petri dishes and allow solidifying. Store the prepared plates at 4 °C and use within 1 month.

2.2 Molecular Evolution of Cystine-Stabilized Proteins

2.2.1 Molecular Evolution Cycle

1. PEG/NaCl solution: 20 % (w/v) polyethylene glycol 8000 (J.T. Baker) and 2.5 M sodium chloride (Sigma) in ddH_2O, sterilized with autoclave. Store at room temperature.
2. Phosphate buffer: 50 mM NaH_2PO_4, 300 mM NaCl, pH 8.0, filter-sterilized. Store at room temperature.
3. Guanidine hydrochloride (GdnHCl, Sigma).
4. Dithiothreitol (DTT) (Sigma): 500 mM freshly prepared in ddH_2O, filter-sterilized.
5. N-ethylmaleimide (NEM) (Sigma): 1 M freshly prepared in 70 % EtOH.
6. Bovine factor Xa (fXa) reaction buffer: 20 mM Tris–HCl, 50 mM NaCl, and 1 mM $CaCl_2$, pH 6.8.
7. Bovine factor Xa protease (fXa) (QIAGEN).
8. Ni-NTA binding buffer: 50 mM NaH_2PO_4, 300 mM NaCl, pH 8.0.

9. Ni-NTA wash buffer: 10 mM imidazole in Ni-NTA binding buffer.
10. Ni-NTA elution buffer: 200 mM imidazole in Ni-NTA binding buffer.
11. Ni-NTA agarose beads (QIAGEN).
12. Micro-filter column, 0.5 μm pore size (Millipore).

2.2.2 Intact Fraction Measurement

1. Transfer 200 μL phage solution (from each round of the previous selection cycle) to a centrifugal filter and spin at 6,600 × *g* for 5 min to remove phage culture medium.
2. Wash the filter with 200 μL phosphate buffer. Remove the wash buffer by centrifuging at 6,600 × *g* for 5 min.
3. Add 100 μL fXa reaction buffer containing GdnHCl (0–2 M) and 10 mM DTT to the filter and incubate for 20 min.
4. Add NEM (final concentration 80 mM) to the filter and incubate for 30 min.
5. Remove the reaction mixture from phage particles by centrifugation, followed by washing the filter two times, each with 200 μL fXa reaction buffer.
6. Add 35 μL of fXa reaction buffer to the filter and collect the solution by centrifuging the filter upside down at 2,000 × *g* for 2 min.
7. Add 10 μL of the collected phage solution to a well of a non-absorbing plate containing 0.2 U fXa in 90 μL fXa reaction buffer. Add another 10 μL of the collected phage solution to 90 μL fXa reaction buffer without fXa in a second well of the same plate as negative control. Incubate at room temperature for 2 h.
8. Add 50 μL of 1.5 M GdnHCl to all wells to terminate the enzymatic reaction.
9. Transfer 100 μL solution from each well to a corresponding well in a His-sorb plate. Incubate at room temperature for 1 h.
10. Wash the His-sorb plate three times with 250 μL PBST and once with 250 μL PBS buffer.
11. Add 100 μL anti-M13-HRP antibody (1:3,000) in PBST + skim milk (3 %) to each well and shake for 1 h.
12. Clear the anti-M13-HRP from the His-sorb plates with three washes of 250 μL PBST and one wash of 250 μL PBS buffer.
13. Add 100 μL ABTS substrate to each well for 10 min. Add 100 μL 1 % SDS to terminate the chromogenic reaction.
14. Measure the OD_{405nm} with 96-well ELISA Reader (*see* **Note 11**). Calculate intact fraction from the measured absorbance (Fig. 2).

2.2.3 Phage-Displayed Protein Expression Test and Single-Colony Analysis

1. Infect mid-log *E. coli* ER2738 with the selected phage from the evolution procedure (Subheading 3.2.1). Spread serially diluted infected cells to 2YT/ampicillin (100 μg/mL) agarose plates.
2. Pick well-separated single colonies into 96-well deep-well culture plate. Each well contains 1 mL 2YT medium, 100 μg ampicillin, and 10^8 cfu helper phage M13KO7. Incubate for 1 h at 37 °C with shaking.
3. Add 200 μL 2YT medium with 300 μg/mL kanamycin and 600 μg/mL ampicillin to each well. Incubate at 37 °C with vigorous shaking overnight.
4. Centrifuge the overnight culture at 3,000 × *g* for 30 min at 4 °C.
5. Transfer 100 μL supernatant from each well to a corresponding well of a 96-well His-sorb plate to perform the ELISA as described in Subheading 3.2.2.
6. Colonies with OD_{405nm} values greater than 0.5 are selected for intact fraction measurement following the procedure in Subheading 3.2.2.
7. Determine the phagemid sequence of selected phage following the procedure described in Subheading 3.2.3 of the chapter on substrate phage display in this volume.

2.3 Proteinaceous Binder Engineering with Cystine-Stabilized Mini-Protein Libraries

2.3.1 Constructions of pC-6X, pC-7X, and pC-8X Libraries

1. DNA primers for pC-6X, pC-7X, and pC-8X libraries (IDT, USA) (*see* **Note 2**):

   6×: 5′ CGG CCG CAC CAC AAA AAC CMN NMN NMN NAC AAA CCC TAC CTT CGA TAC AAT CMN NGT CMN NMN NAC ATC GCA TTA GTC TAG AGC C 3′.

   7×: 5′ CGG CCG CAC CAC AAA AAC CMN NMN NMN NMN NAC AAA CCC TAC CTT CGA TAC AAT CMN NGT CMN NMN NAC ATC GCA TTA GTC TAG AGC C 3′.

   8×: 5′ CGG CCG CAC CAC AAA AAC CMN NMN NMN NMN NMN NAC AAA CCC TAC CTT CGA TAC AAT CMN NGT CMN NMN NAC ATC GCA TTA GTC TAG AGC C 3′.

   Where M = A, C, 50 % each; N = A, G, T, C, 25 % each.

2.3.2 VEGF–Bead Conjugation

1. Ultralink Biosupport resin (Pierce).
2. Quench solution: 3 M ethanolamine, pH 9.0 (Sigma).

2.3.3 Biopanning of Phage Library to VEGF-Conjugated Bead

1. Glycine buffer: 0.2 M glycine, pH 2.0.
2. PBST–milk: 3 % skim milk in PBST.
3. Neutralization buffer: 1 M Tris–HCl, pH 9.1.

2.3.4 VEGF ELISA Test

1. Maxisorb 96 well plate (Nunc).
2. 3 % gelatin (Sigma): 3 % (w/v) gelatin in PBS or PBST.

3 Methods

3.1 Phage Display Library Construction

3.1.1 Single-Strand Template DNA Preparation

1. Construct pC-Mx2_TAA1 and pC-Mx2_TAA3 phagemids (Fig. 3 shows the sequences of the constructs): Anneal Mx2-TAA1_F and Mx2-TAA1_R (or Mx2-TAA3_F and Mx2-TAA3_R for pC-Mx2_TAA3 construction) following IDT manufacturer's instruction. Digest the annealed dsDNA fragments with *Xba*I and *Eag*I. Digest double-stranded pCANTAB5E-derived phagemid (pC-YV1 (*see* **Note 3**)) with *Xba*I and *Eag*I. Ligate the digested insert DNA fragment and

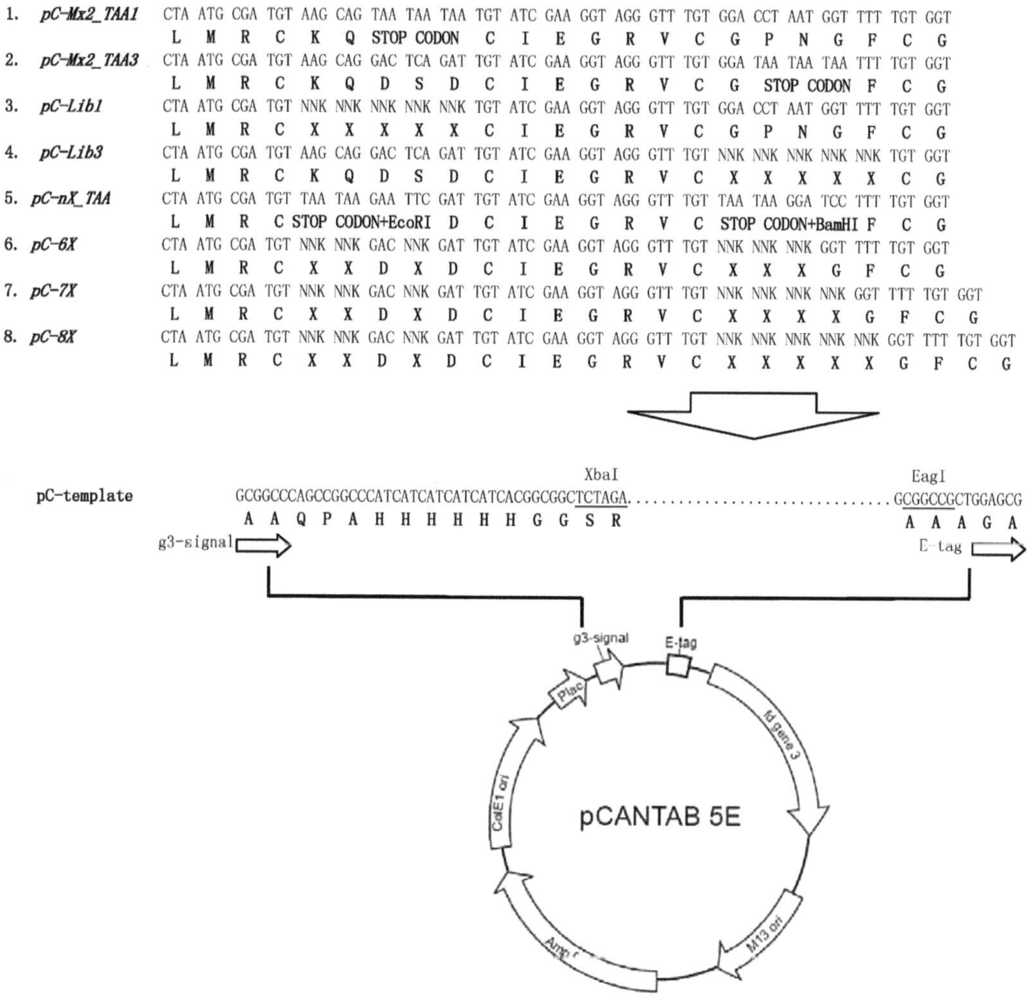

Fig. 3 Constructs of pCANTAB5E-derived phagemids

the digested dsDNA phagemid. Details of the procedure are documented in Subheading 3.1.2 of the accompanied chapter on substrate phage display.

2. Electroporate the recombinant phagemids into *E. coli* strain ER2738 followed by phage rescue (*see* Subheading 3.1.2 of the accompanied chapter on substrate phage display) and sequence confirmation (*see* Subheading 3.2.3 of the chapter on substrate phage display in this volume).
3. Infect *E. coli* strain CJ236 with the phage-harboring phagemid with confirmed sequence (*see* Subheading 3.1.3 of the chapter on substrate phage display in this volume).
4. Rescue phages harboring dU-ssDNA phagemid from CJ236 culture (*see* Subheading 3.1.3 of the chapter on substrate phage display in this volume).
5. Extract dU-ssDNA phagemid from the rescued phages with Qiagen-Qiaprep spin M13 kit. The yield of dU-ssDNA is about 40μg/30 mL phage supernatant. Check the purity of the dU-ssDNA with agarose gel electrophoresis (1.2 %): pCANTAB5E dU-ssDNA should appear at about 3,000 bp (*see* **Note 4**) (*see* Subheading 3.1.3 of the chapter on substrate phage display in this volume).

3.1.2 Library Construction

The following procedure is modified from the Sidhu and Weiss protocol [8].

1. Primer phosphorylation mixture:

27 μL	ddH_2O
3 μL	Primer DNA fragment (1 μg/μL; Subheading 2.1.2)
4 μL	10× T4 polynucleotide kinase reaction buffer
2 μL	10 mM ATP
4 μL	T4 polynucleotide kinase (10 U/μL)
40 μL	Total volume

2. Incubate at 37 °C for 2 h.
3. Annealing reaction mixture:

   8 μg template dU-ssDNA (from Subheading 3.1.1, **step 5**).

   3–24-fold molar ratio of phosphorylated primer (from **step 1** of this section).

   10 μL 10×TM buffer.

   Adjust the total volume to 100 μL with ddH_2O in a PCR tube.
4. Put the PCR tube containing the annealing reaction mixture in a PCR machine with the following annealing protocol: 90 °C for 150 s, followed by cooling to 20 °C in 90 min.

5. Extension reaction mixture:

   100 μL annealing mixture from **step 3**.

   10 μL 10 mM dNTP.

   4 μL 10 mM ATP.

   6 μL 100 mM DTT.

   1.5 μL T4 DNA ligase.

   1.5 μL T7 DNA polymerase.

   Total volume to 123 μL, incubate at 20 °C overnight.

6. Purify the DNA product from the extension reaction mixture with the QIAquick gel extraction kit following the manufacturer's instruction, except for the last elution step. Instead of using EB elution buffer, add 35 μL 50 °C ddH_2O to QIAquick column. Incubate for 10 min. Centrifuge the QIAquick column at 10,000 × *g* for 1 min. Use 2 μL of the product for agarose gel (1.2 %) electrophoresis analysis (*see* **Note 5**).

3.1.3 Electroporation

1. Inoculate 2 mL 2YT medium containing 20 μg/mL tetracycline with one colony of *E. coli* strain ER2738 from a plate with colonies less than 1 week old. Incubate at 37 °C with shaking overnight.
2. Inoculate 500 mL LB medium with 2 mL ER2738 overnight culture above. Incubate at 37 °C with shaking until OD600 reaches 0.8 (approximately 2.5–3 h).
3. Chill the cells on ice for 30 min.
4. Pour the cell culture into pre-chilled centrifuge bottle, and harvest the cells by centrifugation at 3,000 × *g* for 15 min.
5. Wash the pellet with 150 mL pre-chilled ddH_2O three times.
6. Wash the pellet with 150 mL pre-chilled 10 % glycerol twice.
7. Resuspend the pellet with 1 mL 10 % glycerol.
8. Check the washed cells for electro-competency and contamination. The electro-competency of the cells is evaluated by transforming 10 ng of pUC18 to 80 μL of the electro-competent cells. This test should generate ~10^8 ampicillin-resistant colonies with the electroporation protocol below. Blank test should also be carried out to control contamination in the electro-competent cells.
9. Add the purified library DNA (33 μL from **step 6**, Subheading 3.1.2) to the electro-competent cells (1 mL). Incubate on ice for 2 min. Add 80 μL of the DNA/electro-competent cell mixture to an electroporation cuvette (2 mm gap); there should be about 16 cuvettes for electroporation.
10. Electroporate the DNA/electro-competent cell mixture with the following setting: 3 kV, 25 μF, 200 Ω, for cuvette with

2 mm gap between the electrode plates. Add pre-warmed 1 mL SOC (37 °C) to each of the cuvettes immediately after electroporation.

11. Pool the SOC-rescued cells and incubate the culture at 37 °C, 225 rpm, for 30 min. Plate 10 μL of the SOC-rescued cells on 2 YT agar plate with 100 μg/mL ampicillin and 2 % glucose for titering the electroporation efficiency. The electroporation efficiency should be higher than 10^9 cfu for the combined 16 electroporations.
12. Add ampicillin to the pooled SOC-rescued cells for a final concentration of 20 μg/mL. Incubate at 37 °C for 1 h.
13. Add M13KO7 helper phage (~10^{11} cfu, *see* **Note 6**) into the pooled SOC-rescued cells. Incubate at 37 °C for 60 min.
14. Adjust the final volume to 100 mL with 2YT medium. Add ampicillin and kanamycin to a final concentration of 100 and 50 μg/mL, respectively; incubate overnight at 37 °C.
15. Pellet the cells by centrifuging at 4 °C, 10,000 × *g*, for 10 min. Centrifuge the supernatant again with the same setting. The supernatant contains the phage-displayed library. Titer the phage with freshly prepared log-phase (OD_{600} = 0.4–0.6) *E. coli* strain ER2738. The phage titer should be around 10^{10}–10^{11} cfu/mL. The phage library solution should be stable for weeks at 4 °C. For longer storage, add sterilized glycerol to a final concentration of 50 % and store the phage library at −20 °C.

3.2 Molecular Evolution of Cystine-Stabilized Proteins

3.2.1 Molecular Evolution Cycle

1. Add 4 mL PEG/NaCl to 10 mL of phage library solution to precipitate the phage particles (*see* Subheading 3.1.1).
2. Resuspend the phage pellet in 100 μL of phosphate buffer (pH 8.0) with a concentration of GdnHCl ranging from 0 to 2 M (*see* **Note 7**).
3. Add dithiothreitol (DTT) to the solution to make the final DTT concentration of 10 mM. Incubate for 20 min at room temperature.
4. Add *N*-ethylmaleimide (NEM) to the solution to make the final concentration of 80 mM. Incubate for 30 min at room temperature (*see* **Note 8**).
5. Precipitate the phage particles with 20 % PEG/NaCl. Resuspend in 60 μL bovine factor Xa (fXa) reaction buffer containing 2 U of fXa for 1 h at room temperature (*see* **Note 9**).
6. Add the reaction mixture to 1 mL of Ni-NTA binding buffer containing 50 μL of Ni-NTA agarose beads for 1 h with gentle rotation.
7. Transfer the Ni-NTA beads to a micro-filter and wash 21 times, each with 500 μL wash buffer (*see* **Note 10**).

8. Elute the Ni-NTA-bound phage particles from the Ni-NTA beads with 500 μL elution buffer.
9. Amplify the eluted phage particles following the procedure described in Subheading 3.2.1 of the accompanied chapter on substrate phage display. The phage particles are used for the next round of selection and for intact fraction measurement to characterize the stability of the phage-displayed proteins (Fig. 2 and Subheading 3.2.2).

3.2.2 Intact Fraction Measurement

1. Transfer 200 μL phage solution (from each round of the previous selection cycle) to a centrifugal filter and spin at 6,600×*g* for 5 min to remove phage culture medium.
2. Wash the filter with 200 μL phosphate buffer. Remove the wash buffer by centrifuging at 6,600×*g* for 5 min.
3. Add 100 μL fXa reaction buffer containing GdnHCl (0–2 M) and 10 mM DTT to the filter and incubate for 20 min.
4. Add NEM (final concentration 80 mM) to the filter and incubate for 30 min.
5. Remove the reaction mixture from phage particles by centrifugation, followed by washing the filter two times, each with 200 μL fXa reaction buffer.
6. Add 35 μL of fXa reaction buffer to the filter and collect the solution by centrifuging the filter upside down at 2,000×*g* for 2 min.
7. Add 10 μL of the collected phage solution to a well of a nonabsorbing plate containing 0.2 U fXa in 90 μL fXa reaction buffer. Add another 10 μL of the collected phage solution to 90 μL fXa reaction buffer without fXa in a second well of the same plate as negative control. Incubate at room temperature for 2 h.
8. Add 50 μL of 1.5 M GdnHCl to all wells to terminate the enzymatic reaction.
9. Transfer 100 μL solution from each well to a corresponding well in a His-sorb plate. Incubate at room temperature for 1 h.
10. Wash the His-sorb plate three times with 250 μL PBST and once with 250 μL PBS buffer.
11. Add 100 μL anti-M13-HRP antibody (1:3,000) in PBST+skim milk (3 %) to each well and shake for 1 h.
12. Clear the anti-M13-HRP from the His-sorb plates with three washes of 250 μL PBST and one wash of 250 μL PBS buffer.
13. Add 100 μL ABTS substrate to each well for 10 min. Add 100 μL 1 % SDS to terminate the chromogenic reaction.
14. Measure the OD_{405nm} with 96-well ELISA Reader (*see* **Note 11**). Calculate intact fraction from the measured absorbance (Fig. 2).

3.2.3 Phage-Displayed Protein Expression Test and Single-Colony Analysis

1. Infect mid-log *E. coli* ER2738 with the selected phage from the evolution procedure (Subheading 3.2.1). Spread serially diluted infected cells to 2YT/ampicillin (100 μg/mL) agarose plates.
2. Pick well-separated single colonies into 96-well deep-well culture plate. Each well contains 1 mL 2YT medium, 100 μg ampicillin, and 10^8 cfu helper phage M13KO7. Incubate for 1 h at 37 °C with shaking.
3. Add 200 μL 2YT medium with 300 μg/mL kanamycin and 600 μg/mL ampicillin to each well. Incubate at 37 °C with vigorous shaking overnight.
4. Centrifuge the overnight culture at 3,000 × *g* for 30 min at 4 °C.
5. Transfer 100 μL supernatant from each well to a corresponding well of a 96-well His-sorb plate to perform the ELISA as described in Subheading 3.2.2.
6. Colonies with OD_{405nm} values greater than 0.5 are selected for intact fraction measurement following the procedure in Subheading 3.2.2.
7. Determine the phagemid sequence of selected phage following the procedure described in Subheading 3.2.3 of the chapter on substrate phage display in this volume.

3.3 Proteinaceous Binder Engineering with Cystine-Stabilized Mini-Protein Libraries

3.3.1 Constructions of pC-6X, pC-7X, and pC-8X Libraries

1. Construct a phagemid template pC-nX_TAA (Fig. 3) following the procedure described in Subheading 3.1.1.
2. Construct phage-displayed protein libraries pC-6X, pC-7X, and pC-8X (Fig. 3) with the primers shown in Subheading 2.3.1 following the procedures described in Subheadings 3.1.2 and 3.1.3. The design of the three libraries makes use of the structural determinant information obtained from the experiments described in Subheading 3.2. In particular, the sequence motifs—XXDXD and XXXGF—are expected to stabilize the disulfide bonding configuration of the Min23 structure scaffold.

3.3.2 VEGF–Bead Conjugation

1. Dialyze the refolded VEGF protein solution against PBS buffer.
2. Adjust the VEGF concentration to about 100 μg/mL.
3. Add the VEGF protein solution to 10 mg Ultralink Biosupport resin. Rotate at 4 °C overnight for protein conjugation.
4. Centrifuge the beads at 1,000 × *g* for 10 min at room temperature. Remove the supernatant carefully.
5. Add 1.5 mL of quench solution to the beads to block nonreacted sites. Gently rotate the beads in solution at room temperature for 2.5 h.
6. Centrifuge the beads at 1,000 × *g* for 5–10 min. Discard supernatant.

7. Resuspend the beads with 1.5 mL PBS. Incubate at room temperature for 15 min with gentle rotation.
8. Centrifuge the beads at 1,000 × *g* for 5–10 min. Discard supernatant.
9. Resuspend beads in 1.0 M NaCl to remove non-specifically bound protein for 15 min with rotation.
10. Centrifuge the beads at 1,000 × *g* for 5–10 min. Discard supernatant.
11. Resuspend the beads with 1 mL PBS for 15 min with gentle rotation.
12. Centrifuge the beads at 1,000 × *g* for 5–10 min. Discard supernatant. Repeat this wash twice.
13. Resuspend the beads in 100 μL PBS.

3.3.3 Biopanning of Phage Library to VEGF-Conjugated Bead

1. Concentrate the phage library by PEG/NaCl precipitation (Subheading 3.1.1).
2. Dissolve the phage particles in PBST–milk (~10^{12} cfu/mL). Incubate at room temperature for 30 min.
3. Incubate phage libraries from Subheading 3.3.1 with ethanolamine-blocked Ultralink Biosupport beads (without VEGF conjugation) at 4 °C for 2 h to remove phage that bind to the bead matrix.
4. Transfer the supernatant to the VEGF-conjugated Ultralink Biosupport bead at 4 °C for 2 h with gentle rotation.
5. Wash the beads ten times with 1 mL PBST and four times with 1 mL PBS.
6. Add 1 mL of glycine buffer to the beads and incubate for 10 min to elute the phage from the beads.
7. Neutralize the eluted phage with 200 μL neutralization buffer.
8. Centrifuge the beads at 1,000 × *g*. Amplify the eluted phage particles in the supernatant following the procedure described in Subheading 3.2.1 of the chapter on substrate phage display in this volume.

3.3.4 VEGF ELISA Test

1. Add 1 μg/100 μL VEGF to a Nunc maxisorb 96-well plate for target protein coating. Store at 4 °C overnight.
2. Incubate the ELISA plate at room temperature for 1 h with shaking.
3. Remove the VEGF solution from the plate. Add 300 μL of 3 % PBS gelatin to each well for blocking. Shake the plate at room temperature for 1 h.
4. Mix the phage solution with the same volume of 3 % PBST gelatin. Add 100 μL of the mixture to each well of the VEGF-coated ELISA plate. Incubate at room temperature for 1 h.

5. Wash the plate three times with 250 μL PBST and one time with 250 μL PBS.
6. Add anti-M13-HRP (1:3,000, in 3 % PBST gelatin) to each well and bind at room temperature for 1 h.
7. Wash the plate three times with 250 μL PBST and one time with 250 μL PBS.
8. Add 100 μL ABTS substrate to each well for 10 min. Add 100 μL 1 % SDS to terminate the chromogenic reaction.
9. Measure the OD_{405nm} with 96-well ELISA reader.

4 Notes

1. It is preferable to extend 6–8 extra residues at the 3′ and 5′ ends with high G/C content before/after restriction enzyme digestion site.
2. The primer regions flanking the noncomplementary part encoded with degenerate codons need to extend roughly ~20 base pairs each. The estimated melting temperatures of both ends of the complementary regions need to be balanced and to be near 55 °C. 2–3 G/C bases at the 3′ and 5′ ends are preferable.
3. pC-YV1 is derived from pCANTAB5E by inserting a hexa-HIS tag followed by a multiple cloning site between the g3-signal sequence and the E-tag, where substrate sequences and peptide libraries can be conveniently inserted. The details of the construction of pC-YV1 are described in Subheading 3.1.2 of the chapter on substrate phage display in this volume.
4. If extra band appears at 7,000 bp, the dU-ssDNA extract might contain the single-stranded genome of the helper phage. In this case, repeat **steps 4** and **5**, reduce the amount of helper phage added, and reduce the helper phage incubation time in **step 4**. In principle, the dU-ssDNA phagemid extract needs to be pure.
5. The electrophoresis pattern of the DNA products consists of mostly 2–3 bands. None of these bands are lower than the ssDNA template. If the ssDNA template band remains visible, redo **steps 1–4** in this section. Completing the extension reaction is a prerequisite for library construction with high complexity.
6. M13KO7 helper phage preparation can be found in Subheading 3.1.1 of the chapter on substrate phage display in this volume.
7. This protocol is not suitable for GdnHCl concentration higher than 3 M because few phage particles remain infective after

the GdnHCl/DTT/NEM treatment above the 3 M GdnHCl/10 mM DTT threshold concentration. The pH of the phosphate buffer needs to be readjusted after adding GdnHCl. This pH readjustment is necessary because the optimum pH for the DTT reduction reaction is around 8.

8. NEM can block free thiol groups in proteins, preventing the peptide refolding, and also block free thiol groups in unreacted DTT to stop the reduction reaction.
9. Both DTT and GdnHCl inhibit fXa activity. Thus, the supernatant containing these chemicals should be removed as completely as possible.
10. Repeated washing of the beads with wash buffer containing 10 mM imidazole can prevent nonspecific phage binding on the Ni-NTA beads.
11. Saturated signal of OD_{405nm} above 1.5 does not reflect linearly with the phage concentration bound to the plate well. Repeat the ELISA measurement with diluted input phage.

Acknowledgement

This work is supported by grants from NHRI and NSC of Taiwan.

References

1. Gracy J, Le-Nguyen D, Gelly JC, Kaas Q, Heitz A, Chiche L (2008) KNOTTIN: the knottin or inhibitor cystine knot scaffold in 2007. Nucleic Acids Res 36:D314–D319
2. Chang H-J, Hsu H-J, Chang C-F, Peng H-P, Sun Y-K, Yu H-M, Shih H-C, Song C-Y, Lin Y-T, Chen C-C, Wang C-H, Yang A-S (2009) Molecular evolution of cystine-stabilized mini-proteins as stable proteinaceous binders. Structure 17(4):620–631
3. Heitz A, Le-Nguyen D, Chiche L (1999) Min-21 and min-23, the smallest peptides that fold like a cystine-stabilized beta-sheet motif: design, solution structure, and thermal stability. Biochemistry 38:10615–10625
4. Gatto B, Cavalli M (2006) From proteins to nucleic acid-based drugs: the role of biotech in anti-VEGF therapy. Anticancer Agents Med Chem 6:287–301
5. Kajander T, Cortajarena AL, Regan L (2006) Consensus design as a tool for engineering repeat proteins. Methods Mol Biol 340: 151–170
6. Hsu HJ, Tsai KC, Sun YK, Chang HJ, Huang YJ, Yu HM, Lin CH, Mao SS, Yang AS (2008) Factor Xa active site substrate specificity with substrate phage display and computational molecular modeling. J Biol Chem 283:12343–12353
7. Chen CT, Yang EW, Hsu HJ, Sun YK, Hsu WL, Yang AS (2008) Protease substrate site predictors derived from machine learning on multilevel substrate phage display data. Bioinformatics 24:2691–2697
8. Sidhu SS, Weiss GA (2004) Constructing phage display libraries by oligonucleotide-directed mutagenesis. In: Clackson T, Lowman HB (eds.) Phage display. Oxford University Press, New York, pp 27–41

Chapter 2

Construction of a Filamentous Phage Display Peptide Library

Annette Fagerlund, Astrid Hilde Myrset, and Mari Ann Kulseth

Abstract

The concept of phage display is based on insertion of random oligonucleotides at an appropriate location within a structural gene of a bacteriophage. The resulting phage will constitute a library of random peptides displayed on the surface of the bacteriophages, with the encoding genotype packaged within each phage particle. Using a phagemid/helper phage system, the random peptides are interspersed between wild-type coat proteins. Libraries of phage-expressed peptides may be used to search for novel peptide ligands to target proteins. The success of finding a peptide with a desired property in a given library is highly dependent on the diversity and quality of the library. The protocols in this chapter describe the construction of a high-diversity library of phagemid vector encoding fusions of the phage coat protein pVIII with random peptides, from which a phage library displaying random peptides can be prepared.

Key words Phage display, Phagemid, pVIII, Peptide library, Filamentous phage

1 Introduction

The phage display technology, originally established by Smith et al. in 1985 [1], may be used for generating libraries of random peptides displayed on the surface of bacteriophage (*reviewed in* refs. 2, 3). Such libraries are created by inserting random DNA sequences at an appropriate location within a structural gene of a phage particle. Peptides with affinity for a target protein may then be selected through cycles of selection and amplification of phages.

The most commonly used vectors for phage display are Ff filamentous *Escherichia coli* bacteriophage, which infects *E. coli* using the F-pilus as the primary receptor. Without undergoing lysis, infected bacteria produce and secrete the phage particles, which are about 65 Å in diameter and nearly 1 μm in length. The phage particle consists of a circular single-stranded DNA genome encoding 11 genes, covered (in wild-type phage) by approximately 2,700 copies of the major coat protein pVIII. The phage is an excellent cloning vehicle since different lengths of the packaged genome are

Andrew E. Nixon (ed.), *Therapeutic Peptides: Methods and Protocols*, Methods in Molecular Biology, vol. 1088, DOI 10.1007/978-1-62703-673-3_2, accounted for by simply adjusting the number of pVIII copies. The filament is enclosed at each end by four minor coat proteins, in which pIII, present in about five copies at one end of the phage, is required for adsorption to the F-pilus.

Phage display peptide libraries are usually constructed with the random peptides fused at or near the amino terminus of pIII or pVIII. The libraries can be constructed so that all copies of the phage protein contain the inserted peptide, by inserting the random DNA sequences into phage vectors derived from the wild-type genome. Such polyvalent display is commonly used for display on pIII of short peptides that do not interfere with the essential function of the pIII protein [1].

Alternatively, phages containing a mixture of recombinant and wild-type coat proteins may be generated. This is often accomplished by using a phagemid vector encoding the fusion coat protein. A phagemid is a plasmid carrying the phage-derived origin of replication enabling efficient packaging of the phagemid into the phage, in addition to a plasmid origin of replication. Phage production is accomplished through superinfection of phagemid-carrying cells with helper phage, carrying the genes for wild-type coat protein, the other proteins required for phage propagation, and a compromised phage origin or packaging signal leading to inefficient packaging of the helper phage genome. By placing the gene encoding the fusion coat protein under the control of an inducible promoter, it is possible to regulate the density of random peptides displayed on the phage.

This chapter presents protocols for construction of a library of *E. coli* cells carrying the pA2 phagemid [4] encoding fusion proteins of pVIII with random 9-mer peptides, in which the recombinant gene is under the control of the arabinose-inducible P_{BAD} promoter. The protocols may easily be adapted for use with other phagemid vectors and other configurations of displayed peptides.

The success of screening for peptide binders to a chosen target using a phage display library is highly dependent on the quality of the library, and the chance of finding a peptide with a desired property in a given library is proportional to its diversity [5]. Therefore, it is important to focus on generating high-quality and high-diversity libraries. This involves purifying high-quality phagemid vector and random insert DNA and ensuring that high ligation and transformation efficiencies are achieved. Quality analysis of the library is performed by polymerase chain reaction (PCR) analysis and DNA sequencing of individual clones. The procedure for helper phage infection and subsequent production of phage particles is described in Chapter 10 in this volume, along with protocols for selection and amplification of phages displaying peptides with affinity towards a chosen target.

2 Materials

2.1 Preparation of Phagemid Vector and Random Oligonucleotide Insert

1. Phagemid vector pA2 (GenBank Accession EU481505) [4].
2. *Eco*RI, *Bgl*II, and 10× NEBuffer *Eco*RI (New England Biolabs).
3. SeaKem GTG Agarose and NuSieve GTG Agarose (Fisher Scientific).
4. TAE buffer (50×): 2 M Tris–acetate pH 8.0, 0.05 M EDTA pH 8.0. Store at room temperature.
5. Gel casting trays for agarose gel electrophoresis of approximate size 20 cm × 30 cm and 10 cm × 10 cm, with combs that contain one broad well for application of large samples flanked by small wells for molecular weight marker DNA.
6. Gel loading buffer (6×): 0.25 % (w/v) bromophenol blue, 30 % (v/v) glycerol. Store at room temperature.
7. Ethidium bromide: 10 mg/mL stock solution. Store at 4 °C. Working concentration: 0.5 μg/mL. Ethidium bromide is a mutagen. Wear gloves when handling.
8. Molecular weight marker DNA: GeneRuler 1 kb DNA Ladder and GeneRuler Ultra Low Range DNA Ladder (MBI Fermentas) or equivalent.
9. Elutrap ElectroElution System with BT1 and BT2 Membranes (Whatman) (*see* **Note 1**) [6].
10. Ammonium acetate: 7.5 M.
11. Ethanol: 96 and 70 % (v/v).
12. Tris–HCl pH 8.0: 10 mM.
13. Degenerated oligonucleotide: 5′-cctttgct<u>agatct</u>ccgca(mnn)$_9$g ca<u>gaattc</u>accctcagcag and primer for synthesis: 5′-ctgctgagggt<u>gaattc</u>tgc. Restriction sites are underlined, and N is A, T, C, or G (equimolar) and M is A or C (equimolar) (*see* **Note 2**). Order these from a commercial company (e.g., Invitrogen or Eurogentech), produced with a purification step that removes incomplete oligos. Dilute to 100 pmol/μL in dH_2O and store at −20 °C.
14. SOH buffer (5×): 200 mM Tris–HCl pH 7.5, 50 mM $MgCl_2$, 250 mM NaCl.
15. dNTP solution: Illustra Solution PCR Nucleotide Mix dNTP Set (10 mM each dATP, dCTP, dGTP, dTTP) (GE Healthcare). These are high-purity nucleotides.
16. DNA Polymerase I, Large (Klenow) Fragment: 5 units/μL (New England Biolabs).
17. Phenol, chloroform, and isoamyl alcohol, mixed in the ratio of 25:24:1 (v/v/v). Use phenol saturated with 100 mM Tris–HCl pH 8.0. Store at 4 °C. Phenol can cause severe burns and penetrates skin. Wear gloves when handling.

18. Chloroform and isoamyl alcohol, mixed in the ratio of 24:1 (v/v). Store at room temperature.
19. Glycogen: UltraPure Glycogen, 20 µg/µL (Invitrogen).
20. Bio-Spin 6 Chromatography Columns (BioRad).

2.2 Ligation and Transformation Protocols

1. T4 DNA Ligase, 400 units/µL, and 10× T4 DNA Ligation Buffer (New England Biolabs).
2. Adenosine 5′-triphosphate (ATP), 100 mM (GE Healthcare).
3. Qiaex II Gel Extraction Kit (Qiagen).
4. *E. coli* ElectroMax DH12S cells (Invitrogen) (*see* **Note 3**). pUC19 plasmid at 10 pg/µL is provided with the cells.
5. Gene Pulser Cuvettes, 0.2 cm gap (Bio-Rad).
6. Gene Pulser II apparatus (Bio-Rad) or equivalent electroporator.
7. SOC medium: 2 % (w/v) tryptone, 0.5 % (w/v) yeast extract, 0.5 mM NaCl, 25 M KCl. Autoclave. Shortly before use, add 20 mL 1 M glucose (sterile-filtered) and 5 mL 2 M $MgCl_2$ (autoclaved).
8. Ampicillin: For stock solution, dissolve to a concentration of 100 mg/mL in dH_2O, sterile-filtrate, and store in aliquots at −20 °C.
9. SOC-Amp medium: SOC medium with ampicillin added to 60 µg/mL.
10. SOC-Amp agar plates: Add 1.5 % (w/v) Bacto-agar prior to autoclavation. Add glucose and $MgCl_2$ as described above. Cool to ~50 °C, add 60 µg/mL, and dispense into Petri dishes.
11. Glycerol: 50 % (v/v). Autoclave.

2.3 Quality Analysis of Library by PCR and DNA Sequencing

1. Forward primer 5′-accctcgttccgatgctg and reverse primer 5′-tcgctattacgccagctg. These primers anneal up- and downstream of the cloning site in the pVIII gene on the pA2 phagemid. For stock solutions, dilute to 100 pmol/µL in dH_2O and store at −20 °C.
2. DyNAzyme II DNA Polymerase and 10× Reaction Buffer (Finnzymes).
3. dNTP solution: 10 mM each (Finnzymes).

3 Methods

The presented procedure assumes the use of the pA2 phagemid (Fig. 1), but can easily be adapted for use with other phagemid vectors. The degenerate oligo used (Subheading 3.2) will result in the construction of a library of random 9-mer peptides flanked by

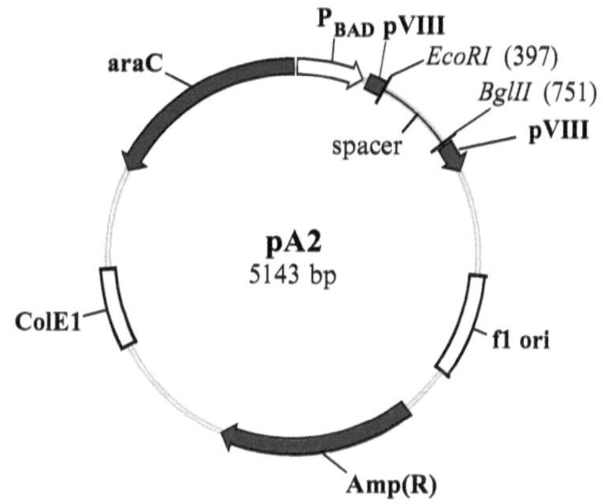

Fig. 1 The pA2 phagemid vector, used for phage display. The vector contains the P_{BAD} promoter/operator region, the pVIII gene encoding the phage major coat protein, spacer fragment to be exchanged with replication origin, and Amp(R) β-lactamase gene (reproduced from ref. 4 with permission from Springer). Features of the pA2 phagemid vector for phage display: *P_{BAD}* P_{BAD} promoter/operator region, *pVIII* gene encoding the M13 major coat protein, *spacer* spacer fragment resistance; *ColE1*; the; and *araC* gene encoding the P_{BAD} transcriptional regulator

cysteine codons (Fig. 2). By designing different synthetic random oligonucleotides, the procedure can be adapted for production of unconstrained as well as constrained libraries with peptides of different lengths.

3.1 Preparation of Phagemid Vector

1. Set up the following restriction digest and incubate at 37 °C for 2 h (or overnight): 100 μg (*see* **Note 4**) pA2 phagemid, 15 μL *Eco*RI (20 units/μL), 30 μL *Bgl*II (10 units/μL), 50 μL 10× NEBuffer *Eco*RI, and dH_2O to a total volume of 500 μL.
2. Separate the 4789 bp pA2 vector from the 354 bp spacer fragment (*see* **Note 5**) by preparative agarose gel electrophoresis. Cast a 250 mL 1 % (w/v) SeaKem GTG agarose gel in TAE buffer, without adding intercalating dyes (*see* **Note 6**) in an appropriately sized gel casting tray (approximately 20 cm × 30 cm), using three combs, each containing one broad well flanked by small wells for molecular weight marker DNA. Place the tray with the gel in a horizontal gel electrophoresis tank and fill with TAE buffer.
3. Mix 100 μL 6× gel loading buffer with the pA2 phagemid restriction digest and load 200 μL into each of the three large wells. Molecular weight marker DNA may be added to the small flanking wells.
4. Run the gel at 5 V/cm until the bromophenol blue has migrated about 5 cm.

a

```
5′-CTGCTGAGGGTGAATTCTGC-3′
3′-GACGACTCCCACTTAAGACG(NNM)9 ACGCCTCTAGATCGTTTTCG-5′
```

⇓ Anneal and extend primer using Klenow fragment.

```
5′-CTGCTGAGGGTGAATTCTGC(NNK)9 TGCGGAGATCAAGCAAAAGC-3′
3′-GACGACTCCCACTTAAGACG(NNM)9 ACGCCTCTAGTTCGTTTTCG-5′
```

⇓ Digest double-stranded oligo using BglII and EcoRI.

```
5′-AATTCTGC(NNK)9 TGCGGA-3′
    3′-GACG(NNM)9 ACGCCTCTAG-5′
```

b

```
ATGAAAAAGTCTTTAGTCCTCAAAGCCTCTGTAGCCGTTGCTACCCTCGTTCCGATGCTGTCTTTC
 M  K  K  S  L  V  L  K  A  S  V  A  V  A  T  L  V  P  M  L  S  F

                EcoRI       Degenerated oligonucleotide insert      BglII
GCTGCTGAGGGTGAATTCTGCNNKNNKNNKNNKNNKNNKNNKNNKNNKTGCGGAGATCTCGCAAAA
 A↑A  E  G  E  F  C  X  X  X  X  X  X  X  X  X  C  G  D  L  A  K

GCGGCCTTTGACTCCCTGCAAGCCTCAGCGACCGAATATATCGGTTATGCGTGGGCGATGGTTGTT
 A  A  F  D  S  L  Q  A  S  A  T  E  Y  I  G  Y  A  W  A  M  V  V

GTCATTGTCGGCGCAACTATCGGTATCAAGCTGTTTAAGAAATTCACCTCGAAAGCAAGCTAA
 V  I  V  G  A  T  I  G  I  K  L  F  K  K  F  T  S  K  A  S STOP
```

Fig. 2 (**a**) Schematic illustration of the procedure for preparation of random oligonucleotide insert. The cysteine codons flanking the random NNM-codons are *underlined*, and the *Eco*RI and *Bgl*II restriction sites are in *bold*. (**b**) The sequence of the pVIII gene and protein encoded on pA2, with the random oligonucleotide correctly inserted. The *arrow* indicates the site of signal peptide cleavage sequence upon export from the *s* cell. Thus the amino terminal AEGDDP of the mature wild-type protein is replaced by AEGEFCXXXXXXXXXCGDL in the recombinant pVIII with inserted peptide

5. Cut off the right and left parts of the gel using a clean scalpel so that the excised pieces contain the lanes containing size marker DNA and ~0.5 cm of the large lanes containing the digested pA2 sample.
6. Immerse the end pieces in TAE buffer containing ethidium bromide, and incubate with gentle agitation for 30 min.
7. Place the stained gel pieces in a UV light box, and mark the position of the 4,789 bp pA2 vector fragment with a scalpel. Reorient the stained gel with the unstained gel, and excise the gel pieces containing the completely digested 4,789 bp pA2 vector fragment from the unstained gel using a clean scalpel. Transfer these gel pieces to a clean container. Stain the remainder of gel with ethidium bromide, and view under UV light to ensure that the bands have been correctly excised.

8. To elute the pA2 vector fragment from the unstained gel using electroelution in the Elutrap ElectroElution System (*see* **Note 1**), assemble the BT1 and BT2 membranes in three Elutrap device elution chambers according to the manufacturer's instructions.
9. Cut the excised gel containing the band of interest in pieces of lengths of 1–2 cm, and place them next to one another in the three Elutrap devices, in contact with the inner membrane at the anode (+).
10. Place the Elutrap device in an electrophoresis chamber, and fill with enough TAE buffer to cover the gel slices. Run the electroelution at 140 V for 6 h (*see* **Note 7**).
11. Reverse the polarity of the electrophoresis apparatus and apply 200 V for 20 s in order to remove any sample DNA attached to the BT1 membrane. Collect each elute with a pipette, by reaspirating a few times to collect sample material attached to the surface of the chamber. Transfer the combined samples to a clean Eppendorf tube.
12. Ethanol-precipitate the elute by adding 0.5 volumes of 7.5 M ammonium acetate and 3.5 volumes of 96 % ethanol to 1 volume of elute. Mix and incubate for minimum 1 h at −20 °C. Recover the DNA by centrifugation at 16,000 × *g* for 30 min at 4 °C. Remove the supernatant and wash the pellet in 1 mL 70 % ethanol at 4 °C. Vortex briefly and centrifuge as before for 10 min. Remove the supernatant and allow the pellet to dry completely at room temperature. Dissolve the pellet in 100 μL 10 mM Tris–HCl pH 8.0. Store at −20 °C.
13. Determine the concentration of purified pA2 vector and calculate the molar yield using the molecular weight of the pA2 vector, 4,789 bp × 649 = 3.11×10^6 g/mol. A total of 22.5 pmol purified vector DNA is required for one initial test ligation with control and the final library construction. Repeat the procedure in Subheading 3.1 until sufficient material is obtained (*see* **Notes 4** and **8**).

3.2 Preparation of Random Oligonucleotide Insert

1. Mix 800 pmol each of degenerated oligo and primer for synthesis (*see* **Note 9** and Fig. 2a) with 200 μL 5× SOH buffer and add dH_2O to a total volume of 952 μL. Anneal the two oligonucleotides by incubation at 65 °C for 5 min followed by 5 min at room temperature.
2. To synthesize double-stranded degenerated oligonucleotide, add 40 μL dNTP solution (10 mM each) and 8 μL DNA Polymerase I, Large (Klenow) Fragment, mix, and incubate at 37 °C for 2 h.
3. Remove the Klenow Fragment by phenol:chloroform extraction (*see* **Note 10**), as follows: Add 1 mL phenol:chloroform:isoamyl

alcohol (25:24:1) to the reaction mixture, vortex for 1 min, and centrifuge at 12,000 × *g* for 2 min. Transfer the upper aqueous phase to a fresh tube. Repeat the extraction once with 1 mL phenol:chloroform:isoamyl alcohol (25:24:1) and once with 1 mL chloroform:isoamyl alcohol (24:1).

4. As a carrier molecule, add 8 μL glycogen (20 μg/μL) to the reaction and mix. Subject the double-stranded oligo to ethanol precipitation as described above (**step 12** in Subheading 3.1), but dissolve the dried pellet in 405 μL 1× NEBuffer *Eco*RI.
5. To digest the double-stranded oligo, add 15 μL *Eco*RI (20 units/μL) and 30 μL *Bgl*II (10 units/μL), mix, and incubate for 2 h at 37 °C.
6. Separate the 41 bp central random oligo from the flanking fragments by preparative agarose gel electrophoresis as follows: Cast a 75 mL 4 % (w/v) NuSieve GTG agarose gel in TAE buffer, without adding intercalating dyes (*see* **Note 6**) in a gel casting tray of approximately 10 cm × 10 cm using one comb with one broad well flanked by small wells for molecular weight marker DNA. Place the tray with the gel in a horizontal gel electrophoresis tank and add TAE buffer.
7. Mix 90 μL 6× gel loading buffer with the purified random oligo restriction digest and load the entire sample into the large well of the gel. Low-range molecular weight marker DNA may be added to the small flanking wells.
8. Run the gel at 5 V/cm until the bromophenol blue has migrated about 5 cm (*see* **Note 11**).
9. Excise the 41 bp digested random oligo insert from the unstained gel as described in **steps 5–7** in Subheading 3.1.
10. Elute the random oligonucleotide insert from the unstained gel using electroelution in the Elutrap ElectroElution System, as described in **steps 8–11** in Subheading 3.1, but use only one Elutrap device and run the electroelution for approximately 2 h (*see* **Note 11**).
11. Use two Bio-Spin 6 Chromatography Columns according to the manufacturer's instructions to change the buffer from TAE to 10 mM Tris–HCl, pH 8.0, which is suitable for the subsequent ligation reaction. Store at −20 °C.
12. Determine the concentration of purified random oligonucleotide insert and calculate the molar yield using the molecular weight of 41 bp × 649 = 2.66×10^4 g/mol. A total of 19.5 pmol purified insert DNA is required for one test ligation and the final library construction.

3.3 Small-Scale Test Ligation and Transformation

Before performing the actual large-scale ligation and transformation of the phage display library, it is strongly advisable to perform the small-scale experiment described here to ensure that the individual preparations of vector and insert are of adequate quality for generating a library of the desired diversity. Controls are also performed to ensure that the ligation and transformation conditions are optimized for high-efficiency cloning.

1. Mix 1.5 pmol pA2 vector with 1.5 pmol random oligo insert and add dH_2O to a total volume of 213 μL. Also prepare one control reaction with 1.5 pmol pA2 vector only. Mix and incubate at 45 °C for 5 min and then place the reactions on ice.
2. Add 1.5 μL (600 units) T4 DNA ligase, 24 μL 10× ligase reaction buffer, and 1.2 μL ATP (100 mM) to each reaction and incubate overnight at 16 °C.
3. Purify each ligation reaction using Qiaex II Gel Extraction Kit (Qiagen), according to the Qiaex II Protocol for Desalting and Concentrating DNA Solutions, using 10 μL resuspended Qiaex II suspension in each reaction. Elute the DNA from each dried Qiaex II pellet using 20 μL 10 mM Tris–HCl pH 8.0. Following centrifugation, approximately 15 μL supernatant may be collected from each reaction. Store on ice until transformation in **steps 5–7**.
4. Run 1 μL each of the ligation reaction and control reaction on a 1 % (w/v) agarose gel containing ethidium bromide. Bands that are present in the ligation reaction but absent in the control reaction should be visible and contain the ligated product.
5. Transfer 4 μL purified ligation mixture, 4 μL purified control reaction, and 1 μL pUC19 (10 pg/μL) to each of the three Eppendorf tubes, and place on ice.
6. For each reaction to be transformed thaw 80 μL ElectroMax DH12S cells (*see* **Note 3**) and place the cells and one 0.2 cm electroporation cuvette on ice.
7. Perform the following transformation procedure individually for each sample: Transfer 80 μL ElectroMax DH12S cells to the tube containing DNA to be transformed, mix by gently pipetting up and down once, and incubate on ice for 1 min. Carefully transfer the sample to the bottom of a chilled cuvette without trapping air bubbles. Electroporate the reaction in the Gene Pulser II electroporation apparatus using settings of 25 μF, 200 Ω, and 2.5 kV. Immediately resuspend the cells in the cuvette using two volumes of 960 μL ice-cold SOC medium, and pool the cell suspension in a 50 mL Nunc-tube, placed on ice.
8. Incubate the cell suspensions at 37 °C for 60 min with shaking (230 rpm).
9. Prepare duplicate tenfold serial dilutions of each culture in SOC medium, and plate suitable dilutions onto SOC-Amp agar plates. After overnight incubation at 37 °C, score the

number of independent transformants by counting the number of colony-forming units (cfu) on each plate (*see* **Note 12**). Save the plates at 4 °C for analysis as described in Subheading 3.4.

3.4 Quality Analysis of Library by PCR and DNA Sequencing

A high-quality phage display library is characterized by minimal occurrence of inserts with frameshifts and stop codons, multiple inserts, or empty vectors. The following procedure can be used both for analysis of the clones obtained from the initial testing (**step 9** in Subheading 3.3) and for quality analysis of the final library (**step 12** in Subheading 3.5) (*see* **Note 13**). In the PCR, the fragment of the pA2 phagemid containing pVIII with random oligonucleotide insert is amplified and analyzed.

1. Pick >100 single colonies containing the pA2 phagemid (from **step 9** in Subheading 3.3 or **step 12** in Subheading 3.5), and resuspend each colony in 10 μL SOC medium.
2. Set up PCR reactions for each resuspended colony according to standard laboratory protocols. As template, use 2 μL of each resuspended colony in a 50 μL reaction and primers that flank the insert:pVIII cloning site (*see* Subheading 2.3).
3. Run the PCR for 30 cycles using 60 °C annealing temperature.
4. Analyze the PCR products on a 2 % (w/v) agarose gel containing ethidium bromide. The correct construct containing one random oligonucleotide insert will give one band of size 320 bp. Larger bands indicate the presence of multiple inserts or empty vectors containing the spacer fragment (*see* **Note 13**).
5. To determine the frequency of clones which contain stop codons or frameshift mutations in the random sequence (*see* **Note 13**), the PCR products may be DNA sequenced using the reverse primer from the PCR reaction as the sequencing primer.

3.5 Ligation and Transformation of Phagemid Library

It is advisable to wait to perform the final large-scale ligation and transformation of the phage display library until the results from testing as described in Subheadings 3.3 and 3.4 have indicated that production of a phage display library of the desired diversity and quality will be achieved.

1. Ligate pA2 vector and random oligonucleotide insert as described in **steps 1–3** in Subheading 3.3, but set up 12 identical reactions, requiring a total of 18 pmol each of pA2 vector and random oligonucleotide insert. Also prepare one control reaction with 1.5 pmol pA2 vector only. Following elution from the Qiaex II silica particles as described in **step 3** in Subheading 3.3, combine the 12 ligation samples to obtain approximately 180 μL elute, and place on ice.
2. Run the reactions on an agarose gel as described in **step 4** in Subheading 3.3 to ensure that the ligation reaction was successful.

3. Transfer 4 μL purified ligation mixture into each of the ten Eppendorf tubes and place on ice.
4. Perform the transformation for the ten samples individually as described in **steps 6** and 7 in Subheading 3.3, but pool the cell suspensions for all ten transformation reactions in the same 50 mL Nunc-tube on ice, to give a total volume of 20 mL.
5. Incubate the combined cell suspension at 37 °C for 60 min with shaking (230 rpm). Immediately proceed to **step 6** after initiating this incubation (*see* **Note 14**).
6. Repeat **steps 3–5** three additional times for a total of 40 electrotransformation reactions.
7. Perform the following procedure for each of the four series of electroporation reactions: After exactly 60-min incubation at 37 °C (**step 5**), prepare duplicate tenfold serial dilutions of the culture in SOC medium and plate suitable dilutions onto SOC-Amp agar plates. Proceed to **step 11** after incubating the plates overnight at 37 °C.
8. To amplify the library, dilute each 20 mL culture into 1 L SOC-Amp medium (*see* **Note 15**) and divide between five dry-sterilized (*see* **Note 16**) 2-L baffled Erlenmeyer bottles (or two 5-L bottles). Incubate at 37 °C for 5–7 h in a rotary-shaking incubator at 230 rpm until the optical density at 600 nm (OD_{600}) is approximately 2.8.
9. Centrifuge in phage-decontaminated centrifuge tubes (*see* **Note 16**) at 11,000 × *g* for 10 min at 4 °C. Discard the supernatant and resuspend the pellet in 5 mL SOC medium. Wash out the tubes with an additional 5 mL SOC medium and pool the cells. Place the tube on ice until cells from all four series of electroporation reactions have been collected, combine in one tube, and mix.
10. Transfer 1 mL aliquots of resuspended bacteria into prechilled cryotubes, each containing 450 μL sterile 50 % glycerol. Mix gently, snap-freeze on a dry ice/ethanol bath, and store at −80 °C. This is the bacterial pA2$(NNK)_9$ library.
11. Following overnight incubation at 37 °C, count the number of cfu on each plate prepared in **step** 7. Calculate the number of independent transformants from each of the four electrotransformation series as follows: number of colonies on plate × total transformation volume (20 mL) × dilution factor. Summarize the numbers obtained in this calculation for the four series of ten transformations to determine the size or titre of the library. Save the plates at 4 °C for quality analysis as described in Subheading 3.4.
12. To determine the number of viable bacteria per glycerol stock after freezing, thaw one of the vials prepared in **step 11**, dilute to

100 mL in SOC-Amp medium, and incubate in a dry-sterilized 1 L Erlenmeyer bottle at 37 °C with shaking at 230 rpm for 30 min. Prepare duplicate tenfold serial dilutions of the culture in SOC medium and plate suitable dilutions onto SOC-Amp agar plates. After overnight incubation at 37 °C, count the number of colonies on the plates and calculate the number of viable cfu per vial.

4 Notes

1. Electroelution is a method that allows the recovery of high-quality DNA from agarose gels. We have used the Elutrap ElectroElution System, but other electroelution systems available can also be used.
2. Due to the degenerate genetic code, certain amino acids will be present at a higher frequency than others in the displayed random peptides. To reduce this effect, the library is constructed with only 32 codons of the form NNK, where K = G or T. This will result in no amino acid being encoded by more than three codons and additionally eliminate two of the three stop codons. The remaining amber stop codon (TAG) can be suppressed by propagating the phage in strains of *E. coli* containing amber suppressor mutations.
3. *E. coli* DH12S cells contain an F′ episome (allowing expression of the F-pilus) and are therefore suitable for use in generating phage libraries. They are also suitable for induction of the P_{BAD} promoter on the pA2 phagemid using L-arabinose, since they carry the *ara*D− mutation which blocks metabolism of arabinose but does not affect its transport into the cell. Finally, the ElectroMax DH12S cells have high transformation efficiency, which is necessary for generation of high-diversity libraries. Transformation with 10 pg pUC19 should yield $>10^{10}$ transformants per μg plasmid.
4. The recovery yield of the 4,789 bp fragment of pA2 vector using this procedure might be as low as 25 %. However, the main concern is the purity and the quality of the preparation. Since a minimum of 22.5 pmol = 70 μg purified vector is required, the procedure described in Subheading 3.1 needs to be repeated about three times to obtain sufficient material. The indicated 100 μg plasmid starting material is chosen since it is a suitable amount to process in 1 day on one large preparative gel and using one Elutrap ElectroElution apparatus.
5. The pA2 phagemid contains a 354 bp spacer fragment between the unique restriction sites used for insertion of random oligonucleotides into the pVIII gene (*see* Fig. 1). The presence of this large spacer fragment instead of a small fragment simplifies

isolation of the pure 4,789 bp vector fragment digested with both restriction enzymes (**step 2**, Subheading 3.1), and reduces the fraction of empty vector clones in the library.

6. The phagemid vector and insert samples are excised from gels without added ethidium bromide or other intercalating dyes in order to omit exposure to UV light, which may have a significant detrimental impact on the downstream cloning efficiency [7].
7. After electroelution, staining one gel piece from the Elutrap device with ethidium bromide as described in **step 6** in Subheading 3.1 will reveal whether all the DNA has been eluted from the gel. If not, continue the electroelution of the remaining unstained gel pieces for a further 1–2 h.
8. Different batches of vector preparations might be of different quality, thus generating varying number of transformants per μg DNA when the final library is prepared. The different preparations of vector should therefore not be pooled until they have been individually assessed in the small-scale test ligation and transformation protocol described in Subheading 3.3.
9. The degenerated oligo used in the present procedure will result in a library of random 9-mer peptides flanked by cysteine-codons (Fig. 2) which will form disulfide bonds, resulting in cyclic random peptide sequences. In some cases, this more defined conformation will promote high-affinity binding to a target [8]. By designing different synthetic random oligonucleotides, linear peptide libraries or libraries with peptides of different lengths can be produced. If a phagemid vector other than pA2 is used, ensure that correct flanking restriction sites enable insertion of the degenerated oligonucleotide in frame with the pVIII (or pIII) gene.
10. Any residual active Klenow Fragment may fill in the 4 bp overhangs created during the subsequent restriction digest (**step 5** in Subheading 3.2). The manufacturer indicates that Klenow Fragment will be inactivated by incubation at 75 °C for 20 min. For the purpose of generating a high-diversity library, however, we rather recommend removal of the Klenow Fragment by phenol:chloroform extraction, as we noted a 100-fold increase in library size by this approach.
11. The migration of the bromophenol blue in the 4 % agarose gel is almost equivalent to that of the digested random oligonucleotide. Thus, the migration of the oligo during both gel electrophoresis and electroelution can be followed using the blue dye.
12. For the ligation of pA2 phagemid and the random oligo insert, it should be possible to obtain 2.5×10^8 cfu per transformation reaction, which would result in a library of approximately 10^{10} independent transformants in 40 transformation reactions. The number of transformants from the control ligation should be less than 1 % of that from the ligation mixture.

13. The effective size of the library is the library titre as determined by the number of individual transformants, minus the fraction of clones without the random oligo insert, with multiple random oligo inserts, or with stop codons or frameshift mutations in the random oligo sequence. If the frequency of clones with multiple inserts is high (>5 %) in the test reaction, the ratio of vector and insert used in the ligation reaction should be optimized. DNA sequencing of single clones will detect stop codons and frameshift mutations, and also reveal whether there is a bias imposed on the frequency of individual codons in the random oligos of the library. However, keep in mind that bias against individual amino acid residues or peptide sequences is mainly imposed during the phage production step (described in Chapter 10 in this volume).
14. It is advisable that two persons carry out the transformation procedure described in **steps 3–8** in Subheading 3.5, in which four series of ten successive transformation reactions are performed in 1 day. One person may perform the electroporation reactions of the second to fourth series (**step 6**) at the same time as the second person performs the initial processing of the resulting cultures of transformed cells (**steps** 7 and **8**).
15. SOC medium contains glucose, which will repress expression of pVIII-fusion proteins from the P_{BAD} promoter during propagation of *E. coli* clones containing pA2 phagemid. This will reduce the bias caused by different growth rates between clones encoding different peptides.
16. Phage particles are able to withstand surprisingly harsh conditions, such as low pH and autoclaving. If the glass- and plasticware used for phage display library construction has been previously used for phage preparation experiments, it is important to take measures to ensure that the library is not contaminated with previously isolated phage. Any glassware used should be dry-sterilized at 180 °C overnight. Bacterial cultures containing phages should be destroyed by addition of sodium hypochlorite. All plastic centrifuge tubes should be incubated overnight in sodium hypochlorite followed by washing, autoclaving, and sterilization under UVC light for 4 h (e.g., in a laminar air flow [LAF]-bench). Other measures for avoiding contamination include the use of aerosol-barrier pipet tips, regular UVC radiation of the lab, and use of disposable pipettes and containers whenever possible.

Acknowledgement

This work was performed at and supported by Medical Diagnostics Research, GE Healthcare.

References

1. Smith GP (1985) Filamentous fusion phage: novel expression vectors that display cloned antigens on the virion surface. Science 228:1315–1317
2. Cesareni G, Castagnoli L, Cestra G (1999) Phage displayed peptide libraries. Comb Chem High Throughput Screen 2:1–17
3. Paschke M (2006) Phage display systems and their applications. Appl Microbiol Biotechnol 70:2–11
4. Fagerlund A, Myrset AH, Kulseth MA (2008) Construction and characterisation of a 9-mer phage display pVIII-library with regulated peptide density. Appl Microbiol Biotechnol 80:925–936
5. Noren KA, Noren CJ (2001) Construction of high-complexity combinatorial phage display peptide libraries. Methods 23:169–178
6. Göbel U, Maas R, Clad A (1987) Quantitative electroelution of oligonucleotides and large DNA fragments from gels and purification by electrodialysis. J Biochem Biophys Methods 14: 245–260
7. Gründemann D, Schömig E (1996) Protection of DNA during preparative agarose gel electrophoresis against damage induced by ultraviolet light. Biotechniques 21: 898–903
8. Bonnycastle LL, Mehroke JS, Rashed M, Gong X, Scott JK (1996) Probing the basis of antibody reactivity with a panel of constrained peptide libraries displayed by filamentous phage. J Mol Biol 258:747–762

Chapter 3

Engineering Bioactive Peptide-Based Therapeutic Molecules

Jong Sang Ryu, A. Yeon Cho, Sang Won Seo, and Hosung Min

Abstract

Peptides are increasingly emerging as human therapeutic drugs. By screening very large phage display libraries, novel bioactive peptides that bind to the target of interest with desired biological properties can be identified. Peptides that are obtained in this fashion become the basis for therapeutic molecule development. However, naked peptides are usually not sufficient to be therapeutic molecules by themselves. They need to be chemically modified or conjugated to other molecules to obtain desired physicochemical and in vivo properties. In this chapter, we describe a general methodology of identifying bioactive peptides by biopanning of peptide phage libraries. As an example of therapeutic peptide modifications, we also describe a method for fusing the peptides to the Fc portion of antibody molecule to increase in vivo stability and activity.

Key words Phage display, Target-specific peptide, Bioactive peptides, Biopanning, Fc-peptide fusion, Targeted therapeutics

1 Introduction

With the advent of recombinant combinatorial library technologies, identification of potential peptide therapeutic molecules can be relatively effortless [1–3]. In phage display technology, libraries of peptides are displayed as viral coat protein fusions on the outer surface of bacteriophage [4, 5]. The fusion gene products are encoded within the genome of filamentous phage and can be identified by sequencing, linking the genetic information to functional peptides [6, 7]. Phage display offers a powerful tool for identification of peptides that can bind and regulate the function of target proteins [8].

In general, the biggest advantage of peptide-based therapeutics is their flexibility. Peptides can be modified with nonnatural amino acids or chemically conjugated to design molecules with improved therapeutic properties, such as bioactivity, stability, and pharmacokinetics [9, 10]. Therapeutic peptides are generally mass-produced chemically for flexible modification purposes, while

Andrew E. Nixon (ed.), *Therapeutic Peptides: Methods and Protocols*, Methods in Molecular Biology, vol. 1088, DOI 10.1007/978-1-62703-673-3_3, © Springer Science+Business Media, LLC 2014

peptides with natural amino acids can also be synthesized using the recombinant DNA and fermentation technologies [11].

Perhaps the most challenging issue that faces peptide-based therapeutics is their in vivo metabolic stability [12]. Mainly due to their small sizes, peptides are generally rapidly filtered through kidney. Unstructured linear peptides are also easily degraded by various proteases in circulation. Another disadvantage for chemically produced peptides is high cost of goods [13]. This is especially true for longer peptides (>50 amino acids) and peptides with extensive modifications since manufacturing process can become very inefficient and cost prohibitive. Recently, therapeutic molecules fused with Fc portion of an antibody were developed [14]. These molecules show good in vivo stability as well as pharmacokinetic/pharmacodynamic (PK/PD) profiles and yet are cost-effective to produce since they are recombinantly produced in cells.

This chapter focuses on the general method of identifying bioactive peptides from Dyax's phage libraries. In addition, a strategy for generating therapeutic agents by fusing bioactive peptides to the Fc portion of an immunoglobulin constant region is discussed.

2 Materials

2.1 Target Protein Preparation

2.1.1 Biotinylation of Target Protein

1. Slide-A-Lyzer 10K MWCO dialysis cassettes (PIERCE, Rockford, IL).
2. Phosphate-buffered saline (PBS): 10× stock of PBS can be prepared by dissolving 80 g NaCl, 2 g KCl, 14.4 g $Na_2HPO_4{\cdot}2H_2O$, and 2.4 g KH_2PO_4 in 800 ml of distilled water and topping up to 1 l (adjust pH with HCl or NaOH if necessary). Store at room temperature after autoclaving and prepare working solution by dilution of one part 10× stock with nine parts autoclaved distilled water.
3. Bio-Rad Protein Assay kit (Bio-Rad, Hercules, CA).
4. EZ-Link® Sulfo-NHS (*N*-Hydroxysuccinimide)-Biotinylation kit (PIERCE).
5. EZ™ Biotin Quantization Kit (PIERCE).

2.1.2 Target Protein Immobilization

1. Dynabead® M-270 streptavidin (Invitrogen Dynal, Oslo, Norway).
2. Blocking solution: 0.1 % (w/v) Bovine Serum Albumin (BSA; Sigma, St. Louis, Mo) in PBS.
3. Sodium azide (Sigma) is a useful preservative.
4. Magnetic particle concentrator (MPC; Invitrogen Dynal).

2.1.3 Validation of Protein Immobilization

1. Washing solution: 0.1 % (v/v) Tween 20 in PBS (PBS-T).
2. Blocking solution: 0.1 % (w/v) BSA in PBS (BSA/PBS).
3. TMB (3,3′,5,5′-tetramethylbenzidine) Substrates (PIERCE): A chromogen that yields a blue color when oxidized with hydrogen peroxide. The color changes to yellow with the addition of 5 N of H_2SO_4 with maximum absorbance at 450 nm.

2.2 Screening High-Affinity Phage Clones

2.2.1 Affinity Selection: Biopanning

1. Library dilution solution: 0.05 % (v/v) PBS-T.
2. Washing solution: 0.1 % (v/v) PBS-T.
3. Elution solution: 100 mM Triethylamine (TEA; Sigma). Dissolve TEA in autoclaved distilled water immediately before use.
4. Neutralizing solution: 1 M Tris–HCl (pH 7.5).
5. TE buffer: 10 mM Tris, 1 mM EDTA in distilled water (pH 7.5).

2.2.2 Phage Amplification and Plaque Formation Assay

1. Luria-Bertani (LB) medium: 5 g yeast extract, 10 g sodium chloride, and 10 g tryptone are dissolved in 1 l of distilled water. Autoclave and store at 4 °C.
2. NZCYM media (2×): 10 g NZ amine (casein hydrolysate), 10 g yeast extract, 10 g sodium chloride, and 4 g $MgSO_4 \cdot 7H_2O$ are dissolved in 1 l of distilled water. Autoclave and store at 4 °C.
3. Antibiotics: Tetracyclin (Sigma) powder is dissolved in ethanol at 50 mg/ml stock. Ampicillin (Sigma) powder is dissolved in distilled water at 100 mg/ml stock. Store at −20 °C.
4. NZCYM agar plate: 15 g bacto agar is dissolved in 1 l of 1× NZCYM media. Autoclave and cool at room temperature. Add 500 μl of ampicillin stock solution, pour in 23 cm × 23 cm large square dish, and store at 4 °C.
5. PEG solution: 20 % (w/v) polyethyleneglycol 8000 (PEG8000) and 3.5 M ammonium acetate are dissolved in autoclaved distilled water.
6. Phage resuspension buffer: 200 mM NaCl, 100 mM Tris pH 8.0, 1 M EDTA.
7. LB agar plate: 15 g bacto agar is dissolved in 1 l of LB media. Autoclave and cool at room temperature. Pour in appropriate dish and store at 4 °C.
8. Top agar: 7 g of bacto agar is dissolved in 1 l of LB media. Autoclave and aliquot 20 ml into polypropylene tube and store at 4 °C.

2.3 Phage Clonal Analysis: Phage ELISA and DNA Sequence Analysis

2.3.1 Phage ELISA

1. LB media, TE buffer (pH 7.5), and PBS (pH 7.2).
2. 96-well microtiter plate (NUNC, Roskilde, Denmark).
3. MaxiSorp™ flat-bottom 96-well plate (NUNC).
4. Plate sealing film (Nunc).
5. Antibiotics: Ampicillin and tetracycline stock solution.
6. Blocking solution: 5 % (w/v) skimmed milk in PBS.
7. Pure streptavidin dissolved in PBS at 2 mg/ml concentration.
8. Binding solution: 4 % (w/v) skimmed milk in 0.05 % (v/v) PBS-T.
9. Washing solution: 0.01 % (v/v) PBS-T.
10. HRP-conjugated anti-M13 antibody (GE Healthcare, Pittsburgh, PA).
11. Antibody dilution solution: 2 % (w/v) skimmed milk in 0.01 % (v/v) PBS-T.
12. TMB substrates and 5 N H_2SO_4.

2.3.2 DNA Sequence Analysis

1. 10× PCR buffer, $MgCl_2$, dNTP, and Taq polymerase: Expand High Fidelity PCR system (Roche, Indianapolis, IN, USA).
2. PCR primers:

   5′-CGGCGCAACTATCGGTATCAAGCTG-3′.

   5′-CATGTACCGTAACACTGAGTTTCGTC-3′.
3. PCR purification kit (QIAGEN, Valencia, CA).

2.4 Fc-Peptide Fusion Proteins

2.4.1 Expression and Purification of Fc-Peptide Fusion Proteins

1. Gel purification kit (QIAGEN).
2. Human embryonic kidney cell line (HEK) 293 (ATCC No. CRL-1573).
3. Tissue culture plate (Corning).
4. Growth media: DMEM (Invitrogen Gibco, Gaithersburg, MD) supplemented with 100 units/ml penicillin G sodium, 100 μg/ml streptomycin, 4 mM L-glutamine, and 10 % fetal bovine serum.
5. Reduced serum media: Opti-MEM® I (Invitrogen Gibco).
6. Lipofectamine™ 2000 (Invitrogen Gibco).
7. Serum-free media: DMEM (Invitrogen Gibco) supplemented with 100 units/ml penicillin G sodium, 100 μg/ml streptomycin, and 4 mM L-glutamine.
8. HiTrap™ Protein A HP (GE Healthcare).
9. HiTrap™ Sepharose™ HP Ion Exchange Columns (GE Healthcare).
10. 1 M Tris–HCl, pH 8.0.
11. 20 mM NaOAc, 50 mM–500 nM NaCl.

2.5 Molecular Assays to Evaluate Fc-Peptide Fusion Proteins

2.5.1 Affinity ELISA

1. MaxiSorp™ flat-bottom 96-well plate (NUNC).
2. Blocking solution: 0.5 % (w/v) BSA/PBS.
3. Washing solution: 0.1 % (v/v) PBS-T.
4. Binding solution: 0.1 % (w/v) BSA/PBS.
5. Goat anti-human IgG(Fc)-HRP (Alpha Diagnostic International, San Antonio, TX).
6. TMB substrates and 5 N H_2SO_4.

2.5.2 Affinity Assay: BIAcore

1. BIAcore X100 (GE Healthcare).
2. Human antibody capture kit (GE Healthcare).
3. CM5 sensor chip (GE Healthcare).
4. Running buffer: HBS-EP (GE Healthcare): contains 0.01 M HEPES pH 7.4, 0.15 M NaCl, 3 mM EDTA, 0.005 % (v/v) Surfactant P20.

3 Methods

3.1 Target Protein Preparation

Oriented immobilization of biologically active proteins is a useful tool to facilitate the identification of therapeutically effective binding partners. This technique allows for increased accessibility of active binding sites and increases the stability of the target molecules. There have been various techniques developed for protein immobilization including the use of polyclonal/monoclonal antibodies, boronate affinity gel, tag moiety and metal complexes, and affinity interaction of biotin–avidin/streptavidin [15]. This section focuses on the high-affinity interaction of streptavidin bead with biotinylated proteins.

3.1.1 Biotinylation of Target Protein

1. These directions assume the use of a Sulfo-NHS-Biotinylation kit (PIERCE). If the target protein is originally in a different buffer to PBS (pH 7.2), it is necessary to dialyze target protein in PBS (*see* **Note 1**).
2. The concentration of target protein is measured using the Bio-Rad Protein Assay kit with diluted protein samples before biotinylation (1:1, 1:10, and 1:100) according to the manufacturer's instructions.
3. Immediately prior to use, 1 mg/ml of Sulfo-NHS-Biotin solution is prepared by dissolving it in ultrapure water. If the Sulfo-NHS-Biotin has been stored cold, it is necessary to equilibrate to room temperature before use.
4. 20-fold excess molar of Sulfo-NHS-Biotin solution is added to the target protein and the mixture is incubated on ice for 2 h or at room temperature for 30–60 min. To determine the

amount of Sulfo-NHS-Biotin required, refer to the equation below:

$$\frac{X \text{ mg of Sulfo} - \text{NHS} - \text{Biotin}}{443} = \left(\frac{\Upsilon \text{ mg of target protein}}{\text{Molecular weight of target protein}} \right) \times 20$$

X = amount of Sulfo-NHS-Biotin.

Υ = amount of target protein.

443 = molecular weight of Sulfo-NHS-Biotin.

5. The reaction mixture is transferred into slide-A-Lyzer 10K MWCO dialysis cassette and dialyzed with ice-cold BupH PBS overnight at 4 °C to remove unincorporated Sulfo-NHS-Biotin.

   The level of biotin incorporation is determined using EZ™ Biotin Quantitation Kit. The HABA/avidin solution is prepared (*see* **Note 2**) and OD_{500} of 900 μl HABA/avidin solution is measured and recorded as A_{500} HABA/avidin (A_{500} H\A). 100 μl of biotinylated target protein sample is added to 900 μl of HABA/avidin solution and OD_{500} was measured and recorded as A_{500} HABA/avidin/biotin sample (A_{500} H\A\B).

6. Biotin incorporation rate (moles of biotin per moles of protein) can be calculated as below. These calculations are based on the Beer–Lambert Law (Beer's Law): $\Delta A_\lambda = \varepsilon_\lambda b \Delta C$.

ΔA_λ: Absorbance change of the sample at a particular wavelength (λ). For the HABA assay at 500 nm, $\Delta A_\lambda = (0.9 \times A_{500}$ H\A) − (A_{500} H\A\B).

ε_λ: Extinction coefficient at the wavelength (λ). For HABA/avidin samples at 500 nm, pH 7.0 extinction coefficient is 34,000 M^{-1} cm^{-1}.

b: Cell path length expressed in centimeters (cm). A 10 mm square cuvette has a path length of 1.0 cm.

ΔC: Concentration change of the HABA/avidin complex expressed in molarity (= mol/L = mmol/ml), equal to the concentration of incorporated biotin.

3.1.2 Target Protein Immobilization

1. The streptavidin dynabead vial is resuspended by rotating or vortexing. The beads are transferred to a new tube to the amount required based on their binding capacity. Typically, 100 μl of beads are required for all three rounds of panning. The beads are resuspended with blocking solution containing 0.02 % sodium azide to produce 1×10^9 beads/ml.
2. To wash the beads, an appropriate amount (~0.6 ml) of resuspended beads is transferred to a new tube. The tube is placed on a magnetic particle concentrator (MPC) for 1–2 min until all beads are grouped to the wall of the tube.

3. Using a micropipette, the supernatant is discarded while the tube is on an MPC. Beads are resuspended by vortexing after adding 1 ml of PBS. **Steps 2** and **3** are repeated three to six times.
4. After final washing, the beads are resuspended with 0.6 ml of PBS and divided into six 100 μl aliquots.
5. To immobilize target protein to streptavidin beads, an equal volume of biotinylated target protein is added to the tube containing beads. Prior to the addition of the target protein the supernatant of the beads should be removed on an MPC as quickly as possible (do not let the beads dry) (*see* **Note 3**).
6. The beads and target protein mixture are resuspended by tapping or inverting the tubes and incubated on a gently rocking platform at room temperature for 30–60 min.
7. The supernatant is discarded and 100 μl of blocking solution is added to each tube. Tubes are incubated on a gently rocking platform at room temperature for 30–60 min.
8. The beads are then washed five times with 500 μl of PBS before being subjected to the panning procedures. After final washing, the beads are resuspended with 100 μl of PBS.
9. Additional beads should be prepared for negative selections. For each panning condition, 500 μl of the bead stock is subjected to the above procedure except the incubation step with biotinylated target protein. In the last washing step, the beads are divided into five 100 μl aliquots.

3.1.3 Validation of Target Protein Immobilization

1. 50 μl of beads coated with target protein are aliquoted into each well of a v-bottom 96-well plate. Typically, two wells are used per each immobilization reaction: one is used to validate the immobilization and the other is used as a negative control without antibody.
2. The beads are washed by adding 100 μl of washing solution three times on a magnetic separator.
3. After final washing, 100 μl of primary antibody diluted in blocking solution is added to each well. For negative control beads, blocking solution without primary antibody is added. The plate is sealed with film and incubated on a gently rocking platform for 1 h at room temperature.
4. On a magnetic separator, supernatant in the plate is discarded. The beads are washed five times with 100 μl of washing solution.
5. Appropriate amount of HRP-conjugated secondary antibody diluted in blocking solution is added to each well. The plate is sealed and incubated on a gently rocking platform for 1 h at room temperature.

6. The beads are washed by adding 100 μl of washing solution five times using a magnetic separator.
7. During the washing, TMB substrate solution is equilibrated to room temperature. To validate target protein immobilization, 100 μl of TMB substrate solution is added to each well and stored in a dark place. Air bubbles are removed by gentle tapping.
8. When the color of the reaction mixture turns blue, 50 μl of 5 N H_2SO_4 solution is added to stop the reaction. Absorbance of reaction mixture is measured at 450 nm wavelength.

3.2 Selection of Target-Binding Phage Clones

This instruction assumes the use of Dyax's phage peptide library system (Dyax Corp.). Three filamentous phage libraries, TN8-IX (5×10^9 independent transformants), TN12-I (1.4×10^9 independent transformants), and Linear (2.3×10^9 independent transformants), are used to select for target-binding peptides. To efficiently identify target protein-binding peptides, it is recommended to eliminate the nonspecific binding phage population. Optimizing detailed conditions, such as protein concentration, immobilization on a solid phase, phage elution methods, and stringency of washing, is the route to success. In addition, various alternative approaches to elute phage can be combined for more effective selection [16–19]. In general, three rounds of panning are performed to obtain relevant peptide candidates and all conditions in the panning procedure should be determined with great circumspection, in consideration for the properties of target protein and the phage titer resulted from every round of panning. In this section, the procedures for first-round panning are focused.

3.2.1 Affinity Selection: The First-Round Panning

1. For each panning condition, the phage library is aliquoted from the library stock and diluted to give 400 μl of 10^{11} pfu/ml.
2. The 300 μl of diluted library stock is added to 100 μl of the beads prepared for negative selection. The resulting mixture is incubated for 10 min at room temperature with rotation.
3. The phage supernatant is collected using an MPC and is added to the second 100 μl of the beads prepared for negative selections for another negative selection step. Four more negative selection steps are performed.
4. The pre-adsorbed phage supernatant after the negative selection steps represents the depleted phage population. The depleted phage population is transferred to a new tube and target protein-coated beads were added. This mixture is incubated for 2 h at room temperature on a gently rocking platform, allowing specific phage to bind to the target protein. The beads are then collected using an MPC.

5. After the supernatant is discarded, the beads are washed about ten times with washing solution followed by twice with PBS.
6. The bound phages are eluted with 1 ml of elution solution for 10 min on a rotator. The pH of the phage-containing solution is immediately neutralized with 0.5 ml of neutralizing solution.
7. After elution, the titer is determined and eluted phages are amplified for the next round of panning.

3.2.2 Amplification

1. 20 ml of fresh *E. coli* (XL-1 Blue MRF') is cultured up to OD_{600} 0.5 in LB media containing 12.5 μg/ml tetracycline. Cells are centrifuged and resuspended in 1 ml of PBS. For each amplification step, fresh *E. coli* culture should be prepared.
2. 1 ml of phage eluates are added to a concentrated 1 ml *E. coli* culture. After being incubated at 37 °C for 15 min, 2 ml of NZCYM (2×) media containing 50 μg/ml of ampicillin is added to the mixture and incubated at 37 °C for 15 min. After incubation, the mixture is plated on a large NZCYM agar plate containing 50 μg/ml ampicillin and incubated overnight at 37 °C.
3. The large NZCYM agar plate is scraped off using 35 ml of LB media, and the agar plate is further rinsed with additional 35 ml of LB media. The resulting *E. coli*/phage mixture in LB media is centrifuged at 5,000 ×*g* for 15 min to pellet the *E. coli* and the resulting supernatant collected.
4. 12.5 ml of ice-cold PEG solution is added to 50 ml of phage solution and the tube is incubated on ice for 2 h.
5. After centrifugation at 10,000 ×*g*, 4 °C, for 30 min, the phage pellet is resuspended in 6 ml of the phage resuspension buffer. 1.5 ml of the PEG solution is added after centrifugation to remove any residual *E. coli* or its debris and the solution centrifuged again at 10,000 ×*g*, 4 °C, for 30 min.
6. The resulting phage pellet is then resuspended in 400 μl of PBS. This solution is subjected to a final centrifugation to get rid of any remaining *E. coli* debris. The resulting phage preparation is used as the input phage in the next round of panning.

3.2.3 Determining Phage Titer: Plaque Formation Assay

1. 10 ml of fresh *E. coli* culture (XL-1 Blue MRF′) is grown to OD_{600} 1.0 in LB media containing 12.5 μg/ml tetracycline.
2. Input phage, depleted phage, and output phage population are diluted to count titers based on estimates of expected phage numbers (*see* **Note 4**).
3. 300 μl of *E. coli* culture at **step 1** is aliquoted to 15 ml polypropylene tubes. The diluted phages are added to the *E. coli* culture and the infections are allowed to proceed for 15 min at room temperature. To verify the cells are not contaminated by

any of the phage population, one tube containing only *E. coli* culture without phage infection should be prepared.

4. After adding 3 ml of melted top agar (maintained below 50 °C and cooled just before hardening), the tube is vortexed gently and the mixture is quickly poured onto dried LB agar plate.
5. The plate is tilted around to spread top agar over plate evenly, stood on level area for at least 5 min to let top agar harden, and incubated overnight at 37 °C.
6. The resulting plaques that appear on the surface of the plate are counted and the phage titer in plaque-forming units (pfu) is calculated as follows:

$$\text{number of pfu per unit volume} = \frac{\text{number of plaques}}{\text{phage volume}\,(10\,\mu l)} \times \text{dilution fold.}$$

3.3 Phage Clone Analysis

The phage clones are analyzed by phage ELISA to select hits and DNA sequence analysis to select unique clones. Based on the combined results from these two assays, candidates are chosen for further use as bioactive peptides.

3.3.1 Phage ELISA

1. After the elution steps of the third panning, a small fraction of the phage eluate is plated out as in the plaque formation assay. Individual plaques are picked and placed into 96-well microtiter plates containing 100 μl of TE buffer in each well. These master plates are incubated at 4 °C overnight to allow phage to elute into the TE buffer.
2. A fresh culture of *E. coli* is grown to OD_{600} 0.5 in LB media containing 12.5 μg/ml tetracycline. A 30 μl aliquot of this culture is loaded into each well of a 96-well microtiter plate and 10 μl of eluted phage is added to each well and allowed to infect the bacteria for 15 min at room temperature. At least one well containing only *E. coli* cells without phage infection should be included.
3. After adding 130 μl of LB media containing 12.5 μg/ml of tetracycline and 50 μg/ml of ampicillin to each well, the plate is sealed with a plate-sealing film and incubated while shaking at 37 °C overnight. For control culture without phage, no antibiotics are added.
4. 100 μl of target protein (1 mg/ml in PBS) is added to each well of a 96-well Maxisorp plate and allowed to incubate overnight at 4 °C. As a control, 100 μl of pure streptavidin (2 mg/ml in PBS) is added to each well of a separate Maxisorp plate.

5. On the following day, the target protein or pure streptavidin is discarded. Nonspecific protein-binding sites on the plate are blocked by the addition of 300 μl of blocking solution per well and the plate is incubated at 4 °C overnight (alternatively, 1 h at room temperature). After discarding the blocking solution, the plate is washed three times with washing solution.
6. 50 μl of binding solution is added to plates and 50 μl of overnight cultures from each well in the 96-well microtiter plate is transferred to the corresponding wells of both the target protein-coated plates as well as the control, streptavidin-coated, plates.
7. The two kinds of plates are incubated for 1 h at room temperature. After incubation, the plates are washed three times with washing solution.
8. HRP-conjugated anti-M13 antibody is diluted to 1:7,500 with antibody dilution solution. 100 μl of diluted antibody solution is added to the Maxisorp plates and allowed to incubate for 1 h at room temperature. The plates are then washed five times with washing solution.
9. 100 μl of TMB substrate is added and the plates are placed in dark place for the color reaction to occur. When the mixture turns blue, the reaction is stopped with 50 μl of 5 N H_2SO_4. Any air bubbles are removed by gentle tapping. The resulting absorbance of the reaction mixture is measured at 450 nm.

3.3.2 DNA Sequence Analysis

1. Plaque eluates, phage culture supernatant, or precipitated phage are all available as PCR template (*see* **Note 5**).
2. For each phage clone, the specified PCR primer pair is used to amplify about 500 nucleotide fragment:

   Primer 1 (5′-CGGCGCAACTATCGGTATCAAGCTG-3′).

   Primer 2 (5′-CATGTACCGTAACACTGAGTTTCGTC-3′).
3. The following program can be used for amplification of candidate sequences: 94 °C for 5 min [94 °C for 30 s, 55 °C for 30 s, 72 °C for 45 s] × 35 cycles; 72 °C for 7 min; cool to 4 °C. The PCR products from each reactions are purified using the PCR purification kit.
4. The purified PCR products are sequenced and analyzed using standard methodologies.
5. From the sequences of selected candidates, consensus sequences and core amino acid sequences can be determined.

3.4 Fc-Peptide Fusion Proteins

For the development of peptide therapeutics, it is useful to express the selected peptide as an Fc fusion protein. This instruction assumes the use of a mammalian expression vector and HEK 293 cell line.

3.4.1 Expression of Fc-Peptide Fusion Protein

1. A linker sequence such as five glycine residues, eight glycine residues, or one lysine residue can be linked to oligonucleotide sequence encoding the peptides of interest.
2. The linked oligonucleotide sequences are ligated into the vector containing the human IgG_1(Fc) gene, which has been previously digested with corresponding restriction endonucleases (*see* **Note 6**).
3. The resulting ligation mixture is transformed into *E. coli* cells using standard procedures. A correct clone is used as a recombinant vector source to express Fc-peptide fusion protein described below.
4. One day before transfection, $0.5–2 \times 10^5$/ml of healthy proliferating HEK 293 cells that have been regularly passaged are plated in tissue culture plate with 10 ml of growth medium. Cells should be 90–95 % confluent at the time of transfection.
5. For each transfection sample, 20–25 μg of sterile plasmid vector is diluted in 1.5 ml of reduced serum media by gently mixing and the appropriate amount of lipofectamine is also diluted gently in 1.5 ml of reduced serum media before use (*see* **Note 7**).
6. After a 5-min incubation, the diluted plasmid DNA is combined with diluted lipofectamine and mixed gently. The mixture is incubated for 20 min at room temperature (solution may appear cloudy).
7. For each transfection the lipofectamine/DNA mixture (3 ml) is added to a plate containing cells and medium. After the plate is mixed gently by rocking back and forth, the cells are incubated at 37 °C in a CO_2 incubator for 18–48 h prior to testing for transgene expression. The growth media may be changed after 4–6 h.
8. For stable cell lines, the cells are passaged at a 1:10 (or higher) dilution into fresh growth media 24 h after transfection. Selective medium is added, if desired, the following day.

3.4.2 Purification of Fc-Peptide Fusion Proteins

1. The positive clones resulting from the transfection are further screened under increasing selective pressures. One cell line with the highest expression level of Fc-peptide fusion protein is selected and gradually adapted to serum-free media.
2. The Fc-fusion molecules contained in the supernatants collected from the serum-free media are isolated at first by protein A affinity chromatography according to the manufacturer's instruction. The eluted Fc fusion protein is adjusted to a pH that ranges from 5 to 7.0 using Tris–HCl, pH 8.0.

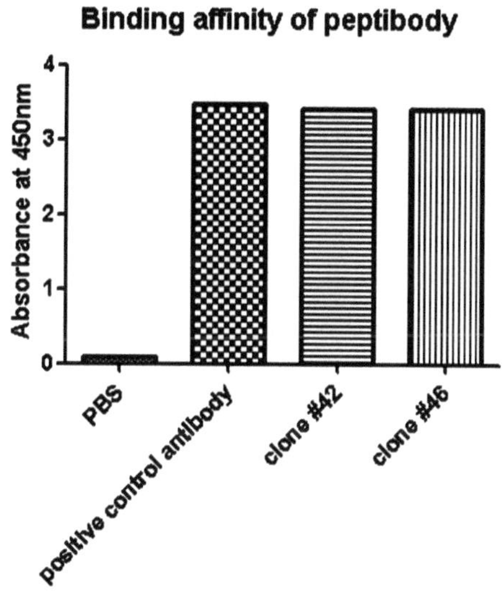

Fig. 1 Affinity ELISA of peptibody clones. Target protein is immobilized to immunoplate and incubated with peptibody and control antibody. PBS is used for negative control

3. If necessary, the Fc fusion protein is further purified with other chromatography methods according to the manufacturer's instruction.

3.5 Molecular Assays to Evaluate Fc-Peptide Fusion Proteins

Affinity ELISA and BIAcore assays are used to assess binding affinity and kinetic parameters of the Fc-peptide fusion proteins.

3.5.1 Affinity ELISA (See Fig. 1)

1. For initial screening of candidate Fc-peptide fusion proteins, purified native target proteins are used.
2. 100 μl of target protein (1 mg/ml in PBS) is added to each well of a 96-well Maxisorp plate and allowed to incubate at room temperature for 2 h. The target protein is then discarded and the plates blocked with 250 μl of blocking solution, followed by an incubation at room temperature for 2 h.
3. After the blocking solution is discarded, 100 μl of each Fc-peptide fusion protein is added to each well (starting at a concentration of 40 nM and then serially diluting fourfold in binding solution). The plates are then incubated overnight at room temperature.
4. The plates are washed with washing solution four times. Add 100 μl of goat anti-human IgG(Fc)-HRP previously diluted 1:5,000 in binding solution to each well. The plates are incubated for 1 h at room temperature.
5. The plates are then washed five times in washing solution, after which 100 μl per well of TMB substrate is added. The plates

are incubated for 5–15 min in the dark until a blue color develops. When the mixture turns blue, the reaction is stopped with 50 μl of 5 N H_2SO_4 solution. Absorbance of the reaction mixture is measured at 450 nm.

3.5.2 Affinity BIAcore

1. This instruction assumes the use of BIAcore X100 system and detailed procedures should be performed according to the manufacturer's instructions.
2. Following the standard procedures described in the human antibody capture kit, a monoclonal mouse anti-human IgG (Fc) antibody is immobilized on CM5 sensor chip.
3. Each candidate Fc-peptide fusion protein is injected and captured by the monoclonal mouse anti-human IgG (Fc) antibody immobilized on the CM5 sensor chip.
4. Various concentrations (0–100 nM) of target protein are then injected to allow measurement of the association reaction and of the dissociation which occurs once injection of the target protein ends and the injection of running buffer begins.
5. Affinity (K_D) is calculated from the rate constants (K_d/K_a) by BIAcore software.
6. Lower affinity indicates greater affinity of the Fc-peptide fusion product for target protein.

4 Notes

1. Target protein is transferred into a 10K MWCO slide-A-Lyzer and then the cassette is put into 2 l of PBS (pH 7.2) at 4 °C by gently stirring overnight. The PBS buffer is changed once after 2 h of dialysis.
2. The HABA/avidin solution is prepared by adding 10 mg of avidin and 600 μl of 10 mM HABA to 19.4 ml of PBS. This solution is stable for 2 weeks at 4 °C.
3. The concentration of target protein immobilized on the beads can be altered to eliminate nonspecific phage.
   (a) For the low-stringency first-round selection, the target protein is immobilized on the beads at a concentration of 4 μg (or higher) of the protein per 100 μl of the bead stock from the manufacturer.
   (b) For a higher stringency selection in the second round, 2 μg (or lower) of target protein is immobilized on 100 μl of the beads. The coating concentration is then reduced to 1 μg (or lower) of target protein per equal volume of beads for the third round of selection.
4. The supernatant resulting from a fresh phage infection may have a titer ranging from 10^8 to 10^9 pfu/ml, and a titer of a purified

phage can be expected to be between 10^9 and 10^{11} pfu/ml. Phage dilutions should be made to cover these pfu ranges.

5. Phage culture supernatant and precipitated phage should be diluted to 1:100 prior to use.
6. The peptides selected as potentially bioactive can be used to construct fusion proteins in which either a monomer of each peptide or a tandem dimer of each peptide (with a linker between the monomer units) is fused in-frame to DNA encoding a linker followed by the Fc region of human IgG_1. For optimal expression in mammalian cells and extracellular secretion of the Fc-peptide fusion protein, appropriate sequences encoding Kozak consensus and signal peptide are also required.
7. To obtain the highest transfection efficiency and low cytotoxicity, transfection conditions should be optimized by varying cell density as well as DNA and lipofectamine concentrations; make sure that cells are greater than 90 % confluent and vary in DNA (μg):lipofectamine ratios from 1:0.5 to 1:5.

Acknowledgement

This work was supported by Samsung Electronics Co., Ltd.

References

1. Scott JK, Craig L (1994) Random peptide libraries. Curr Opin Biotechnol 5:40–48
2. Houghten RA (1994) Combinatorial libraries. Finding the needle in the haystack. Curr Biol 4:564–567
3. Smith GP, Petrenko VA (1997) Phage display. Chem Rev 97:391–410
4. Cortese R, Monaci P, Nicosia A, Luzzago A, Felici F, Galfré G et al (1995) Identification of biologically active peptides using random libraries displayed on phage. Curr Opin Biotechnol 6:73–80
5. Lowman HB (1997) Bacteriophage display and discovery of peptide leads for drug development. Annu Rev Biophys Biomol Struct 26: 401–424
6. Markland W, Roberts BL, Saxena MJ, Guterman SK, Ladner RC (1991) Design, construction and function of a multicopy display vector using fusions to the major coat protein of bacteriophage M13. Gene 109:13–19
7. Li Z, Zhang J, Zhao R, Xu Y, Gu J (2005) Preparation of peptide-targeted phagemid particles using a protein III-modified helper phage. Biotechniques 39:493–497
8. Lien S, Lowman HB (2003) Therapeutic peptides. Trends Biotechnol 21:556–562
9. Veronese FM, Mero A (2008) The impact of PEGylation on biological therapies. BioDrugs 22:315–329
10. Takahashi-Nishioka T, Yokogawa K, Tomatsu S, Nomura M, Kobayashi S, Miyamoto K (2008) Targeted drug delivery to bone: pharmacokinetic and pharmacological properties of acidic oligopeptide-tagged drugs. Curr Drug Discov Technol 5:39–48
11. Houghten RA (1985) General method for the rapid solid-phase synthesis of large numbers of peptides: specificity of antigen–antibody interaction at the level of individual amino acids. Proc Natl Acad Sci U S A 82: 5131–5135
12. Ladner RC, Sato AK, Gorzelany J, de Souza M (2004) Phage display-derived peptides as therapeutic alternatives to antibodies. Drug Discov Today 9:525–529
13. Krumpe L, Mori T (2006) The use of phage-displayed peptide libraries to develop tumor-targeting drugs. Int J Peptide Res Ther 12: 79–91

14. Sergeeva A, Kolonin MG, Molldrem JJ, Pasqualini R, Arap W (2006) Display technologies: application for the discovery of drug and gene delivery agents. Adv Drug Deliv Rev 58: 1622–1654
15. Oliner J, Min H, Leal J, Yu D, Rao S, You E et al (2004) Suppression of angiogenesis and tumor growth by selective inhibition of angiopoietin-2. Cancer Cell 6:507–516
16. Turková J (1999) Oriented immobilization of biologically active proteins as a tool for revealing protein interactions and function. J Chromatogr B Biomed Sci Appl 722:11–31
17. Hetian L, Ping A, Shumei S, Xiaoying L, Luowen H, Jian W et al (2002) A novel peptide isolated from a phage display library inhibits tumor growth and metastasis by blocking the binding of vascular endothelial growth factor to its kinase domain receptor. J Biol Chem 277:43137–43142
18. Kelley BD, Booth J, Tannatt M, Wu QL, Ladner R, Yu J et al (2004) Isolation of a peptide ligand for affinity purification of factor VIII using phage display. J Chromatogr A 1038:121–130
19. Hills R, Mazzarella R, Fok K, Liu M, Nemirovskiy O, Leone J et al (2007) Identification of an ADAMTS-4 cleavage motif using phage display leads to the development of fluorogenic peptide substrates and reveals Matrilin-3 as a novel substrate. J Biol Chem 282:11101–11109

Chapter 4

T7 Lytic Phage-Displayed Peptide Libraries: Construction and Diversity Characterization

Lauren R.H. Krumpe and Toshiyuki Mori

Abstract

In this chapter, we describe the construction of T7 bacteriophage (phage)-displayed peptide libraries and the diversity analyses of random amino acid sequences obtained from the libraries. We used commercially available reagents, Novagen's T7Select system, to construct the libraries. Using a combination of biotinylated extension primer and streptavidin-coupled magnetic beads, we were able to prepare library DNA without applying gel purification, resulting in extremely high ligation efficiencies. Further, we describe the use of bioinformatics tools to characterize library diversity. Amino acid frequency and positional amino acid diversity and hydropathy are estimated using the REceptor LIgand Contacts website http://relic.bio.anl.gov. Peptide net charge analysis and peptide hydropathy analysis are conducted using the Genetics Computer Group Wisconsin Package computational tools. A comprehensive collection of the estimated number of recombinants and titers of T7 phage-displayed peptide libraries constructed in our lab is included.

Key words T7, Lytic, M13, Filamentous, Phage, Display, Peptide, Library, Diversity, Construction

1 Introduction

Filamentous phages (M13, fl, and fd, Fig. 1a) have been frequently utilized in the construction of phage-displayed peptide libraries [1, 2], and these libraries have successfully yielded binding ligands for various targets [3, 4]. However, earlier studies demonstrated specific and positional amino acid biases in populations of peptides randomly selected from M13 libraries, biases likely a result of the *Escherichia coli* (*E. coli*) membrane-dependent processes of filamentous phage morphogenesis (Fig. 2a) [5]. The popularity of filamentous phage-displayed peptide library usage may be, in part, due to the initial development of the technology with filamentous bacteriophage and/or the commercial availability of filamentous phage-displayed peptide libraries.

Other phage display systems have been developed using lytic phages such as T7, T4, and λ [6] (Fig. 1b), and these types of

Andrew E. Nixon (ed.), *Therapeutic Peptides: Methods and Protocols*, Methods in Molecular Biology, vol. 1088,
DOI 10.1007/978-1-62703-673-3_4, © Springer Science+Business Media, LLC 2014

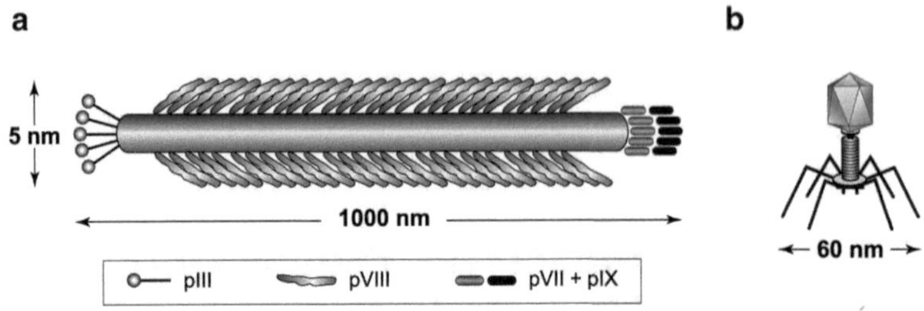

Fig. 1 Morphology of filamentous phage and T7 lytic phage. Filamentous phage (**a**) consists of an approximately 1 μm protein-encapsulated DNA genome, and is approximately 5 nm in width. Filamentous phage proteins pIII and pVIII, and less commonly pVII and pIX, have been utilized for peptide and/or protein display. T7 lytic phage (**b**) is approximately 60 nm in width. The icosahedral capsid of T7 contains the viral DNA genome and consists of proteins 10A and 10B. The gene for protein 10B has been manipulated for display of peptides/proteins. For further discussion of bacteriophage morphology *see* ref. 16

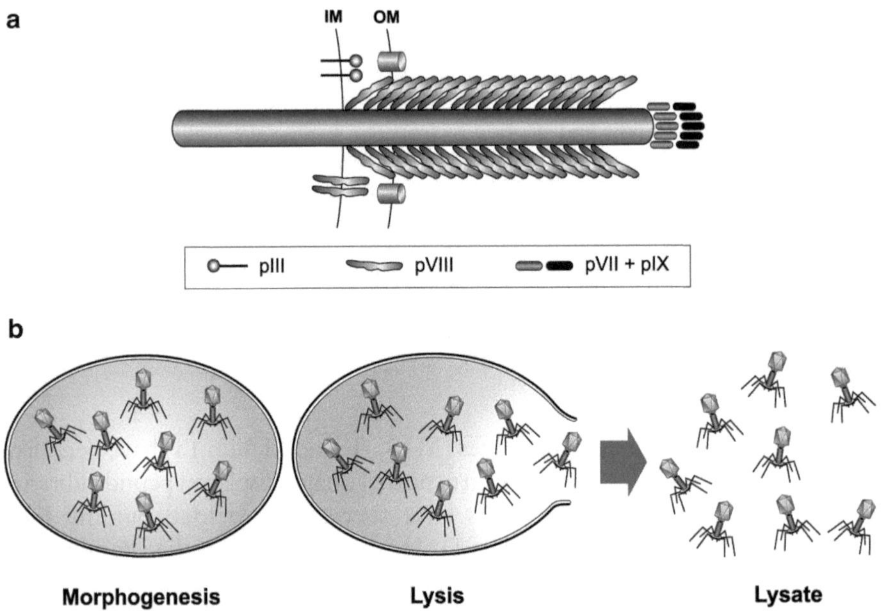

Fig. 2 Morphogenesis of filamentous and lytic phage. Filamentous phage (**a**) assembles at the *E. coli* inner membrane (IM) and is secreted through the outer membrane (OM). Filamentous phage-displayed peptides/proteins must be compatible with the secretory processes, and therefore filamentous phage morphogenesis can impart bias on the repertoire of displayed amino acid sequences. T7 lytic phage (**b**) assembles independently within the *E. coli* cytoplasm. Mature T7 virions are released by host cell lysis, and therefore are not subjected to bias imparted by morphogenesis. For further discussion of bacteriophage morphogenesis *see* ref. 16

libraries have also been used to successfully identify binding ligands [1, 5, 7, 8]. Peptides displayed on lytic phage do not have to be compatible with host cell synthesis and secretion complexes because lytic phage morphogenesis takes place within the *E. coli* cytoplasm, and mature phage virions are released by cell lysis (Fig. 2b). Our research revealed that libraries produced with lytic phage could surpass the diversity of filamentous phage-displayed peptide libraries [5]. Using a bioinformatics-assisted computational approach, diversity characteristics of random amino acid sequences obtained from T7 phage-displayed peptide libraries were compared with diversity characteristics of populations obtained from M13 libraries. Peptide libraries constructed with the T7 system have fewer amino acid biases, increased peptide diversity, and more normal distributions of peptide net charge and hydropathy than the M13 libraries. The superior diversity of T7-displayed libraries may afford the opportunity of successfully isolating peptides that bind to molecular targets of interest.

T7 phage-displayed peptide libraries can be easily constructed due to the commercial availability of necessary reagents and protocols (T7Select System, Novagen [9]). One simply needs to acquire the desired T7Select vector(s), *E. coli* host strains, and molecular biology reagents. However, special attention must be applied to the design and synthesis of the DNA that will be used to construct the phage library. Available protocols do not include procedures for the design and construction of the library insert DNA. We developed, and herein disclose, a system of producing high-quality library insert DNA. Library DNA is prepared using streptavidin-coupled magnetic beads for immobilization and enzymatic digestion, and subsequently purified by standard precipitation techniques. The purified DNA is then ligated into T7Select vectors, packaged into phage particles, and amplified in *E. coli*. Once amplified, the library is ready for diversity characterization and biopanning. Biopanning procedures are not included in this chapter but can be found in the T7Select System Manual [10]. In addition to phage-displayed peptide library construction, this chapter also outlines how to obtain populations of phage-displayed amino acid sequences, and ultimately how to use bioinformatics tools to characterize the diversity of the libraries based on a small subset of amino acid sequences.

2 Materials

2.1 T7 Phage-Displayed Peptide Library Construction

2.1.1 T7 Phage Vector and Library Oligonucleotide DNA

1. T7Select Cloning Kit (*see* Subheading 3.1.1): T7Select *Eco*RI/*Hin*dIII Vector Arms, ligation control insert DNA, T7Select Packaging Extracts, packaging control DNA, *E. coli* host strain glycerol stocks, and T7SelectUP and T7SelectDOWN sequencing primers (Novagen, Gibbstown, NJ).
2. Synthetic randomized library oligonucleotide DNA (IDT, Coralville, IA; Glen Research, Sterling, VA).
3. Synthetic biotinylated extension primer (IDT).

2.1.2 Annealing, Extension, Enzymatic Digestion, and Purification of Library Oligonucleotide DNA

1. 10 mM Tris–HCl, 1 mM ethylenediamine tetraacetic acid (EDTA) pH 8.0 (TE Buffer, Quality Biological, Gaithersburg, MD).
2. TE buffer, 100 mM NaCl, pH 8.0 (*see* **Note 1**).
3. 10 mM deoxyribonucleotide triphosphates (dNTPs, New England BioLabs (NEB), Ipswich, MA).
4. Klenow fragment DNA polymerase 1 (NEB).
5. EcoPol buffer, 10×: 100 mM Tris–HCl, 50 mM $MgCl_2$, 75 mM dithiothreitol (DTT), pH 7.5.
6. Dynabeads M-280 streptavidin (SA)-coupled magnetic beads (DYNAL, Invitrogen, Carlsbad, CA).
7. B + W buffer, 2×: 10 mM Tris–HCl, 1 mM EDTA, 2 M NaCl, pH 7.5.
8. Restriction endonucleases *Hin*dIII, *Eco*RI, NEBuffer2, 100× bovine serum albumin (NEB).

2.1.3 Tris–Borate–EDTA Polyacrylamide Gel Electrophoresis (TBE-PAGE) Analysis of Digested and Purified Library DNA

1. TBE running buffer: 89 mM Tris base, 89 mM boric acid, 2 mM EDTA, pH 8.3 (Invitrogen).
2. Novex High-Density TBE Sample Buffer, 5×: 18 mM Tris base, 18 mM Boric acid, 0.4 mM EDTA, 3 % Ficoll Type 400, 0.02 % Bromophenol Blue, 0.02 % Xylene Cyanol (Invitrogen).
3. Novex Pre-Cast 20 % TBE polyacrylamide gels (Invitrogen).

2.1.4 Ligation of Library Insert DNA into T7Select EcoRI/HindIII Vector Arms

1. DNA ligation kit: Sterile water, 10× ligase buffer, 100 mM DTT, 10 mM ATP, T4 DNA ligase, ligase dilution buffer (Novagen).
2. T7Select *Eco*RI/*Hin*dIII Vector Arms (T7Select Cloning Kit, Novagen).
3. *Eco*RI/*Hin*dIII-digested and purified library oligonucleotide insert DNA.

2.2 In Vitro Packaging, Phage Titering by Plaque Assay, and Library Amplification

1. T7Select Packaging Extracts and packaging control DNA (Novagen).
2. T7Select *Eco*RI/*Hin*dIII Vector Arms and library insert DNA ligation reaction.
3. T7Select system-appropriate *E. coli* host strain, see T7Select System Manual.
4. Luria Bertani (LB) broth, per liter: 10 g Bacto Tryptone (Difco, Becton Dickinson, Franklin Lakes, NJ), 5 g yeast extract, 10 g NaCl, sterilized by autoclaving at 120 °C for 20 min.

5. Top agarose: LB formulation with 0.6 g agarose per 100 ml, sterilized by autoclaving.
6. LB agar plates: LB formulation with 15 g/l agar, sterilized by autoclaving.
7. T7Select vector and *E. coli* host strain appropriate antibiotics, see T7Select System Manual. Stock solutions of antibiotics should be frozen in aliquots at −20 °C.
8. 500 mM stock solution of isopropyl β-D-1-thiogalactopyranoside (IPTG), frozen in aliquots at −20 °C.
9. M9 Salts, 20×, per liter: 20 g NH_4Cl, 60 g KH_2PO_4, 120 g $Na_2HPO_4 \cdot 7H_2O$, sterilized by autoclaving.
10. T7 phage growth medium: 100 ml LB with the following sterile components added, 5 ml 20× M9 salts, 2 ml 20 % glucose, 0.1 ml 1 M $MgSO_4$.
11. Chloroform (J.T. Baker, Phillipsburg, NJ).
12. 5 M NaCl stock solution, sterilized by autoclaving.
13. 80 % glycerol solution, sterilized by autoclaving.

2.3 Phage-Displayed Peptide Library Diversity Characterization

2.3.1 Phage Clone Amplification and DNA Sequencing

1. *See* Subheading 2.2, **items 3–10**.
2. Sterilized inoculation needles or sterilized toothpicks.
3. Megatiter plates: Sterilized 96×2.2 ml wells, and gas-permeable plate sealing tape (Continental Lab Products (CLP), SanDiego, CA).
4. Minitube racks: Sterilized 96×1.1 ml minitube strips and caps (CLP).
5. Platinum PCR SuperMix High Fidelity (Invitrogen).
6. T7SelectUP and T7SelectDOWN sequencing primers (Novagen).
7. Agarose.
8. Ethidium bromide solution, 10 mg/ml (Sigma).
9. 50× TAE buffer solution: Tris base 242.28 g/l 2 M, disodium EDTA·$2H_2O$ 18.61 g/l 0.05 M, glacial acetic acid 57.1 ml/l (Quality Biological).

2.3.2 Diversity Analysis of Phage-Displayed Amino Acid Sequences

1. List of displayed amino acid sequences in text (.txt) format.
2. REceptor LIgand Contacts (RELIC) bioinformatics web server: http://relic.bio.anl.gov [11].
3. GCG Wisconsin Package computational tools. Version 10.3 Accelrys was utilized.

3 Methods

3.1 T7 Phage-Displayed Peptide Library Construction Using Streptavidin-Coupled Magnetic Beads

3.1.1 T7 Phage Vector and Library Oligonucleotide DNA

1. T7 phage vector DNA can be manipulated to vary the number of displayed peptides. Choose T7Select 415-1b for 415 peptide copies per phage particle (high-copy number). Choose T7Select 10-3b for 5–15 peptide copies per phage particle (mid-copy number). Choose T7Select 1-1 or 2 for 0.1–1 peptide copies per phage particle (low-copy number). As the number of peptide copies per phage particle increases, the size limit of the displayed molecule decreases. In addition, vector selection dictates which strain of *E. coli* is used for phage propagation. For more details, see the T7Select System Manual.
2. Obtain synthetic randomized library oligonucleotide DNA and the biotinylated extension primer (*see* **Note 2**). For an example of a peptide library DNA construct, *see* Fig. 3.

Fig. 3 Construction of a random 12-mer T7 phage-displayed peptide library. Library oligonucleotide DNA is synthesized as the antisense strand, and the biotinylated extension primer is synthesized as the sense strand. "N" represents a hand-mixed equimolar ratio of guanine, adenine, cytosine, and thymine nucleotides, and "M" is restricted to an equimolar mixture of adenine or cytosine nucleotides. *Eco*RI and *Hin*dIII restriction endonuclease recognition sites are shown in *boxes*. The diagram shows the three main processes of library construction: (*1*) extension of the DNA duplex using Klenow DNA polymerase and dNTPS, (*2*) enzymatic digestion and purification of library insert DNA using streptavidin (SA)-coated magnetic beads, and (*3*) ligation of library insert DNA into the T7Select vector. The final ligated vector sequence is shown, along with the amino acid translation of the construct, Gly-Asp-Pro-Asn-Ser-Gly-Gly-Gly-(NNK; a random amino acid)$_{12}$, STOP

3.1.2 Annealing of Extension Primer and Randomized Library DNA

1. Centrifuge the lyophilized oligonucleotides in a tabletop microcentrifuge at 16,000 × *g* for 1 min at room temperature.
2. Dissolve the oligonucleotides in TE buffer at a concentration of approximately 1 μg DNA per μl TE buffer. Incubate the solutions with orbital rotation at room temperature for 15 min to ensure dissolution.
3. Measure the concentration of the dissolved oligonucleotides by ultraviolet light absorbance spectrophotometry.
4. In a screw-top microcentrifuge tube, combine 4 μg each of the biotinylated extension primer and the library oligonucleotide DNA. Bring the final volume of the solution to 50 μl with TE buffer, 100 mM NaCl, pH 8.0.
5. Heat 1 l of water to boiling in a 1 l beaker. Add the microcentrifuge tube containing the library oligonucleotide DNA and biotinylated extension primer. Remove the beaker from the heat source and allow the water to cool to 35 °C or less.

3.1.3 Extension of the Annealed Duplex DNA

1. Centrifuge the annealing reaction in a tabletop microcentrifuge at 16,000 × *g* for 1 min at room temperature.
2. In a new screw-top microcentrifuge tube, combine the following reagents, mix gently by pipetting, and incubate for 10 min in a 37 °C circulating water bath:

   116 μl water.

   20 μl 10× EcoPol buffer.

   50 μl annealing reaction (from Subheading 3.1.2).

   8 μl 10 mM dNTPs.

   6 μl Klenow fragment DNA polymerase I (5 U/μl)* (*see* **Note 3**).
3. Inactivate the enzyme by incubation for 1 h in a 65 °C circulating water bath.
4. Remove 5 μl of the extension reaction to a new microcentrifuge tube for later analysis.
5. Store the extension reactions at −20 °C or incubate on ice until proceeding with digestion.

3.1.4 Restriction Endonuclease Digestion of Extended DNA Using SA-Coupled Magnetic Beads

The following procedure is modified from the DYNAL Dynabeads M-280 SA product insert [12].

1. Resuspend Dynabeads M-280 SA-coupled magnetic beads, aliquot 1 ml to microcentrifuge tube, apply magnet, and remove supernatant.
2. Wash beads with 1 ml of 1× B + W buffer, vortex, apply magnet, and remove supernatant.
3. Add the following components to the washed beads:

314.7 μl water.

490.3 μl 2× B+W buffer.

195.0 μl extended duplex DNA.

(*See* **Note 4**.)

4. Incubate the binding reaction on a rotating wheel for 30 min at room temperature.
5. Apply magnet, remove, and save supernatant.
6. Wash beads with 1 ml 1× NEBuffer2, 1× BSA, vortex, apply magnet, and remove supernatant.
7. Resuspend beads in 1 ml 1× NEBuffer2, 1× BSA, 800 U/ml *Hin*dIII* solution. It is critical to first digest with the restriction enzyme farthest from the immobilized end of the DNA duplex. Incubate at 37 °C for 3 h with gentle rotation (*see* **Note 5**). Apply magnet, remove, and save supernatant.
8. Wash beads with 1 ml 1× NEBuffer2, 1× BSA, vortex, apply magnet, and remove supernatant.
9. Resuspend beads in 1 ml 1× NEBuffer2, 1× BSA, 800 U/ml *Eco*RI* solution. Incubate at 37 °C for 3 h with gentle rotation (*see* **Note 5**). Apply magnet, remove, and save supernatant. This supernatant is the doubly digested library DNA to be used for cloning into the T7Select *Eco*RI/*Hin*dIII vector arms.
10. Purify and concentrate the DNA from the supernatants generated in Subheading 3.1.4 by phenol extraction and ethanol precipitation [13] (*see* **Note 6**).
11. Estimate the concentration and purity of the DNA populations generated in Subheading 3.1.4 by ultraviolet light absorbance spectrophotometry (*see* **Note 7**).
12. Store the digested DNA at −20 °C.

3.1.5 TBE-PAGE Analysis of Digested and Purified Library DNA

Prior to proceeding with library DNA ligation optimization, confirm restriction digestion by TBE-PAGE analysis.

1. Separately combine up to 200 ng of the following DNA populations with Novex High-Density TBE Sample Buffer: supernatants generated in Subheading 3.1.4, **steps 5** and **7**, biotinylated extension primer, library oligonucleotide DNA, undigested extended DNA duplex (Subheading 3.1.3, **step 4**), and dual-restriction enzyme-digested library DNA (Subheading 3.1.4, **step 9**).
2. Assemble the Novex electrophoresis apparatus, add 20 % TBE gel, and cover with TBE running buffer according to the manufacturer's instructions [14].
3. Load the DNA samples into the gel and include several quantitative DNA molecular weight ladders.

4. Run the gel at 200 V, constant current, for approximately 1 h.
5. Visualize the DNA fragments by staining the gel with ethidium bromide (0.5 μg/ml) in TBE buffer.

3.1.6 Ligation of Library DNA into T7Select EcoRI/HindIII Vector Arms

Follow the section entitled "Ligation of Inserts and Vector Arms" within the T7Select System Manual, with the following exceptions:

1. Ensure proper storage of T7Select Cloning Kit reagents. T7Select Vector Arms are temperature sensitive and should be stored at 4 °C.
2. The suggested starting ratios, per 5 μl ligation, are 0.5 μg cut vector (0.02 pmol) and 2:1 and 4:1 molar excess of cut duplex. Dilute the insert DNA with TE buffer as necessary (*see* **Note 8**).
3. We recommend performing the ligation-positive control and the negative ligation control concurrently with each ligation reaction.
4. The reaction is temperature sensitive. Ensure a constant incubation temperature of 16 °C using a refrigerated water bath.

3.2 In Vitro Packaging, Phage Titering by Plaque Assay, and Library Amplification

3.2.1 Small-Scale In Vitro Packaging and Ligation and Packaging Efficiency Optimization

The following protocol was adapted from the "In Vitro Packaging" section of the T7Select System Manual.

1. On ice, aliquot 5 μl of T7Select Packaging Extract to five prechilled, screw-top microcentrifuge tubes.
2. Add 1 μl of the four ligation reactions described in Subheading 3.1.6 to tubes 1 through 4 containing 5 μl packaging extract. To the fifth tube of packaging extract, for the packaging positive control, add 0.5 μl (0.2 μg) T7Select Packaging Control DNA. To mix the reactions, only stir gently with the pipet tip. Do not vortex or pipet up and down.
3. Incubate the reactions for 2 h at room temperature.
4. Stop the reaction by adding 294 μl sterile LB broth. This will bring the final volume to 300 μl.
5. Proceed immediately to plaque assay as described in the T7Select System Manual, or add 20 μl chloroform to the packaging reactions, stir with pipet tip, and store at 4 °C (*see* **Note 9**).
6. Count the number of plaques (plaque-forming units, pfu) and calculate the phage titer (pfu/ml) and packaging efficiency (pfu/μg vector arms) as described in the T7Select System Manual (*see* **Note 10**).

3.2.2 Large-Scale In Vitro Packaging

For large-scale library construction, repeat the ligation reactions using the molar ratio conditions that resulted in the highest packaging efficiencies, as described in Subheadings 3.1.6 and 3.2.1, and in the T7Select System Manual. We typically performed five to ten 5 μl individual ligation reactions for each library produced, which were subsequently packaged using 25 μl T7Select Packaging Extract as described in the T7Select System Manual.

1. After the 2-h packaging reaction is complete, stop the reaction with sterile LB broth, which will bring the final volume of each reaction to 300 μl.
2. Pool the packaged library DNA into one sterile tube and remove 1.5 μl to a sterile microcentrifuge for titering by plaque assay. Use the titer of the packaging reaction to estimate the number of recombinants generated in the library (*see* **Note 11**). *See* Table 1 for values of the number of recombinants generated per library constructed in our laboratory.
3. Proceed to amplification in the appropriate *E. coli* host strain, or store as directed in Subheading 3.2.1.

3.2.3 Library Amplification

T7 phage-displayed libraries must be amplified once in *E. coli* prior to biopanning for surface display of random peptides. It is extremely important to provide proper aeration of bacterial cultures when growing T7 phage. Use baffled Erlenmeyer flasks and fill to 20 % or less of stated flask capacity.

1. Infect an *E. coli* culture with the remaining packaged library DNA (Subheading 3.2.2) as described in the section entitled "Liquid Lysate Amplification (500 ml scale)" within the T7Select System Manual.
2. Check the culture for lysis at 30-min intervals. When lysis has occurred, the culture will appear much clearer than before it was infected, and bacterial debris can be observed. It may be helpful to incubate a non-infected culture of *E. coli* alongside the infected culture for comparison of turbidity.
3. After lysis occurs, add NaCl to a final concentration of 0.5 M.
4. Centrifuge the lysate–NaCl solution at 8,000 × *g* at 4 °C for 20 min to pellet the bacterial debris.
5. Prepare the library for cryopreservation by transferring the clarified lysate to a sterile container, adding 0.1 volume 80 % glycerol, and mixing thoroughly. Save 1.5 μl for titering by plaque assay (*see* **Note 12**). *See* Table 1 for the titers of libraries constructed in our laboratory.
6. Prepare 1 ml aliquots of the library in sterile screw-top microcentrifuge tubes.
7. Freeze the libraries at −80 °C.

3.3 T7 Phage-Displayed Peptide Library Diversity Characterization

The statistical methods utilized in this section are applicable to amino acid sequence sample sizes greater than 50 sequences [15]. The following procedures are tailored to producing a sample size of approximately 100 amino acid sequences.

Table 1
T7 phage-displayed peptide libraries

Library	Type	Recombinants (pfu)	Titer (pfu/mL)
CX_4C	Constrained	1.2×10^9	4.7×10^{10}
CX_5C	Constrained	1.2×10^9	3.4×10^{10}
CX_6C	Constrained	1.8×10^9	2.4×10^{11}
CX_7C	Constrained	1.4×10^9	1.1×10^{11}
CX_8C	Constrained	1.3×10^9	3.9×10^{11}
CX_9C	Constrained	1.5×10^9	1.1×10^{11}
$CX_{10}C$	Constrained	1.8×10^9	4.3×10^{11}
$X_3CX_4CX_3$	Constrained + tails	1.4×10^9	4.3×10^{11}
$X_3CX_5CX_3$	Constrained + tails	1.7×10^9	1.4×10^{11}
$X_3CX_6CX_3$	Constrained + tails	1.5×10^9	1.8×10^{11}
$X_3CX_7CX_3$	Constrained + tails	1.6×10^9	3.1×10^{11}
$X_3CX_8CX_3$	Constrained + tails	9.0×10^8	2.8×10^{11}
$X_3CX_9CX_3$	Constrained + tails	1.0×10^9	1.1×10^{11}
$X_3CX_{10}CX_3$	Constrained + tails	5.4×10^8	2.9×10^{11}
X_{12}	Linear	7.0×10^8	3.0×10^{10}
X_{16}	Linear	6.3×10^8	4.3×10^{10}
X_{20}	Linear	1.8×10^9	4.0×10^{11}
Trinuc X_{12}	Linear	1.2×10^9	8.5×10^{10}
		Total 2.3×10^{10}, Avg. 1.3×10^9	Avg. 2.1×10^{11}

"X" represents a random amino acid, and "C" represents a fixed cysteine residue

3.3.1 Phage Clone Amplification and DNA Sequencing

1. Plate the libraries at low density (approximately 50 plaques per plate) using the plaque assay method as described in the T7Select System Manual.
2. Inoculate 350 μl *E. coli* cultures in 96-well, 2.2 ml deep-well plates with single well-isolated plaques by using sterile inoculating needles or sterile toothpicks. Collect three 96-well plates of clones (*see* **Note 13**).
3. Seal the plates with gas-permeable tape and incubate the cultures at 37 °C with fast (approximately 250 rpm) orbital rotation until lysis is observed.
4. Add NaCl to 0.5 M and pellet the bacterial debris by centrifugation at 4 °C.

5. Aliquot 200 μl of the clarified lysates to sterile minitube racks and store at 4 °C. Discard the pelleted debris.
6. In sterile, 96-well PCR reaction plates, set up the PCR reactions as follows:
   Per reaction:
   (*See* **Note 14**.)

Platinum PCR SuperMix High Fidelity	45 μl
Sterile molecular biology-grade water	2 μl
10 μM T7SelectUP Primer (diluted in TE buffer)	1 μl
10 μM T7SelectDOWN Primer (diluted in TE buffer)	1 μl
T7 phage clone clarified lysate	1 μl

7. Amplify the T7 phage vector insert DNA for 35 cycles using the protocol outlined in the Platinum PCR SuperMix High Fidelity product insert.
8. Confirm amplification of insert DNA and check for appropriate insert DNA base pair length by 2 % agarose gel electrophoresis using TAE running buffer.
9. Visualize the DNA fragments by staining the gel with ethidium bromide (0.5 μg/ml) in TAE buffer.
10. Sequence the PCR-amplified T7 phage vector insert DNA using both the T7SelectUP and T7SelectDOWN primers (*see* **Note 15**).
11. Translate the obtained DNA sequences into amino acid sequences.
12. Compile the displayed amino acid sequences into text files (*see* **Note 16**).

3.3.2 Diversity Analysis of Phage-Displayed Amino Acid Sequences

Peptide library diversity characterization methods have been described in detail elsewhere [5, 16–19]. Briefly, the text files generated in Subheading 3.3.1 are used as input into statistical Web-interfaced programs available at the RELIC website (http://relic.bio.anl.gov) [11] and into the Genetics Computer Group (GCG) Wisconsin Package programs Isoelectric and PeptideStructure (Version 10.3, Accelrys). The RELIC suite of tools allow for the analysis of overall and position-dependent amino acid frequencies using the Amino Acid Frequency program (AAFREQ) and for the overall and position-dependent amino acid diversities using the Analysis of Amino Acid Diversity in Multiple Aligned Proteins program (DIVAA). The GCG Wisconsin Package tools offer another approach to diversity characterization by analyzing the biophysical properties of population net charge, hydropathy, and predicted secondary structure. For a thorough discussion of how library diversity analyses are conducted, the reader is directed to ref. 16.

4 Notes

1. Unless otherwise stated, all reagents were of molecular biology grade and purchased from Sigma (St. Louis, MO). All solutions, unless otherwise noted, were prepared using molecular biology-grade water purchased from Quality Biological (Gaithersburg, MD) and were sterilized by filtration using a 0.22 μm membrane.
2. The exact sequences of the library DNA constructs will be unique to each library and depend on the desired sequence and length of the displayed peptide and on the amino acid sequence randomization method. The randomized library oligonucleotide DNA construct is synthesized as the antisense strand. For a description of randomization options, the reader is directed to references [5, 16, 17]. A 12-mer peptide library construct randomized by the "NNK" method could be synthesized as such: 5′-AAC TGC AAG CTT TTA $(MNN)_{12}$ ACC ACC ACC AGA ATT CGG ATC CCC GAG CAT-3′, where "N" represents a hand-mixed equimolar ratio of adenine, cytosine, thymine, and guanine nucleotides, and "M" represents a hand-mixed equimolar ratio of both adenine and cytosine nucleotides (*see* Fig. 3). The underlined nucleotides represent the *Hin*dIII restriction endonuclease recognition sequence. A complementary biotinylated extension primer for this construct would be synthesized as such: 5′biotin-ATG CTC GGG GAT CCG AAT TCT GGT-3′. The underlined nucleotides represent the *Eco*RI restriction endonuclease recognition sequence. The biotinylated extension primer is synthesized as the sense strand. Both the randomized library oligonucleotide DNA and the biotinylated extension primer should be purified by polyacrylamide gel electrophoresis and shipped as lyophilizates. Store the lyophilized oligonucleotides at 4 °C until use.
3. The asterisk (*) represents that the reagent is an enzyme. All enzymes should be kept at −20 °C until required. Keep the enzyme on ice when not in freezer. Mix only by gentle pipetting. Do not vortex the solution.
4. Ensure that the final concentration of NaCl in the binding reaction is 1 M. Account for NaCl in the duplex extension reaction, Subheading 3.1.3, **step 2**.

5. Do not allow the magnetic beads to pool at the bottom or on the sides of the microcentrifuge tube. Try using a rotating wheel placed inside an incubator.
6. 10 μg glycogen can be added to the ethanol precipitation step as a carrier molecule to assist in DNA precipitation.
7. An absorbance (A) A_{260}/A_{280} ratio above 1.6 for the supernatant generated in Subheading 3.1.4, **step 9**, indicates sufficient purity for subsequent cloning steps. If a lesser value is obtained, then repeat phenol extraction and ethanol precipitation.
8. For calculating insert molar excess, use the conversion factor that 1 pmol of nucleotide pairs has a mass of 660 pg.
9. A plaque assay performed using 10^4, 10^5, and 10^6 dilutions of the packaged library DNA and packaged ligation-positive control should yield a range of plaques amenable to counting. For the packaged ligation-negative control and packaging control DNA, use 10^2 and 10^3 and 10^6 and 10^7 dilutions for the plaque assays, respectively.
10. The packaging efficiency of phage library DNA should approximate or exceed the packaging efficiency of the ligation-positive control. The packaging efficiency of the negative ligation control DNA indicates the background level of nonrecombinant phage. The packaging efficiency of the positive control ligation reaction, performed at a 4:1 molar excess of control insert DNA to cut duplex (0.08 pmol DNA and 0.5 μg vector arms) was greater than 5×10^7 pfu/μg arms (as stated in a T7Select 10-3 Cloning Kit certificate of analysis, Novagen). We typically obtained packaging efficiencies for the positive ligation control, and for ligated library insert DNA, on the 10^7 order of magnitude in our laboratory. The packaging efficiency of the negative control ligation reaction was less than 2.5×10^5 pfu/μg arms (as stated in a T7Select 10-3 Cloning Kit certificate of analysis, Novagen). We typically obtained packaging efficiencies for the negative ligation control on the 10^4 order of magnitude in our laboratory.
11. Plate 10^4 and 10^5 dilutions of the packaged library DNA. The number of recombinants generated in the library can be calculated by multiplying the titer of the packaged library DNA by the total volume of the pooled packaged library DNA. We typically generated 10^8 or 10^9 recombinants for each library (Table 1).
12. Plate 10^8 and 10^9 dilutions for amplified library lysates.
13. Include one non-infected well per plate for turbidity comparison to assist in lysis determination.

14. Include one negative control per plate during the PCR reaction in which a phage clone lysate is omitted to rule out reagent contamination.
15. The PCR reaction should be purified by G50 chromatography prior to performing the sequencing reaction.
16. Omit any fixed amino residues from diversity analyses. For example, if analyzing the diversity of a disulfide-constrained peptide library, then omit the fixed cysteine residues from analyses.

Acknowledgements

This project has been funded in whole or in part with federal funds from the National Cancer Institute, National Institutes of Health, under contract HHSN261200800001E. The content of this publication does not necessarily reflect the views or policies of the Department of Health and Human Services, nor does mention of trade names, commercial products, or organizations imply endorsement by the U.S. Government. This Research was supported [in part] by the Intramural Research Program of the NIH, National Cancer Institute, Center for Cancer Research.

References

1. Krumpe LRH, Mori T (2007) Potential of peptide library technology to identify functional targeting peptides. Expert Opin Drug Discov 2:525–537
2. Krumpe LRH, Mori T (2006) The use of phage-displayed peptide libraries to develop tumor-targeting drugs. Int J Pept Res Ther 12:79 91
3. Yacoby I, Benhar I (2008) Targeted filamentous bacteriophages as therapeutic agents. Expert Opin Drug Deliv 5:321–329
4. Kehoe JW, Kay BK (2005) Filamentous phage display in the new millennium. Chem Rev 105:4056–4072
5. Krumpe LR, Atkinson AJ, Smythers GW, Kandel A, Schumacher KM, McMahon JB, Makowski L, Mori T (2006) T7 lytic phage-displayed peptide libraries exhibit less sequence bias than M13 filamentous phage-displayed peptide libraries. Proteomics 6(2006):4210–4222
6. Castagnoli L, Zucconi A, Quondam M, Rossi M, Vaccaro P, Panni S, Paoluzi S, Santonico E, Dente L, Cesareni G (2001) Alternative bacteriophage display systems. Comb Chem High Throughput Screen 4:121–133
7. Sakamoto K, Ito Y, Mori T, Sugimura K (2006) Interaction of human lactoferrin with cell adhesion molecules through RGD motif elucidated by lactoferrin-binding epitopes. J Biol Chem 281:24472–24478
8. Sakamoto K, Ito Y, Hatanaka T, Soni PB, Mori T, Sugimura K (2009) Discovery and characterization of a peptide motif that specifically recognizes a non-native conformation of human IGG induced by acidic pH conditions. J Biol Chem 284(15):9986–9993
9. Rosenberg A, Griffin K, Studier W, McCormick M, Berg J, Novy R, Mierendorf R (1996) T7Select phage display system: a powerful new protein display system based on bacteriophage T7. Innovations 6:1–6
10. T7Select System Manual, TB178 Rev. B 0203, Novagen, Gibbstown, NJ. http://www.emdbiosciences.com/docs/docs/PROT/TB178.pdf. Accessed 16 Dec 2008
11. Mandava S, Makowski L, Devarapalli S, Uzubell J, Rodi DJ (2004) RELIC—a bioinformatics server for combinatorial peptide analysis and identification of protein-ligand interaction sites. Proteomics 5:1439–1460

12. DYNAL Dynabeads M-280 Streptavidin product insert. Rev. 014. Invitrogen, Carlsbad, CA. http://tools.invitrogen.com/content/sfs/manuals/112.05D06D%20602.10%20Dynabeads%20M280%20Streptavidin%20%28rev%20014%29.pdf. Accessed 8 Apr 2011
13. Moore D, Dowhan D (2002) Purification and concentration of DNA from aqueous solutions. Curr Protoc Mol Biol Suppl 59:2.1.1–2.1.10
14. Novex Pre-Cast Gel Electrophoresis Guide. Version B. Invitrogen, Carlsbad, CA. http://tools.invitrogen.com/content/sfs/manuals/electrophoresisguide_man.pdf. Accessed 8 Apr 2011
15. Makowski L, Soares A (2003) Estimating the diversity of peptide populations from limited sequence data. Bioinformatics 19:483–489
16. Krumpe LRH (2006) Diveristy comparison of T7 and M13 phage-displayed peptide libraries. Hood College, 2006, 82 pages; AAT 1439197. http://proquest.umi.com/pqdlink?did=1221746261&Fmt=7&clientId=79356&RQT=309&VName=PQD. Accessed 8 Apr 2011
17. Krumpe LR, Schumacher KM, McMahon JB, Makowski L, Mori T (2007) Trinucleotide cassettes increase diversity of T7 phage-displayed peptide library. BMC Biotechnol 7:65
18. Rodi DJ, Mandava S, Makowski L (2004) DIVAA: analysis of amino acid diversity in multiple aligned protein sequences. Bioinformatics 18:3481–3489
19. Rodi DJ, Soares AS, Makowski L (2002) Quantitative assessment of peptide sequence diversity in M13 combinatorial peptide phage display libraries. J Mol Biol 322:1039–1052

Chapter 5

Affinity Selection Using Filamentous Phage Display

Mari Ann Kulseth, Annette Fagerlund, and Astrid Hilde Myrset

Abstract

Display of peptides on filamentous phage, phage display, is an in vitro selection technique well suited for identification of therapeutic peptide binders for a huge variety of protein targets. The peptides are identified in a process where phage libraries are subjected to affinity selection towards a particular protein target. A successful outcome of an affinity selection is dependent on proper surveillance of the phage life cycle, to make sure that the selection is based on affinity for the target, not on bias in phage propagation rate. In this chapter we present two approaches for protein target presentation and a protocol for phage rescue and propagation, which includes several controls to ensure that all phages initially eluted from the protein target are given equal conditions during the following amplification and selection steps.

Key words Phage display, Affinity selection, Peptides, VEGFR2, MPP9

1 Introduction

Novel therapeutic peptides that bind protein targets with high affinity and specificity potentially can be provided by phage display technology. In peptide phage libraries the peptides are displayed on the surface of a filamentous bacteriophage, whereas the DNA encoding the peptide is encapsulated inside the phage. Usually a phagemid vector encoding the recombinant peptide-coat protein and a phage-derived origin of replication for effective encapsulation of DNA is used in combination with a helper phage encoding the entire phage genome with a compromised phage origin of replication or packaging signal leading to inefficient packaging. The combination of phagemid and helper phage increases the flexibility and diversity of the prepared libraries.

In this chapter we describe a step-by-step protocol for identification of peptides by affinity selection of phage-displayed peptide libraries and for surveillance of the phage propagation during the affinity selection process (schematically illustrated in Fig. 1). Various protein targets presented for the phage display libraries in various ways have been successfully used for identification of

Andrew E. Nixon (ed.), *Therapeutic Peptides: Methods and Protocols*, Methods in Molecular Biology, vol. 1088,
DOI 10.1007/978-1-62703-673-3_5, © Springer Science+Business Media, LLC 2014

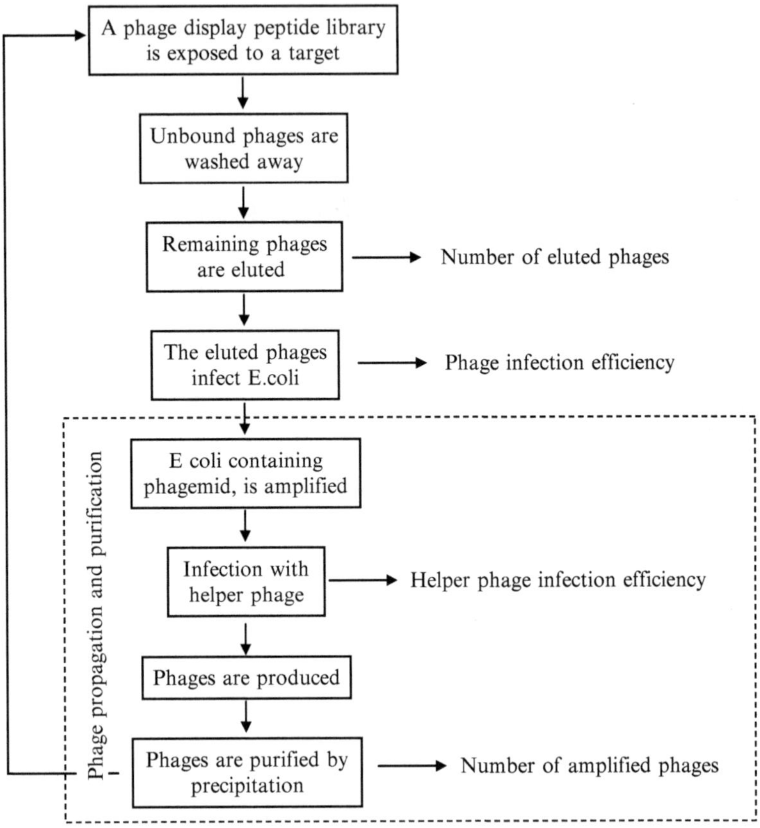

Fig. 1 A schematic illustration of the phage display affinity selection process and the recommended measurements for thorough surveillance of the phage propagation

peptide ligands from phage display libraries. In the present chapter we describe two approaches where purified recombinant proteins are presented in solution or attached to a plastic surface via an antibody. The affinity selection might be divided into five steps which are (1) presentation of the phage library to the protein target and binding of phage to the target, (2) removing of low-affinity or nonspecific binders, (3) elution of bound phage from the protein target, (4) rescue of eluted phage by bacteria infection, and (5) phage propagation and purification.

2 Materials

2.1 Affinity Selection on Target in Solution

1. Phage display peptide library: pA2(NNK)$_9$-library described in ref. 1 and in [Chapter 2]. This library contains 9.4×10^9 unique transformants.
2. Phosphate-buffered saline (PBS): Prepare 10× PBS with 1.37 M NaCl, 27 mM KCl, 100 mM Na_2HPO_4, and 18 mM KH_2PO_4, pH 7.4. Adjust pH with HCl if necessary, and autoclave.

3. Recombinant human vascular endothelial growth factor receptor 2 (VEGFR2): A disulfide-linked homodimeric protein, containing two 988 amino acid monomers where each monomer encodes amino acids 1–764 of the extracellular domain of human VEGFR2 fused to the 6× histidine (His)-tagged Fc of human IgG_1 via the peptide IEGRMD is available from R&D Systems.
4. Ni-NTA magnetic beads (Qiagen).
5. Imidazole (Calbiochem).
6. Tween 20 (Sigma).
7. PBS/Tween/imidazole: 1× PBS, 0.05 % (v/v) Tween 20, 20 mM imidazole.
8. Elution buffer: 0.1 M Glycine–HCl, pH 2.2.
9. Neutralization buffer: 1 M Tris–HCl, pH 9.1.

2.2 Affinity Selection on Immobilized Target

1. Phage display peptide library: *See* Subheading 2.1, **item 1**.
2. MaxiSorp Immunoplate (NUNC).
3. Monoclonal anti-matrix metalloproteinase-9 (MMP-9) antibody (kindly provided by GE Healthcare).
4. 50 mM $NaHCO_3$, pH 9.6.
5. Superblock: Blocking buffer in Tris-buffered saline (TBS) (Pierce).
6. Assay buffer: 50 mM Tris–HCl pH 7.6, 1.5 mM NaCl, 0.5 mM $CaCl_2$, 1 μM $ZnCl_2$, 0.01 % (w/v) BRIJ 35.
7. Recombinant human MMP-9 (R&D Systems).
8. Washing buffer: 0.01 M sodium phosphate buffer pH 7.0, 0.05 % (v/v) Tween 20.
9. *p*-Aminophenylmercuric acetate (APMA) (Sigma). APMA solution is 10 mM APMA dissolved in assay buffer.
10. TBS: 20 mM Tris–HCl pH 7.5, 150 mM NaCl.
11. TBS/Tween: TBS with 0.05 % (v/v) Tween 20.

2.3 Phage Infection and Propagation

1. Bacterial strain: TOP10F' (Invitrogen) Petri dishes (90 mm in diameter).
2. LB medium: 1 % (w/v) tryptone, 0.5 % (w/v) yeast extract, 0.17 M NaCl. Sterilize by autoclaving.
3. LB-agar: LB medium with 1.5 % (w/v) Bacto-agar. Sterilize by autoclaving.
4. SOC medium: 2 % (w/v) tryptone, 0.5 % (w/v) yeast extract, 0.5 mM NaCl, 25 M KCl. Sterilize by autoclaving and add 20 mL 1 M D-glucose (sterile filtrated) per liter medium. Do not add $MgCl_2$ as it inhibits tetracycline.

5. SOC-agar: SOC-medium with 1.5 % (w/v) Bacto-agar. Sterilize by autoclaving and add 20 mL 1 M D-glucose (sterile filtrated) per liter medium.
6. Ampicillin stock solution: 100 mg ampicillin (Sigma) per mL H_2O. Sterilize by filtration and store at −20 °C.
7. Tetracycline stock solution: 5 mg tetracycline (Sigma) per mL ethanol. Store at −20 °C.
8. Kanamycin stock solution: 10 mg kanamycin A (Sigma) per mL H_2O. Sterilize by filtration and store at −20 °C.
9. LB-tet: LB medium with 50 μg/mL tetracycline.
10. LB-amp/kan: LB medium with 50 μg/mL ampicillin and 50 μg/mL kanamycin.
11. SOC-tet: SOC medium with 50 μg/mL tetracycline.
12. SOC-amp/tet: SOC medium with 60 μg/mL ampicillin and 50 μg/mL tetracycline.
13. SOC-agar-amp: SOC-agar with 60 μg/mL ampicillin.
14. SOC-agar-amp/kan: SOC-agar with 60 μg/mL ampicillin and 50 μg/mL kanamycin.
15. M13K07 helper phage (New England Biolabs).
16. L-Arabinose stock: 20 % (w/v) L-arabinose (Sigma) in dH_2O. Sterilize by filtration and store at −20 °C.
17. Glycerol (Sigma): 50 % (v/v) in dH_2O. Sterilize by autoclaving.
18. PEG/NaCl: 16.7 % (w/v) PEG8000 (Sigma), 3.3 M NaCl.
19. TBS/NaN_3: TBS with 0.02 % (w/v) NaN_3.
20. TBS–gelatin: TBS with 1 % (w/v) gelatin (Sigma).

2.4 PCR Amplification of Phagemid DNA

1. AmpliTaqGold, 10× PCR Gold buffer and 25 mM $MgCl_2$ (PE Applied Biosystems).
2. dNTP: dATP, dCTP, dGTP, and dTTP (GE Healthcare), 10 mM each.
3. DMSO (Sigma).
4. Upper primer: 5′-ACCCTCGTTCCGATGCTG-3′.
5. Lower primer: 5′-TCGCTATTACGCCAGCTG-3′.

3 Methods

The outcome of an affinity selection or biopanning of phage display libraries is influenced by several factors, including the following: (1) library size and quality; (2) presentation of peptides on the phage: a monovalent versus polyvalent attachment between phage and target promotes selection based on affinity instead of avidity; (3) selection stringency, target presentation, and washing procedure; (4) equal phage amplification: each phage eluted from the target should be amplified to the same amount, to yield a renormalized library with reduced diversity reflecting the eluted phage and not the phage infection and replication properties. The last factor is as least as important as the others. The outcome of a successful biopanning should be one or several sets of peptides with high consensus, not a few identical peptides. If most of the individual clones isolated after biopanning contain identical DNA, it is most likely an indicator of bias in phage amplification where this phage has outgrown the others. Selection of phage by biopanning should be based on the affinity for the target, not on phage propagation rate.

Ff filamentous phage attaches to *Escherichia coli* via the tip of an F-pilus. Due to low abundance of F-pili (only a few per cell) and the small size of the phage-binding region [2], the efficiency of infection of a bacterial culture is improved with high multiplicity of phage per cell and by high cell density [3]. However, growing the cells past log phase will reduce pilus expression and compromise infectivity. Host strains used for phage display should contain an F-episome (F′) encoding the F-pilus, with a selectable marker to ensure its maintenance.

Upon entry of the single-stranded phagemid into *E. coli*, host RNA and DNA polymerases and topoisomerase will convert it to a double-stranded replicative form (RF). The RF serves as template for transcription of the recombinant peptide-coat protein [3]. Prevention of the recombinant peptide-coat protein transcription has been shown to significantly reduce bias in host cell growth rate [4].

3.1 Laboratory Facilities and Washing Regime

It is essential to keep the *E. coli* cultures and phage preparations free of phage contamination. Filamentous bacteriophage of M13 origin is stable and remains infectious under rather extreme conditions. Wild-type phage and previously isolated phage in the lab might become a serious contamination problem if caution is not exerted.

1. Use aerosol-resistant pipette tips and cotton-plugged pipettes.
2. Phage contamination of glassware is removed by dry-sterilization at 180 °C of all glassware that have been or are going to be in contact with phage.
3. Phage contamination of centrifuge tubes is removed by thorough washing in sodium hypochlorite followed by autoclaving and UV irradiation.
4. The work area and equipment like centrifuges and shakers should be UV irradiated and wiped down with ethanol.

3.2 Affinity Selection on Target in Solution

A selective enrichment for high-affinity peptides might be achieved by allowing binding of the phage display library to a limited amount of target in solution for a prolonged period of time [5]. This strategy favors a monovalent interaction between phage and target and thus increases the likelihood of selecting high-affinity peptides. If the target protein is a dimer, monovalent interaction is dependent on a low-density and well-distributed peptide on the phage particle. A tag present on the target will enable capturing of the target–phage complex at the end of the target presentation period.

Biotin, as used by Barrett et al. [5], is one potential tag. However, biotinylation of a protein might change or hide epitopes on the protein influencing the outcome of the biopanning process. Recombinant proteins with a peptide tag at a location of the protein outside the region of interest represent an excellent alternative. His-tagged proteins can be produced by introducing the cDNA encoding the protein of interest in a commercially available vector and transfect the vector into mammalian cells for production of the recombinant protein. Several His-tagged proteins are also commercially available. His-tagged proteins can be captured from the solution by Ni^{2+}-coated magnetic beads. The following protocol describes an affinity selection to His-tagged recombinant VEGFR2 using the pA2$(NNK)_9$-library described in Chapter 2; however the protocol might easily be adopted for other receptors and phage display libraries.

3.2.1 First Cycle of Affinity Selection on Target in Solution

1. In an Eppendorf tube dilute 3.2 μg (approximately 10 pmol dimeric) VEGFR2 in PBS and add phage in an amount corresponding to approximately 1,000 times the phage library titer (9.4×10^{12} phage of the pA2$(NNK)_9$-library) to a final volume of 1 mL.
2. Rotate the tube at 4 °C for at least 16 h.
3. Add 20 μL 1 M imidazole (*see* **Note 1**) and 30 μL Ni-NTA magnetic beads.
4. Rotate the tube at room temperature for 3 h.
5. Insert the tube in a magnet for approximately 1 min and remove the supernatant.
6. Wash the beads with 1 mL PBS/Tween/imidazole ten times, each time for 5–6 min with rotation at room temperature. Insert the tube in the magnet and remove the supernatant after approximately 1 min. Following the last wash, ensure that all supernatant is removed.
7. Add 500 μL elution buffer and rotate the tube for 30 min at room temperature (*see* **Note 2**).
8. Insert the tube in a magnet for 1 min and transfer the supernatant to a tube containing 80 μL neutralizing buffer.
9. Determine the titer of eluted phage as described in Subheading 3.6.1 and continue from Subheading 3.4.

3.3 Affinity Selection on Immobilized Target

The most common strategy for presentation of a target is coating 96-well microtiter plates with the target and adding the phage display library to the wells. Unless the amount of target protein in the well or the density of peptides displayed on the phage is kept low [5], this approach might favor polyvalent interaction between phage and target, which might also enable low-affinity binders to be selected. However, the simplicity of this approach makes it an attractive alternative.

Therapeutic peptides are often agonists or antagonists of the target protein, and peptides identified by phage display should either bind to the active site of the target or disturb the target enough to deprive its normal activity. If the active site is blocked or covered by a prodomain, this should be cleaved before affinity selection is initiated.

The protocol that follows describes an affinity selection to activated MMP-9 in a 96-well plate, where the MMP-9 is attached to the well in the desired orientation via a monoclonal MMP-9 antibody, which does not disturb the active site. The pA2$(NNK)_{9\text{-library}}$ described in Chapter 2 is used; however the protocol is easily adaptable to other libraries.

3.3.1 First Cycle of Affinity Selection on Immobilized Target

1. Add 200 μL MMP-9 antibody (5 μg/mL) diluted in 50 mM $NaHCO_3$ (pH 9.6) to five wells in a 96-well MaxiSorp immunoplate and incubate overnight at 4 °C.
2. Remove the supernatant and block the wells with 200 μL Superblock for 2 h at room temperature and wash with 300 μL assay buffer three times for 5 min each time.
3. Add 200 μL MMP-9 (1 μg/mL) diluted in assay buffer and incubate overnight at 4 °C.
4. Wash with 300 μL washing buffer three times for 5 min each time.
5. Activate the MMP-9 by adding 200 μL of 10 mM APMA solution and incubate overnight at 4 °C.
6. Wash with 300 μL Superblock three times for 5 min each time.
7. Dilute an amount of phage corresponding to approximately 1,000 times the phage library titer (9.4×10^{12} phage of the pA2$(NNK)_9$-library) to 1 mL in Superblock and incubate at room temperature for 30 min in a rotor wheel. Add 200 μL to the wells containing activated MMP-9 and incubate the plate overnight at 4 °C with gentle shaking.
8. Wash the wells 15 times for 5 min each with 300 μL TBS/Tween.
9. Add 200 μL elution buffer to each well and incubate for 30 min at room temperature with gentle shaking.

10. Combine the eluted phage in an Eppendorf tube and add 160 µL neutralizing buffer.
11. Determine the titer of eluted phage as described in Subheading 3.6.1 and continue from Subheading 3.4.

3.4 Rescue and Amplification of Eluted Phage

1. Inoculate a single colony of TOP10F' in 30 mL LB-tet and incubate at 37 °C with shaking (230 rpm) overnight. Dilute the culture 1:100 in 100 mL SOC-tet and incubate further for approximately 4 h to achieve an optical density at 600 nm (OD_{600}) between 1.7 and 2.0 (*see* **Note 3**).
2. Add the eluted phages and incubate at 37 °C with gentle shaking (70 rpm) for 1 h.
3. Withdraw a sample to measure the infection efficiency as described in Subheading 3.6.2.
4. Dilute the phage-infected culture to 500 mL with SOC-amp/tet, divide the culture into two 2 L Erlenmeyer bottles with baffles, and incubate at 37 °C with vigorous shaking (230 rpm) overnight.
5. Combine the split cultures the following morning. Withdraw 2 mL, dilute to 40 mL using SOC-amp/tet, and incubate the culture at 37 °C with vigorous shaking (230 rpm) for approximately 4 h to achieve an OD_{600} between 1.7 and 2.0 (*see* **Notes 3** and **4**).
6. Add 3×10^{11} M13K07 helper phage (multiplicity of infection [MOI] >6) and incubate at 37 °C with gentle shaking (70 rpm) for 1 h.
7. Withdraw a sample to measure the helper phage infection efficiency as described in Subheading 3.6.3.
8. Centrifuge the culture at $4{,}000 \times g$ for 10 min.
9. Resuspend the bacterial pellet in 20 mL LB medium and transfer to two 2 L Erlenmeyer bottles with baffles, each containing 190 mL LB-amp/kan (*see* **Note 5**).
10. Add the inducer, L-arabinose, to a final concentration of 0.001 % (*see* **Note 6**).
11. Incubate the cultures for phage production at 37 °C with vigorous shaking (230 rpm) for 6–8 h.
12. Combine the cultures and centrifuge once at $4{,}000 \times g$ and the supernatant a second time at $7{,}000 \times g$, both times for 10 min at 4 °C.
13. Transfer the supernatant to a new centrifuge tube and add 0.18× volume PEG/NaCl. Mix by inversion by hands approximately 50 times and incubate at 4 °C overnight.
14. Centrifuge at $20{,}000 \times g$ for 1 h at 4 °C. Discard the supernatant.

15. Resuspend the pellet in 2 mL TBS and heat the samples at 70 °C for 1 h. Precipitate *E. coli* debris by centrifugation at 5,000 × *g* for 10 min at 4 °C.
16. Transfer the supernatant to a new tube and re-precipitate the phage by adding 0.7 mL PEG/NaCl. Mix by inversion and incubate at 4 °C for 1 h.
17. Centrifuge at 20,000 × *g* for 1 h at 4 °C. Discard all the supernatant and resuspend the phage pellet in 1 mL TBS/NaN_3. Store at 4 °C.
18. Determine the phage concentration as described in Subheading 3.6.1.

3.5 Subsequent Cycles of Affinity Selection

Phage capable of binding with various affinities to the target can be detected after one cycle of affinity selection. However, to identify the highest affinity phage, second and third cycles of affinity selection are necessary. These cycles are performed as described for the first cycle of affinity selection using approximately 1.0×10^{12} phage and the target protein presented as described in Subheadings 3.2 or 3.3.

3.6 Surveillance of the Phage Propagation

3.6.1 Determination of the Number of Eluted and Amplified Phage

Infection in small scale is used to monitor the biopanning process. The phage titer is determined by infecting *E. coli* with serial dilutions of phage, and given as colony-forming units (cfu). The number obtained depends on the infection efficiency of the phage. The MOI should be less than 1 (the cells must be in excess) to avoid that the *E. coli* represents a limiting factor. The amount and density of *E. coli* giving the maximum infection efficiency should be optimized for the bacterial strain to be used.

1. Grow an overnight culture of TOP10F' from a fresh colony as described in Subheading 3.4, **step 1**. The next morning, dilute the culture 1:100 in SOC-tet and incubate further for approximately 4 h to achieve an OD_{600} between 1.7 and 2.0.
2. Prepare tenfold serial dilutions of phage in TBS–gelatin in order to obtain a countable number of colonies on the plates (**step 7**).

Suggested dilution ranges	For eluted phage, 10^{-2}–10^{-4}
	For amplified phage, 10^{-8}–10^{-10}

3. Mix 30 μL of the phage dilution with 970 μL of TOP10F′ (prepared in **step 1**) in a 15 mL tube and incubate for 10 min at room temperature.
4. Add 1 mL SOC medium and incubate for 40 min at 37 °C with vigorous shaking (230 rpm).

5. Plate 100 μL on SOC-agar-amp and incubate overnight at 37 °C.
6. Count the colonies and calculate the phage titer according to the following equation:
   Number of phages per μL = Number of colonies × 2/3 × dilution factor (*see* **Note 7**).

3.6.2 Phage Infection Efficiency

A successful affinity selection of phage from a phage library is dependent on an optimal rescue and amplification of eluted phages. The initial step in the phage rescue is the infection of *E. coli* and it is important to monitor this step to ensure that all eluted phages are able to infect *E. coli*.

1. Following infection (Subheading 3.4, **step 2**) withdraw 100 μL and prepare $10–10^{-3}$ dilutions in SOC medium.
2. Plate 100 μL on SOC-agar-amp and incubate overnight at 37 °C.
3. Count the colonies and compare to the number obtained after titration of eluted phage.

3.6.3 Helper Phage Infection Efficiency

Phagemids contain the origin of replication in addition to the gene encoding the chimeric pVIII coat protein. A helper phage is needed to supply in trans all the structural proteins required to generate a complete phage encapsulating the phagemid. The optimal amount of helper phage should be determined to achieve an optimal phage production where each phagemid-infected cell is infected with helper phage and subsequently produces phage. The infection efficiency of helper phage is monitored as follows:

1. Withdraw 100 μL to calculate the number of M13K07-infected bacteria after addition of M13K07 (Subheading 3.4, **step 8**).
2. Prepare a 10^{-6} and 10^{-7} dilution, and spread 100 μL on SOC-agar-amp plates to measure the number of phagemid-infected *E. coli* and 100 μL on SOC-agar-amp/kan plates to measure the number of M13K07-infected *E. coli*. Incubate overnight at 37 °C.

Calculate the M13K07 infection efficiency by comparing the number of colonies obtained on SOC-agar-amp plates and SOC-agar-amp/kan plates (*see* **Notes 8** and **9**).

3.7 Production of a Phage Display Library from a Glycerol Stock

A phage display library is usually stored as glycerol stocks of phagemid-containing *E. coli*. Prior to affinity selection, phage has to be prepared.

1. Thaw glycerol stocks containing a number of phagemid-containing bacteria corresponding to at least ten times the library titer.

2. Resuspend the glycerol stocks in a total of 1 L SOC-amp and divide between five 2 L Erlenmeyer bottles with baffles. Incubate at 37 °C with shaking at 230 rpm until the OD_{600} is 2.25 (*see* **Note 10**), and combine the divided culture in one flask.
3. Withdraw a sample to measure the number of phagemid-containing *E. coli.* as described in Subheading 3.6.2.
4. Transfer 200 mL of the combined culture to a 2 L Erlenmeyer bottle with baffles and add 7×10^{11} M13K07 helper phage (MOI) of >6. Incubate for 1 h at 37 °C with gentle shaking at 70 rpm.
5. Withdraw a sample to measure the helper phage infection efficiency as described in Subheading 3.6.3.
6. Centrifuge the culture at 4,000 × *g* for 10 min at 4 °C. Discard the supernatant.
7. Resuspend the pellet in a total of 1 L LB-amp/kan containing 0.001 % arabinose (*see* **Note 6**) and divide between five 2 L Erlenmeyer bottles with baffles. Incubate at 37 °C with shaking at 230 rpm for 6–8 h.
8. Continue from Subheading 3.4, **step 13**.

3.8 Determination of Phage Display Peptide Sequences

The progress of an affinity selection might be monitored by calculating the enrichment ratio either as the number of phage eluted from a well coated with the target divided by the number of phage eluted from an uncoated well [6] or as the number of phage eluted divided by the number of phage added to the well (Out/In). The enrichment ratio usually increases significantly at one particular cycle (usually cycle 3), indicating an enrichment of phage binding specifically to the target. Individual clones are analyzed after the cycle where the enrichment ratio markedly increases.

DNA sequencing of individual clones might be performed on purified phage or directly on bacteria colonies after an initial PCR amplification of the phagemid DNA.

3.8.1 Purification of Individual Phage for DNA Sequencing

1. Inoculate 30 mL LB-amp/kan in 500 mL Erlenmeyer bottles with baffles, with a single colony from the SOC-amp/kan plates prepared in Subheading 3.6.2, and incubate at 37 °C with shaking (230 rpm) for 8 h or overnight (*see* **Note 11**).
2. Prepare glycerol stocks for the future: Withdraw 700 μL from each culture to a 1.8 mL cryotube and add 300 μL 50 % glycerol. Mix gently and store at −70 °C.
3. Transfer the rest of each culture to new centrifuge tubes and precipitate the bacteria by centrifugation once at 4,000 × *g* and the supernatant a second time at 7,000 × *g*, both times for 10 min at 4 °C.
4. Continue from Subheading 3.4, **step 13**.

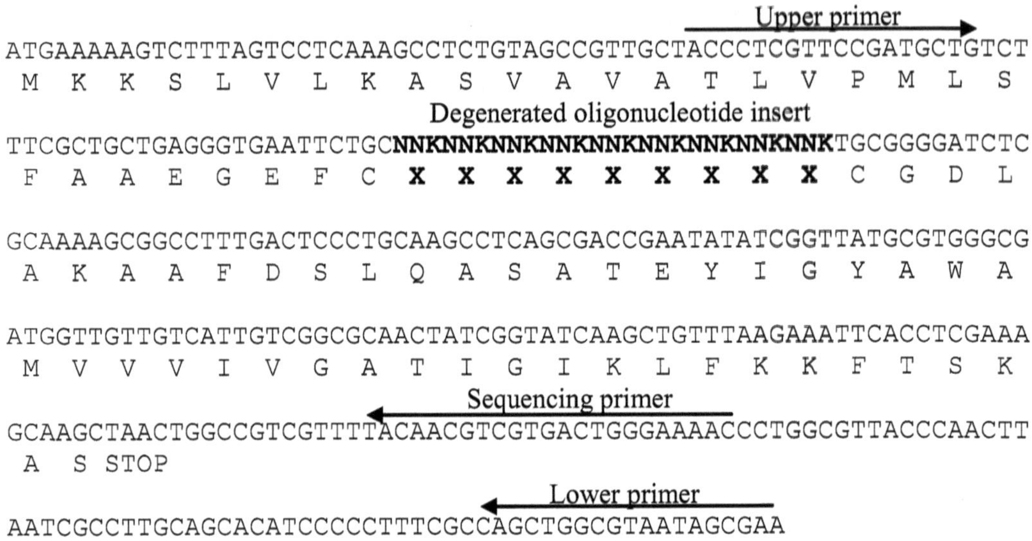

Fig. 2 The DNA sequence of the pVIII gene in pA2 and the encoded amino acids. The inserted degenerated oligonucleotide is included (in *bold*). The PCR primers used for amplification of phagemid DNA in *E. coli* and the primer used for DNA sequencing are indicated by *arrows*

5. Resuspend the phage pellets in 200 μL TBS/NaN_3. Store at 4 °C before DNA sequencing (*see* Subheading 3.8.3).

3.8.2 PCR Amplification of Phagemid DNA from E. coli for Sequencing

Localization of primers used to amplify DNA from the pA2$(NNK)_9$ library is shown in Fig. 2.

1. Pick single colonies from the SOC-agar-amp plates obtained after determination of the number of eluted phage (*see* Subheading 3.6.1, **step 6**) with a small pipette tip and transfer to 10 μL LB medium. Spread an aliquot on SOC-agar-amp/kan plates to maintain the particular colonies.
2. Prepare the following PCR-mix:

   5 μL 10× PCR gold buffer.

   3 μL 25 mM $MgCl_2$.

   1 μL 10 mM dNTP.

   1 μL (25 pmol) upper primer.

   1 μL (25 pmol) lower primer.

   2.5 μL 100 % DMSO.

   0.5 μL (2.5 U) AmpliTaqGold.

   34 μL dH_2O.

   2 μL resuspended colony from **step 1**.

   Include one control sample without any bacterial colony.
3. Subject the PCR-mix to thermal cycling:
   10 min at 95 °C followed by 30 cycles with 1 min at 95 °C, 1 min at 60 °C, and 1 min at 72 °C. Leave at 4 °C until further analysis.

3.8.3 Phage DNA Sequencing

The DNA sequencing might be performed using any commercial DNA sequencing kit and the sequencing primer 5′-GTTTTC-CCAGTCACGACGTTGTA-3′ located 13 bp downstream of pVIII stop codon in the pA2 phagemid (Fig. 2). The amino acid sequence is deduced from the reversed complementary strand obtained after DNA sequencing.

The progress of an affinity selection might also be monitored by performing a DNA sequencing of the mixture of amplified phages after each cycle. Sequencing the amplified phages will reveal if one phage has outgrown the others.

4 Notes

1. Imidazole at a final concentration of 20 mM is added to prevent nonspecific binding to the Ni-NTA beads.
2. The bound phage is recovered by denaturing the target protein by exposure to pH 2.2. M13 bacteriophage remains intact under this condition [7]. pH 12, 15 mM DDT, 1 % DMSO, and 6 M urea or trypsin might also be used.
3. The amount and density of *E. coli*, which will give the maximum infection efficiency, should be optimized for each bacterial strain used. For TOP10F′ both phagemid infection and helper phage infection are performed at OD_{600} between 1.7 and 2.0 to ensure maximal infection efficiency.
4. The amplification of the phagemid-infected *E. coli* prior to helper phage infection is performed in SOC medium containing glucose to reduce bias during the growth of phagemid-infected *E. coli*. d-Glucose represses transcription from the P_{BAD} promoter, which controls expression of the gene encoding the chimeric pVIII on the pA2 phagemid. Repression of chimeric pVIII expression has been shown to reduce the growth bias of different clones due to the presence of different chimeric constructs encoded by the different phagemids [4].
5. Kanamycin is added to select for helper phage M13K07-infected bacteria.
6. 0.001 % Arabinose was shown to give a low amount of peptide displayed on the phage surface [1] and this concentration was chosen to stimulate monovalent contact between phage and target protein.
7. Explanation for "2/3" in the equation: Number of colonies on the plate must be multiplied by 20 to obtain the number of infected *E. coli* in the 2 mL SOC medium and divided by 30 to obtain the number of phages per microliter diluted phage.
8. The bacteria will divide during the 1-h incubation, and a successful M13K07 infection should give a higher number of

colonies after M13K07 infection compared to the number of colonies obtained prior to M13K07 infection.

9. M13K07-infected bacteria will give rise to smaller colonies than phagemid-infected bacteria. Although filamentous phage does not produce a lytic infection in *E. coli*, the growth rate of the phage-producing bacteria is reduced.
10. The pA2-$(NNK)_9$-library was prepared in DH12S strain of *E. coli*. The optimal bacterial density for maximum infection was found to be at OD_{600} between 2.0 and 2.3.
11. The phage production time should in general be 6–8 h. An overnight production might reduce the quality of the phage due to the effect of proteases released by dying bacteria. However, an overnight production does not affect the DNA of the phage and might be used if the phage propagation is performed solely to analyze the phage DNA.

Acknowledgement

This work was performed at and supported by Medical Diagnostics Research, GE Healthcare.

References

1. Fagerlund A, Myrset AH, Kulseth MA (2008) Construction and characterization of a 9-mer phage display pVIII-library with regulated peptide density. Appl Microbiol Biotechnol 80:925–936
2. Jacobson A (1972) Role of F pili in the penetration of bacteriophage fl. J Virol 10:835–843
3. Russel M, Lowmann HB, Clackson T (2004) Introduction to phage biology and phage display. In: Clackson T, Lowmann HB (eds) Phage display: a practical approach. Oxford University Press, New York, pp 1–26
4. Beekwilder J, Rakonjac J, Jongsma M, Bosch D (1999) A phagemid vector using the E. coli phage shock promoter facilitates phage display of toxic proteins. Gene 228:23–31
5. Barrett RW, Cwirla SE, Ackerman MS, Olson AM, Peters EA, Dower WJ (1992) Selective enrichment and characterization of high affinity ligands from collections of random peptides on filamentous phage. Anal Biochem 204:357–364
6. Sidhu SS, Lowman HB, Cunningham BC, Wells JA (2000) Phage display for selection of novel binding peptides. Methods Enzymol 328:333–358
7. Kay BK, Kasanov J, Yamabhai M (2001) Screening phage-displayed combinatorial peptide libraries. Methods 24:240–246

Chapter 6

Bioprospecting Open Reading Frames for Peptide Effectors

Ling Xiong and Charles Scott

Abstract

Recent successes in the development of small-molecule antagonists of protein–protein interactions designed based on co-crystal structures of peptides bound to their biological targets confirm that short peptides derived from interacting proteins can be high-value ligands for pharmacologic validation of targets and for identification of druggable sites. Evolved sequence space is likely to be enriched for interacting peptides, but identifying minimal peptide effectors within genomic sequence can be labor intensive. Here we describe the use of incremental truncation to diversify genetic material on the scale of open reading frames into comprehensive libraries of constituent peptides. The approach is capable of generating peptides derived from both continuous and discontinuous sequence elements, and is compatible with the expression of free linear or backbone cyclic peptides, with peptides tethered to amino- or carboxyl-terminal fusion partners or with the expression of peptides displayed within protein scaffolds (peptide aptamers). Incremental truncation affords a valuable source of molecular diversity to interrogate the druggable genome or evaluate the therapeutic potential of candidate genes.

Key words Incremental truncation, Peptide, Expression library, Target validation, Exonuclease

1 Introduction

One of the most significant challenges to the pharmaceutical industry is the paucity of pharmacologically tractable targets. According to recent estimates, FDA-approved drugs target less than 2 % of the open reading frames in the human genome [1, 2]. Furthermore, despite significant investments in molecular diversity and screening infrastructure, not to mention sequencing of the entire human genome [3] and development and maturation of powerful methods for genetic validation of targets [4, 5], FDA approvals of new molecular entities have declined steadily over the past decade [6–10], with 2007 producing the fewest approvals in a quarter of a century [11]. Clearly, there is considerable need and opportunity for improved methods to discover and validate novel targets for pharmacologic intervention in disease.

Rational design and high-throughput screening are used widely in both the public and private sector as methods to expedite

Andrew E. Nixon (ed.), *Therapeutic Peptides: Methods and Protocols*, Methods in Molecular Biology, vol. 1088,
DOI 10.1007/978-1-62703-673-3_6, © Springer Science+Business Media, LLC 2014

target validation through the generation of lead molecules that act by pharmacologically relevant mechanisms. Rational design exploits atomic resolution structural information to direct the synthesis of specific small-molecule effectors. One of the major obstacles to the use of rational design (and virtual screening, which also relies on atomic resolution target structure) for target validation on the genomic scale is the sparseness of structural databases. Despite significant investment in structural genomics, atomic resolution structural information is available for only 10,000 unique targets, of which approximately 62 % (6,173/10,000) are prokaryotic in origin (*see* http://sbkb.org/tt/). As of September 4th, 2013, atomic resolution structural information was available for only approximately 10 % of the open reading frames in the human genome (2,213 structures). Moreover, classes of targets that have historically proven to be highly druggable (membrane proteins in general and G-protein-coupled receptors in particular) are significantly underrepresented in structural databases.

Limited knowledge of target structure and function has led to extensive use of high-throughput empirical approaches for target validation and drug discovery. High-throughput screening (HTS) of arrayed chemical libraries has been a staple of drug discovery in the pharmaceutical industry for decades, and screening proprietary molecular diversity expedites the pharmaceutical business model. While the limited success of HTS in discovering pharmacologically tractable targets may in part reflect the extent of the druggable genome or limitations associated with high-throughput assay technology, one obvious problem with arrayed library screening is that compound collections designed to engage traditional targets such as enzymes and small-molecule receptors may be poorly suited to interact with unconventional targets [12].

Interactome analysis indicates that the majority of proteins in biological systems function as homo-oligomers and/or as components of macromolecular complexes [13]. The sequence of a potential target may therefore incorporate peptides that display autoregulatory activity, and the sequence of gene products that are known or predicted to have functional interactions with a particular target, or that are genetically implicated in a particular pathway, may also offer rich sources of potential peptide effectors to modulate the function of a target or a pathway of interest. Mining genes implicated in the pathogenesis of disease or involved in pathways of interest is therefore a practical approach to focus discovery efforts in a region of sequence space that is likely to be rich in functional compounds. Indeed, deletion mutagenesis is often employed to uncover functional elements within genes, and atomic resolution structural data is commonly used to identify sub-fragments of larger structures that can regulate function. For example, peptide antagonists of the interaction between p53 and MDM2 [14] and between pro- and antiapoptotic members of the Bcl2 family [15] have been discovered through a combination of deletion mutagenesis and structural studies. These peptides were used to identify sites on their targets that enabled the development of first-in-class therapeutic candidates (Nutlins [16] and ABT-737 [17]). These examples suggest that general, efficient methods to diversify genomic sequence into libraries of constituent peptides could provide a valuable source of molecular diversity to interrogate the druggable genome and evaluate the therapeutic potential of candidate genes.

Here we describe the use of incremental truncation as a tool to diversify open reading frames into libraries of constituent peptides. Incremental truncation has been used extensively to generate libraries of heterodimeric or chimeric gene products for protein engineering applications [18–28], but to our knowledge it has never been used previously as a method to generate molecular diversity for the discovery of peptide effectors or for the validation of druggable targets. Incremental truncation enables production of peptide libraries derived from either continuous or discontinuous sequence elements, as will be described in Subheadings 3.1 and 3.2, respectively.

2 Materials

1. Restriction endonucleases with recommended reaction buffers.
2. Plasmid purification columns (Qiagen, Valencia, CA) and supplied buffers.
3. *E. coli* exonuclease III (Exo III).
4. Mung bean single-stranded endonuclease (MB).
5. dNTP stock solution (10 mM each of dATP, dCTP, dGTP, and dTTP).
6. Klenow fragment from *E. coli* DNA polymerase I (exo+).
7. Absolute ethanol (molecular biology grade) stored at −20 °C.
8. Polyethylene glycol 6000 (50 % solution).
9. Pellet Paint precipitation reagent and supplied buffers (EMD Biosciences, La Jolla, CA).
10. T4 DNA ligase and supplied buffer.
11. DH5α-E electrocompetent cells (Invitrogen Corporation, Carlsbad, CA) stored at −80 °C.
12. Electroporation cuvettes (BIO-RAD Laboratories, Hercules, CA).
13. 1 M $MgCl_2$, filter sterilized.
14. 1 M $MgSO_4$, filter sterilized.
15. 1 M dextrose solution, filter sterilized.

16. SOC bacterial transformation recovery medium: 0.5 % (w/v) yeast extract, 2 % (w/v) tryptone, 10 mM NaCl, and 2.5 mM KCl autoclaved at 121 °C/30 psi for 25 min and allowed to cool to below 60 °C; $MgCl_2$ added to a final concentration of 10 mM, $MgSO_4$ added to a final concentration of 20 mM, and dextrose added to a final concentration of 20 mM.
17. 1.5 % LB-agar plates: 25 g Miller LB Broth (Fisher Scientific, Pittsburgh, PA) and 15 g agar per liter, autoclaved at 121 °C/30 psi for 25 min and cooled to 60 °C prior to addition of appropriate antibiotic(s); 8–10 ml of 1.5 % LB-agar per 100 mm × 15 mm plate; 180–200 ml of 1.5 % LB-agar per 25 cm × 25 cm bioassay dish.
18. 2× YT bacterial cell culture medium: 16 g tryptone, 10 g yeast extract, 5 g NaCl per liter; autoclave at 121 °C/30 psi for 25 min.
19. 50 % (v/v) glycerol solution autoclaved at 121 °C/30 psi for 25 min.
20. Agarose (electrophoresis grade).
21. 0.5 M EDTA, pH 8.0.
22. Tris–acetate–EDTA buffer: 50×: 242 g Tris–HCl base, 100 ml 0.5 M EDTA, 57 ml glacial acetic acid diluted to 1 l.
23. Ethidium bromide (10 mg/ml solution).
24. Oligonucleotide primers complementary to sequence elements 5′ and 3′ to the sequence of interest on the expression plasmid (*X* and *Y* in Figs. 1 and 3).
25. Taq DNA polymerase and supplied buffer.
26. DNA sizing ladder.

3 Methods

Time-based incremental truncation affords a convenient method to diversify open reading frames into comprehensive libraries of constituent peptides derived from either continuous (Subheading 3.1) or discontinuous (Subheading 3.2) sequence elements (*see* **Note 1**). Library members can be expressed as free linear peptides, but they can also be tethered amino- or carboxyl-terminal to a fusion partner, displayed from an internal loop as a peptide aptamer [29–31] or expressed as backbone cyclic peptides [32] through the incorporation of appropriate flanking elements in expression constructs (*see* **Note 2**).

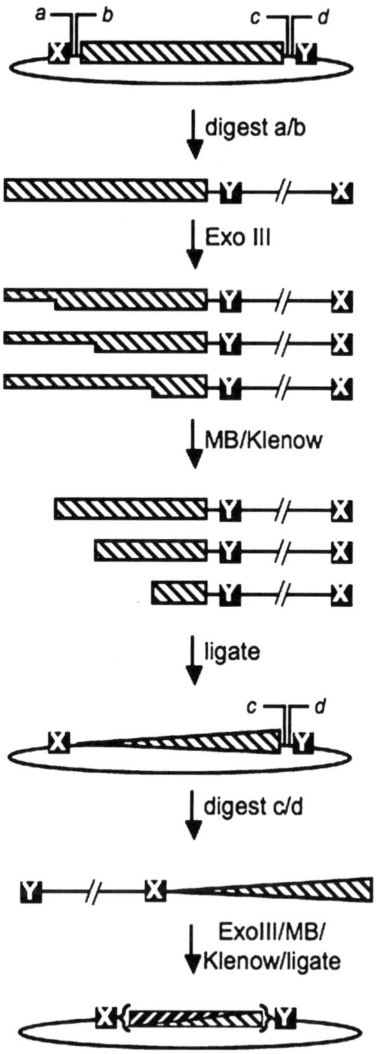

Fig. 1 Generating comprehensive libraries of continuous peptides derived from genomic sequence via successive incremental truncation. An expression vector is created with a sequence of interest (*striped bar*) interposed between two sequence elements (*X* and *Y*) and four unique restriction sites (*a–d*). Restriction sites proximal to the sequence of interest (*b* and *c*) generate Exo III-sensitive ends (blunt or 3′-recessed), while those distal from the sequence of interest (*a* and *d*) generate Exo III-resistant ends (5′-recessed). Linearization via double digestion at sites *a* and *b* directs Exo III activity into the sequence of interest, which generates a 5′-truncation library. Treatment with MB eliminates single-stranded overhangs, and treatment with Klenow produces polished blunt ends for efficient intramolecular ligation. The resulting 5′-truncation library can serve as a substrate for truncation in the opposite direction following linearization via double digestion with restriction endonucleases *c* and *d*. Treatment with Exo III, MB, Klenow, and ligase generates a library consisting of every continuous peptide encoded by the sequence of interest

3.1 Preparation of Peptide Libraries Derived from Continuous Sequence Elements Through Successive Incremental Truncation Reactions

1. Prepare an expression vector with the sequence of interest cloned between two pairs of unique restriction sites (*a/b* and *c/d* in Fig. 1, *see* **Note 3**) embedded between desired flanking elements (*X* and *Y* in Fig. 1).
2. Linearize 10 μg of supercoiled DNA by double (if possible) or sequential digestion with restriction enzymes *a* and *b* (*see* **Notes 4–7**).
3. Column purify the digestion reaction mixture using the QIAquick PCR purification kit, following the manufacturer's suggested protocol, and elute with 100 μl of low-salt (EB) buffer. Determine the concentration of the eluted DNA by absorbance at 260 nm (where a unit of absorbance corresponds to 50 μg/ml of double-stranded DNA).
4. Equilibrate a low-pH/high-ionic-strength quench buffer on ice in a 1.6 ml tube (Qiagen buffer PB, which is supplied with PCR purification kits, is a convenient choice). Five volumes of quench are needed per volume of exonuclease reaction.
5. Prepare the exonuclease reaction in a thin-walled tube on a thermocycler at a final DNA concentration of 30 ng/μl in 1× reaction buffer (NEB1 or buffer recommended by the manufacturer). The total volume of the Exo III reaction mixture is a function of the truncation length and desired resolution (*see* **Notes 8** and **9**). Prior to the addition of Exo III (**step 6**), 1 μl of the reaction mixture is withdrawn and pipetted into the quench tube (**step 4**) to ensure that the parental construct is maintained in the library.
6. Add 100 units of Exo III per μg of DNA to initiate the reaction.
7. Remove 1 μl aliquots at regular intervals (typically every 10–20 s; *see* **Note 10**) and pipette into the chilled quench tube (**step 4**) to generate the truncation library. Note that all time points are pipetted into the same tube.
8. Concentrate and change the buffer of the truncation library with a QIAquick PCR purification column. Elute DNA in $(54 - z)$ μl of EB, where *z* represents the volume of MB added in **step 10**.
9. Add 6 μl of 10× MB reaction buffer and incubate at 30 °C.
10. Add 1 unit of MB per μg of DNA (*see* **Note 11**) and incubate at 30 °C for 30 min.
11. Purify the reaction mixture using a QIAquick column and elute with 10 μl buffer EB per μg of DNA. For volumes less than 30 μl, use MinElute columns (Qiagen, Valencia, CA). Confirm DNA concentration by absorbance at 260 nm.
12. Add 0.1 volumes of T4 DNA ligase buffer (*see* **Note 12**) and dNTPs to a final concentration of 100 μM to DNA eluted in **step 11** and incubate at 37 °C.

13. Add 1 unit of Klenow fragment per μg of DNA, mix gently but thoroughly, and incubate at 37 °C for 5 min.
14. Heat denature the Klenow reaction at 75 °C for 20 min.
15. Dilute the reaction to a final concentration of 3 ng/μl DNA with 1× T4 DNA ligase buffer including 5 % PEG 6000 to promote intramolecular ligation. Equilibrate the solution to 16 °C and initiate ligation with 5 Weiss-units of T4 DNA ligase per μg of DNA. Incubate at 16 °C overnight.
16. Concentrate the DNA by ethanol precipitation using Pellet Paint. To enhance recovery, mix the DNA solution with 2 μl Pellet Paint and 0.1 volumes 3 M sodium acetate, then add 2 volumes of absolute ethanol (at −20 °C), and incubate on ice for 5 min. Spin for 5 min at maximum speed in a microfuge, discard the supernatant, and wash the pellet with 1 volume of room-temperature 70 % ethanol. Allow the pellet to air-dry and resuspend at a concentration of approximately 10 ng/μl in EB.
17. Prewarm SOC to 37 °C. 1 ml of SOC will be needed for each transformation.
18. Transform the ethanol-precipitated truncation library ligation mixture into competent *E. coli* (*see* **Notes 13** and **14**). Recover each transformation in 1 ml of SOC. Pool the transformations and incubate at 37 °C for 45 min with shaking.
19. Plate a dilution series (10^6, 10^4, and 10^2 dilutions on 100 mm plates) on LB-agar containing antibiotic(s) at the appropriate concentration to select for ligation products. Pellet the remainder of the transformation by brief centrifugation at room temperature. Resuspend the pelleted bacteria in 0.5–1 ml of warm SOC and spread on LB-agar plus antibiotic(s) on a 25 cm × 25 cm bioassay dish (*see* **Note 15**). Incubate all plates overnight at 37 °C.
20. Harvest the library from the bioassay dish with 10 ml of harvest medium (6 ml 2× YT, 3 ml 50 % glycerol, 1 ml 1 M dextrose). Freeze 0.5 ml aliquots in liquid nitrogen and store at −80 °C.
21. Prepare supercoiled plasmid DNA encoding the truncation library generated in **steps 1–20** (*see* **Notes 4** and **16**).
22. Linearize 10 μg of supercoiled plasmid by double or sequential digestion with restriction enzymes c and d (*see* **Notes 4–6**).
23. Repeat **steps 3–20**, assuming a truncation length (*L*, *see* **Note 8**) equal to the size of the original sequence of interest in base pairs.
24. Evaluate library coverage using 0.2 μl of harvested library and 20 (or more) randomly selected colonies as template in 25 μl of PCR reaction mixture per amplification (1× reaction buffer, 250 μM dNTP (final), and 0.5 μM oligonucletide primers;

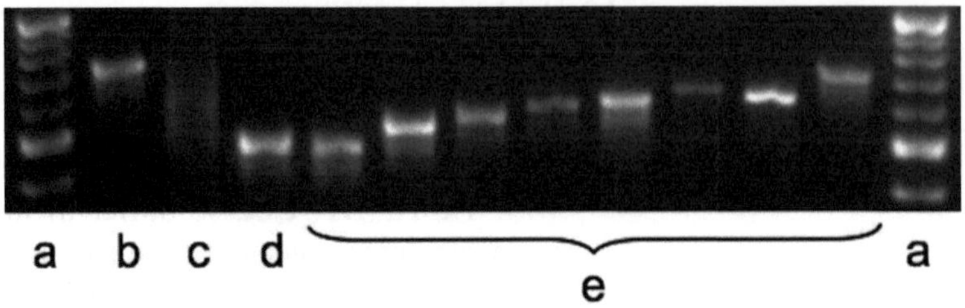

Fig. 2 Evaluating library size and coverage through colony PCR and agarose gel electrophoresis. Small aliquots of harvested library (*c*) and randomly selected colonies (*e*) were used as templates for PCR with outside primers complementary to *X* and *Y* (*see* Fig. 1). DNA products were separated on a 1.5 % agarose gel stained with ethidium bromide. Amplified products from the parental construct prior to truncation (*b*) and a control construct lacking the sequence of interest (*d*) were used to define the upper and lower limits of the desired truncation range. Randomly selected colonies were loaded in order of increasing size to aid in visualization of library coverage

see **Note 17**). Colonies from transformation of the parental plasmid (**step 1**) and the parental plasmid without the sequence of interest (*see* **Note 18**) are useful controls.

25. Visualize the size of library members by agarose gel electrophoresis (agarose and 1× Tris–acetate–EDTA buffer with ethidium bromide (1:25,000 dilution from 10 mg/ml stock), *see* Fig. 2).
26. Prepare plasmid from the colonies selected in **step 24**, and sequence each library member in both directions using the primers from **step 24** to evaluate the quality of the library.

3.2 Preparation of Peptide Libraries Derived from Discontinuous Sequence Elements Through Incremental Truncation

1. Prepare an expression vector including an Exo III-sensitive, unique restriction site within a sequence of interest (*e* in Fig. 3), and a second unique restriction site on the vector outside the sequence of interest (*f* in Fig. 3; *see* **Note 19**).
2. Linearize 10 μg of supercoiled vector by digestion with the restriction enzyme corresponding to site *e* (*see* **Note 6**).
3. Perform **steps 3–11** as described in Subheading 3.1.
4. Add 0.1 volumes of 10× buffer for restriction enzyme *f* and dNTPs to a final concentration of 100 μM to DNA eluted in **step 3** (above) and incubate at 37 °C (*see* **Note 12**).
5. Add 1 unit of Klenow fragment per μg of DNA, mix gently but thoroughly, and incubate at 37 °C for 5 min.
6. Heat denature the Klenow reaction at 75 °C for 20 min.
7. Digest the truncation library with 1 unit of restriction enzyme *f* per μg of DNA (*see* **Notes 6** and **20**) and incubate for 1–3 h at the temperature recommended for optimal enzymatic activity. Monitor the reaction by agarose gel electrophoresis to gauge completion.

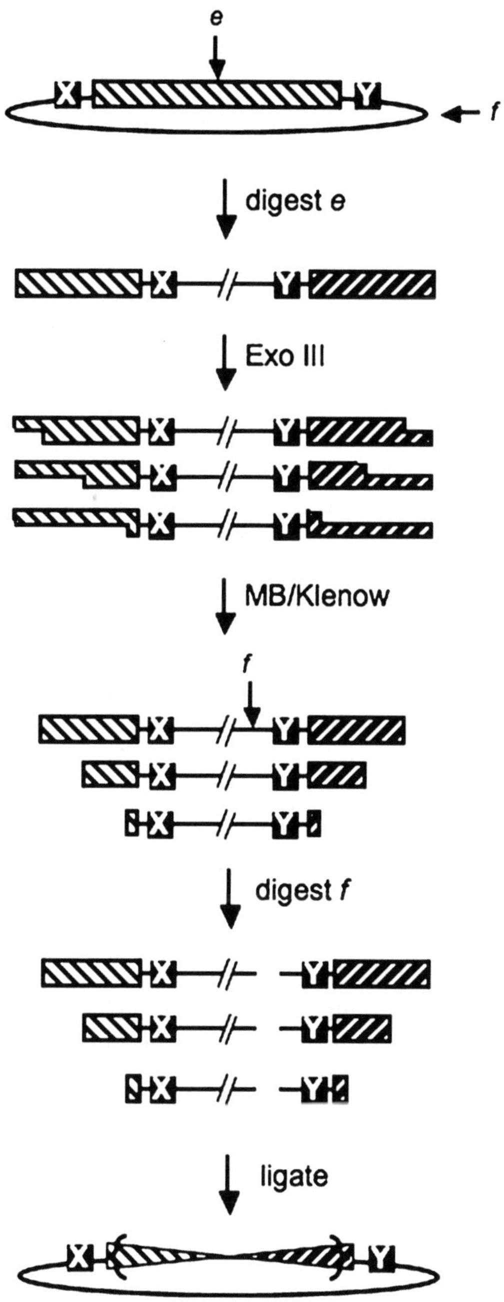

Fig. 3 Generating libraries of discontinuous peptides derived from genomic sequence via bidirectional truncation and independent assortment. An expression vector is created with a sequence of interest (*striped bar*) interposed between two sequence elements (*X* and *Y*, *see* **Note 2**). The vector is linearized via digestion at a unique restriction site within the sequence of interest that generates an Exo III-sensitive end (*e*). Treatment with Exo III, MB, and Klenow generates a library of truncations that are symmetrically displaced around the initiation site. Digestion with a second unique restriction endonuclease at a site outside of the sequence of interest (*f*) enables regions of the truncation library that are 5′- and 3′- from the initiation site to assort independently upon ligation, which generates library members encoding discontinuous peptides with sequences that are both symmetric and asymmetric with respect to the initiation site

8. Concentrate the digestion by ethanol precipitation. Add 2 μl of pellet paint, 0.1 volumes of 3 M sodium acetate, and 2 volumes of absolute ethanol (–20 °C) and incubate on ice for 5 min prior to centrifugation at maximum speed in a microfuge. Discard the supernatent, wash the pellet with 1 volume of room-temperature 70 % ethanol, and suspend in EB at high concentration (>100 ng/μl, if possible).
9. Set up multiple ligations with 150–300 ng of digested, ethanol-precipitated DNA (from **step 8**; *see* **Note 21**) in 10 μl of 1× ligase buffer plus 2 Weiss-units of T4 DNA ligase. Incubate for at least 6 h at 15 °C (preferably overnight).
10. Pool the ligations. Add 2 μl of pellet paint, 0.1 volumes of 3 M sodium acetate, and 2 volumes of absolute ethanol (–20 °C) and incubate on ice for 5 min prior to centrifugation at maximum speed in a microfuge. Discard the supernatant, wash the pellet with one volume of room-temperature 70 % ethanol, and suspend in EB at 10–20 ng/μl.
11. Prewarm SOC to 37 °C. 1 ml of SOC will be needed for each transformation.
12. Chill electroporation cuvettes in an ice bucket. One cuvette is needed for each 10–20 μg (1–2 μl) of ethanol-precipitated ligation mixture (*see* **Notes 13** and **14**).
13. Transform the ethanol-precipitated truncation library from **step 10** into electrocompetent *E. coli* using the transformation protocol recommended by the manufacturer of the transformation apparatus. Recover each transformation in 1 ml of SOC. Pool the transformations and incubate at 37 °C for 45–60 min with shaking.
14. Plate a dilution series (10^6, 10^4, and 10^2 dilutions on 100 mm plates) on LB-agar containing antibiotic(s) at the appropriate concentration to select for ligation products. Pellet the remainder of the transformation by brief centrifugation at room temperature. Resuspend the pelleted bacteria in 0.5–1 ml of warm SOC and spread on LB-agar plus antibiotic(s) on a 25 cm × 25 cm bioassay dish (*see* **Note 15**). Incubate all plates overnight at 37 °C.
15. Harvest the library from the bioassay dish with 10 ml of harvest medium (6 ml 2× YT, 3 ml 50 % glycerol, 1 ml 1 M dextrose). Freeze 0.5 ml aliquots in liquid nitrogen and store at -80 °C.
16. Aliquot 25 μl of PCR reaction mixture (1× reaction buffer, 250 μM dNTP (final), and 0.5 μM oligonucleotide primers) per amplification (*see* **Note 17**). Evaluate library coverage using 0.2 μl of harvested library (from **step 15**) and 20 (or more) randomly selected colonies as template. Colonies from

transformation of the parental plasmid (**step 1**) and the parental plasmid without the sequence of interest (*see* **Note 18**) are useful controls. Evaluate the length range of the library and randomly selected clones by agarose electrophoresis (*see* Fig. 2 for an example).

17. Prepare plasmid from the colonies selected in **step 16**, and sequence each library member in both directions using the primers from **step 16** to evaluate the quality of the library.

4 Notes

1. Random fragmentation of genomic sequence with DNase gives rise to libraries of constituent peptides in a single step [33], but this approach generates fragment libraries that are biased to a particular length based on the amount of DNase used and the incubation time. In addition, fragments encoding short peptides (<20 amino acids), which are of the greatest interest for validating pharmacologic targets, are often selectively depleted during cloning and purification of libraries generated by DNase fragmentation. Sequential truncation generates libraries encoding every constituent peptide, and library sequences remain covalently tethered to expression vectors throughout library generation, which minimizes loss of constructs encoding short peptides because the vector sequence promotes essentially uniform behavior from all library members.
2. For libraries of free linear peptides, *X* in Figs. 1 and 3 corresponds to promoter and start codon and *Y* corresponds to stop codon(s) and termination elements. For peptide libraries tethered amino- or carboxyl-terminal to a protein or displayed from an internal loop as a peptide aptamer [29–31], *X* and/or *Y* corresponds to a fusion partner. For libraries of backbone cyclic peptides, *X* corresponds to the carboxyl-terminal component of an intein [32] protein splicing domain (C-intein or I_C) and Y corresponds to the amino-terminal component (N-intein or I_N).
3. Double or sequential digestion with two unique restriction endonucleases is necessary to generate two types of DNA ends. Proximal sites (*b* or *c* in Fig. 1) must generate Exo III-susceptible blunt or 3′-recessed ends to promote digestion of the sequence of interest. Distal sites (*a* or *d* in Fig. 1) must generate Exo III-resistant ends (5′-recessed) to protect flanking elements from digestion (*X* or *Y* in Fig. 1).
4. Supercoiled DNA is critical for efficient library generation using the incremental truncation method [34, 35]. While there

are commercial reagents that restore supercoiling to nicked DNA (Supercoil-It, Bayou Biolabs, Harahan, LA), plasmid preparation from fresh inocula grown to an optical density (OD_{600}) of less than 2 gives a preponderance of supercoiled DNA in our hands.

5. Column purification steps are a significant contributor to loss of DNA during library production. Often, these steps serve to change buffer composition and/or concentrate DNA. Combining steps (double digestion) or avoiding buffer changes (for example, when performing a sequential digestion, performing the digest requiring the lower salt buffer first, then modifying the buffer for compatibility with the enzyme requiring higher salt to eliminate the need for a column purification step) can reduce loss of DNA.
6. Nicked DNA is incompatible with incremental truncation because it acts as a substrate for Exo III, leading to nonspecific loss of DNA and reduced library sizes. It is important to avoid nicking DNA substrates in all steps prior to Exo III digestion. Restriction digestion can generate nicked products due to star activity, which is often promoted at high enzyme concentration, long incubation times, or non-optimal buffer. However, incomplete digestion leads to libraries biased with the parental construct, which also reduces useful library size. It is therefore worthwhile to optimize DNA linearization steps empirically to determine the minimal concentration of restriction enzymes and minimal incubation times necessary to generate linearized substrate.
7. Figure 1 depicts production of the successive truncation library with 5′→3′ truncation preceding 3′→5′ truncation, but the opposite order of events (3′→5′ truncation preceding 5′→3′ truncation) would in principle produce the same final successive truncation library of constituent polypeptides.
8. The volume of the exonuclease reaction (v) varies as a function of the desired length of truncation and desired resolution based on the empirical formula $v = L/R$ where L represents the length of the sequence of interest (in base pairs) and R represents the desired resolution (in base pairs per μl sampled). For example, to truncate an open reading frame of 300 base pairs at single codon resolution (3 base pairs per μl sampled), a final volume of 100 μl is needed. This corresponds to 3 μg of DNA substrate at a concentration of 30 ng DNA substrate per μl [34]. Truncating an open reading frame of 10 kilobase pairs at 100 base pair resolution also requires a volume of 100 μl and 3 μg of DNA substrate.
9. Resolution is a function of truncation rate and sampling rate based on the relationship $R = r/s$, where r represents rate in

base pairs per minute, and *s* represents sampling rate in μl per minute. While, in principle, either reaction rate or sampling rate can be altered to affect resolution, in practice it is far easier to control the rate of Exo III reaction through changes in temperature and buffer ionic strength [34, 36].

10. It is difficult to manually collect more than 3–5 aliquots per minute, and physically challenging to maintain this rate for more than 45–60 min. Diversifying sequences longer than about 600 base pairs (encoding proteins greater than 200 amino acids) at single codon resolution requires automated sampling, or truncation in two passes at different rates.
11. Removal of single strands by mung bean endonuclease is critical for efficient library generation, so optimizing this step empirically is worthwhile (*see* Ostermeier and Lutz [34] for a detailed protocol to optimize the mung bean endonuclease step).
12. The Klenow fragment of DNA polymerase I from *E. coli* is very tolerant of a wide range of buffer conditions. Performing the Klenow reaction in the buffer needed for the subsequent step saves a column purification step, which increases DNA yield.
13. Either electroporation or chemical transformation may be used for the initial truncation library (Subheading 3.1, **step 18**) because the size of this library is relatively small (equal to the number of base pairs in the sequence of interest ($<10^4$)). Electroporation is recommended for transformation of the library resulting from the second truncation step (Subheading 3.1, **step 23**) because the size of these libraries is approximated by the square of the number of base pairs in the sequence of interest, which can easily exceed 10^6 members. Theoretical sizes for libraries generated by bidirectional truncation and de-symmetrization (Subheading 3.2) are equal to the square of the maximal number of base pairs truncated, which is defined by the distance from the initiation site to the furthest end of the sequence of interest (*see* Fig. 3).
14. Transformation efficiency increases as DNA concentration decreases, particularly for electroporation. However, DNA concentration determines the number of transformations necessary to deliver the library. In our experience 10–20 ng of DNA per transformation provides a reasonable compromise between high transformation efficiency and a manageable number of transformations.
15. Both overly wet and overly dry plates can negatively affect colony yield. This is particularly true for bioassay dishes used for plating libraries. Air drying inverted bioassay dishes at room temperature with the base rotated 45° relative to the lid for 30–60 min prior to plating provides a level of dryness that is conducive to reasonable colony yield.

16. Although plasmid purification from fresh inocula provides the highest quality supercoiled DNA (*see* **Note 4**), bacteria harvested from library plates (Subheading 3.1, **step 20** and Subheading 3.2, **step 15**) can also be used for plasmid preparation directly without the need for growth in liquid culture. The quality of the resulting plasmids is often sufficient for incremental truncation, to transform or transfect into hosts for selection or screening, or as a template for PCR or sequencing.
17. In our hands, Taq polymerase is the best choice for colony PCR. Other polymerases fail to give product with bacteria as template to a much higher degree. With Taq, the efficiency of amplification is about 80–90 % from colonies that harbor target plasmids. Amplifying multiple colonies is worthwhile to ensure that enough amplification products are generated to have confidence that the truncation library affords coverage of the desired truncation range (*see* Fig. 2 for example).
18. Primers for colony PCR and sequencing should be designed based on sequence elements 5′ (*X*) and 3′ (*Y*) to the sequence of interest that are displaced a minimum of 150–200 base pairs from the expected truncation range (*see* Figs. 1 and 3). This way, constructs devoid of the sequence of interest (direct fusions of *X* and *Y*) give visible PCR products that define the smallest desired members of the library (*see* Fig. 2). The additional 300–400 base pairs of sequence appended to each library member improve the efficiency of preferential amplification [37] as a method to enrich for the smallest members of the library following functional selection.
19. Deletion libraries generated by the action of Exo III on plasmid DNA digested at a single, Exo III-susceptible site (*e* in Fig. 3) are symmetrical, but the symmetry can be broken by digestion at a unique site with a second restriction enzyme (*f* in Fig. 3) [34]. Ligation of the resulting digestion mixture enables the deletion libraries 5′ and 3′ from the linearization site to assort independently and generate products corresponding to both symmetrical and asymmetrical deletions from the linearization site. To avoid loss of library members encoding short peptides, it is important that the distance between sites *e* and *f* be substantial (>500 base pairs).
20. Since no additional Exo III steps follow the restriction digestion necessary to de-symmetrize the discontinuous deletion library, longer incubations and higher concentrations of restriction endonuclease are less problematic at this step than when plasmids are being linearized prior to Exo III digestion. Complete digestion is important to avoid biasing the discontinuous deletion library with symmetrical deletion constructs.

21. DNA concentration affects ligation efficiency. For large libraries (>10^6 members) it is worthwhile to determine the optimal DNA concentration per ligation empirically (between 100 and 400 ng/ligation) before ligating the majority of the library DNA.

Acknowledgements

The authors wish to thank Eileen Burke, Shirish Damle, and Alex Smith for material contributions and helpful discussions. This work was supported by National Institutes of Health grant R01-AI053800 (awarded to CPS).

References

1. Imming P, Sinning C, Meyer A (2006) Drugs, their targets and the nature and number of drug targets. Nat Rev Drug Discov 5:821–834
2. Overington JP, Al-Lazikani B, Hopkins AL (2006) How many drug targets are there? Nat Rev Drug Discov 5:993–996
3. International Human Genome Sequencing Consortium (2004) Finishing the euchromatic sequence of the human genome. Nature 431:931–945
4. Chatterjee-Kishore M, Miller CP (2005) Exploring the sounds of silence: RNAi-mediated gene silencing for target identification and validation. Drug Discov Today 10:1559–1565
5. Houdebine LM (2007) Transgenic animal models in biomedical research. Methods Mol Biol 360:163–202
6. Drews J (2006) What's in a number? Nat Rev Drug Discov 5:975
7. Drews J, Ryser S (1997) The role of innovation in drug development. Nat Biotechnol 15:1318–1319
8. Reichert JM (2003) Trends in development and approval times for new therapeutics in the United States. Nat Rev Drug Discov 2:695–702
9. Sams-Dodd F (2005) Target-based drug discovery: is something wrong? Drug Discov Today 10:139–147
10. Swindells MB, Overington JP (2002) Prioritizing the proteome: identifying pharmaceutically relevant targets. Drug Discov Today 7:516–521
11. Marshall A (2008) Raising the game. Nat Biotechnol 26:137
12. Wells JA, McClendon CL (2007) Reaching for high-hanging fruit in drug discovery at protein-protein interfaces. Nature 450:1001–1009
13. Devos D, Russell RB (2007) A more complete, complexed and structured interactome. Curr Opin Struct Biol 17:370–377
14. Kussie PH, Gorina S, Marechal V, Elenbaas B, Moreau J, Levine AJ, Pavletich NP (1996) Structure of the MDM2 oncoprotein bound to the p53 tumor suppressor transactivation domain. Science 274:948–953
15. Petros AM, Nettesheim DG, Wang Y, Olejniczak ET, Meadows RP, Mack J, Swift K, Matayoshi ED, Zhang H, Thompson CB, Fesik SW (2000) Rationale for Bcl-xL/Bad peptide complex formation from structure, mutagenesis, and biophysical studies. Protein Sci 9:2528–2534
16. Vassilev LT, Vu BT, Graves B, Carvajal D, Podlaski F, Filipovic Z, Kong N, Kammlott U, Lukacs C, Klein C, Fotouhi N, Liu EA (2004) In vivo activation of the p53 pathway by small-molecule antagonists of MDM2. Science 303:844–848
17. Oltersdorf T, Elmore SW, Shoemaker AR, Armstrong RC, Augeri DJ, Belli BA, Bruncko M, Deckwerth TL, Dinges J, Hajduk PJ, Joseph MK, Kitada S, Korsmeyer SJ, Kunzer AR, Letai A, Li C, Mitten MJ, Nettesheim DG, Ng S, Nimmer PM, O'Connor JM, Oleksijew A, Petros AM, Reed JC, Shen W, Tahir SK, Thompson CB, Tomaselli KJ, Wang B, Wendt MD, Zhang H, Fesik SW, Rosenberg SH (2005) An inhibitor of Bcl-2 family proteins induces regression of solid tumours. Nature 435:677–681

18. Griswold KE, Kawarasaki Y, Ghoneim N, Benkovic SJ, Iverson BL, Georgiou G (2005) Evolution of highly active enzymes by homology-independent recombination. Proc Natl Acad Sci U S A 102:10082–10087
19. Kawarasaki Y, Griswold KE, Stevenson JD, Selzer T, Benkovic SJ, Iverson BL, Georgiou G (2003) Enhanced crossover SCRATCHY: construction and high-throughput screening of a combinatorial library containing multiple non-homologous crossovers. Nucleic Acids Res 31:e126
20. Luker KE, Smith MC, Luker GD, Gammon ST, Piwnica-Worms H, Piwnica-Worms D (2004) Kinetics of regulated protein-protein interactions revealed with firefly luciferase complementation imaging in cells and living animals. Proc Natl Acad Sci U S A 101:12288–12293
21. Lutz S, Ostermeier M, Benkovic SJ (2001) Rapid generation of incremental truncation libraries for protein engineering using alpha-phosphothioate nucleotides. Nucleic Acids Res 29:E16
22. Lutz S, Ostermeier M, Moore GL, Maranas CD, Benkovic SJ (2001) Creating multiple-crossover DNA libraries independent of sequence identity. Proc Natl Acad Sci U S A 98:11248–11253
23. Ostermeier M, Nixon AE, Shim JH, Benkovic SJ (1999) Combinatorial protein engineering by incremental truncation. Proc Natl Acad Sci U S A 96:3562–3567
24. Ostermeier M, Shim JH, Benkovic SJ (1999) A combinatorial approach to hybrid enzymes independent of DNA homology. Nat Biotechnol 17:1205–1209
25. Park SH, Park HY, Sohng JK, Lee HC, Liou K, Yoon YJ, Kim BG (2009) Expanding substrate specificity of GT-B fold glycosyltransferase via domain swapping and high-throughput screening. Biotechnol Bioeng 102:988–994
26. Tarendeau F, Boudet J, Guilligay D, Mas PJ, Bougault CM, Boulo S, Baudin F, Ruigrok RW, Daigle N, Ellenberg J, Cusack S, Simorre JP, Hart DJ (2007) Structure and nuclear import function of the C-terminal domain of influenza virus polymerase PB2 subunit. Nat Struct Mol Biol 14:229–233
27. Whitehead TA, Je E, Clark DS (2009) Rational shape engineering of the filamentous protein gamma prefoldin through incremental gene truncation. Biopolymers 91:496–503
28. Yadid I, Tawfik DS (2007) Reconstruction of functional beta-propeller lectins via homo-oligomeric assembly of shorter fragments. J Mol Biol 365:10–17
29. Norman TC, Smith DL, Sorger PK, Drees BL, O'Rourke SM, Hughes TR, Roberts CJ, Friend SH, Fields S, Murray AW (1999) Genetic selection of peptide inhibitors of biological pathways. Science 285:591–595
30. Abedi MR, Caponigro G, Kamb A (1998) Green fluorescent protein as a scaffold for intracellular presentation of peptides. Nucleic Acids Res 26:623–630
31. Geyer CR, Colman-Lerner A, Brent R (1999) "Mutagenesis" by peptide aptamers identifies genetic network members and pathway connections. Proc Natl Acad Sci U S A 96:8567–8572
32. Saleh L, Perler FB (2006) Protein splicing in cis and in trans. Chem Rec 6:183–193
33. Gudkov AV, Roninson IB (1997) Isolation of genetic suppressor elements (GSEs) from random fragment cDNA libraries in retroviral vectors. Methods Mol Biol 69:221–240
34. Ostermeier M, Lutz S (2003) The creation of ITCHY hybrid protein libraries. Methods Mol Biol 231:129–141
35. Ostermeier M, Nixon AE, Benkovic SJ (1999) Incremental truncation as a strategy in the engineering of novel biocatalysts. Bioorg Med Chem 7:2139–2144
36. Hoheisel JD (1993) On the activities of *Escherichia coli* exonuclease III. Anal Biochem 209:238–246
37. Kawarasaki Y, Sasaki Y, Ikeuchi A, Yamamoto S, Yamane T (2002) A method for functional mapping of protein–protein binding domain by preferential amplification of the shortest amplicon using PCR. Anal Biochem 303:34–41

Chapter 7

Identification of Ideal Peptides for Heterovalent Ligands

Ajay Shrivastava, Matthew von Wronski, Michael F. Tweedle, and Adrian D. Nunn

Abstract

Recent use of hetero-dimerization to improve the affinity of peptide ligands has made peptides an attractive alternative to small molecules and proteins. Using this strategy, we have developed peptides with affinities comparable to antibodies and specificities often better than small molecules or antibodies. These peptides can be used as a delivery vehicle for drugs or diagnostics, especially in the case of tumor targeting cytotoxic drugs or targeted diagnostics. We describe here an assay to identify an ideal pair of peptides suitable for heterovalent ligands.

Key words Heterodimerization, Heterovalent ligands, Peptide, VEGFR-2, Tetravalent, Neutravidin cross-competition, Cooperativity

1 Introduction

Most peptide usage has been limited to polyvalent vaccine development [1] or peptide hormones directed against G-protein-coupled receptors, GPCRs [2], due to the lower affinity and faster clearance of peptides compared to antibodies and small molecules. Recent developments in targeted cytotoxic drugs (radiotherapy and toxins) and targeted imaging are mainly responsible for revived interest in peptides [3, 4]. Traditionally, peptide usage for even targeted therapy and diagnosis has been limited to peptide receptors (GPCRs) [3]. However, a newer method of using heterovalency to create high affinity peptide ligands for other drug targets like receptor tyrosine kinases (RTKs) and possibly other proteins [4] may expand peptide usage to non-GPCR drug-targeted therapies and diagnostics.

Multivalent interactions are often more potent than monovalent ones [5]. An important new tactic involves creating multivalent "heterovalent" peptide conjugates [4]. By binding to two separate

Andrew E. Nixon (ed.), *Therapeutic Peptides: Methods and Protocols*, Methods in Molecular Biology, vol. 1088,
DOI 10.1007/978-1-62703-673-3_7,

epitopes on a single target molecule such bifunctional molecules can derive the benefit of multivalent binding without needing to span two target molecules, which makes their enhanced binding ability independent of target density. One important aspect of creating high affinity peptide ligands is to identify the peptides best suited for heterovalent ligands. The ideal peptides should bind to two different epitopes on a target molecule and should synergistically enhance each other's binding to the target molecule [4]. We describe here an assay to screen for such peptides. In this assay, tetravalent complexes of biotinylated peptides [6] with Neutravidin HRP (*see* **Note 1**) bound specifically to target (VEGFR-2)-transfected cells but not to mock-transfected cells [4]. Using this assay, the binding behavior of VEGFR-2-binding synthetic heterovalent ligands is predicted using a two step approach. In the first step, monomeric peptides are tested for their ability to cross-compete with tetravalent peptide/Neutravidin HRP (NA-HRP) complexes. In the second step, mixed tetravalent peptide/NA-HRP complexes are created and compared for their specific binding to VEGFR-2-transfected cells with pure tetravalent peptide/NA-HRP complexes (4P-1/NA-HRP or 4P-2/NA-HRP). This process identifies two non-competing peptides capable of cooperating. Subsequent chemical synthesis of heterovalent ligands based on this information provided us several hundred-fold stronger binding than individual peptides or their homovalent ligands [4].

2 Materials

2.1 Transfection Vector

1. Cloning vector, pcDNA6/V5-HisC, was obtained from Invitrogen.
2. As described previously, VEGFR-2 was cloned into pcDNA6/V5-HisC to produce VEGFR-2 expression vector pVEGFR-2 [4].

2.2 Cell Culture

1. 293H cells (Invitrogen).
2. 293H medium (Invitrogen).
3. Lipofectamine 2000 (Invitrogen).
4. Penicillin–streptomycin (Invitrogen).
5. Fetal bovine serum-FetalClone II (HyClone).
6. Poly-D-Lysine-coated 96-well microplates (BD biosciences).
7. Opti-MEM® I reduced serum medium (Invitrogen).
8. Filters (Corning).

a (N-acetyl)-WQPCPWESWTFCWDP-GGGKdi(aminodioxaocta)-Biotin-NH_2

b (N-acetyl)-AGPKWCEEDWYYCMITGT-GGGKdi(aminodioxaocta)-Biotin-NH_2

c (N-acetyl)-GDSRVCWEDSWGGEVCFRYDP-GGGKdi(aminodioxaocta)-Biotin-NH_2

Fig. 1 Peptide structure with linker and biotin (**a**) P-X (negative control) (**b**) P-1 and (**c**) P-2

2.3 M199 Medium

1. 1 N NaOH: 20.4 g of NaOH (Sigma) to 500 mL of distilled water (dH_2O). Store at room temperature.
2. M199 medium: M199 medium packet (Invitrogen), 20 mL of 1 M HEPES (invitrogen) and 2 g of gelatin (DIFCO) to 950 mL of dH_2O (*see* **Note 2**).
3. Adjust the pH of the solution to 7.4 by adding approximately 4 mL of 1 N NaOH.
4. After pH adjustment, the M199 medium is warmed to 37 °C in a water bath for 2 h to dissolve the gelatin, filter sterilized using 0.2 μm filters and stored at 4 °C.

2.4 SoftLink Soft Release Avidin–Sepharose

SoftLink soft release avidin–sepharose (Promega) is prepared by centrifuging the sepharose obtained from Promega at 12,000 rpm (15,300 rcf) for 2 min, washing twice with ice-cold dH_2O (centrifuging in-between the washes) and resuspending the pellet in ice-cold water to make a 50 % slurry in ddH_2O. A fresh 50 % slurry of avidin–sepharose is prepared for each experiment.

2.5 Neutravidin Solutions

1. Neutravidin HRP: 2 mg of Neutravidin HRP in 1 mL of ice-cold distilled water. Aliquots of 50 μL are prepared and stored at −80 °C to be used later (*see* **Note 3**) (Pierce).
2. Neutravidin (Pierce). A fresh blocking solution is prepared for each experiment by adding 20 mL of M199 medium to 10 mg of lyophilized unlabeled Neutravidin. A fresh blocking solution is used for each experiment.

2.6 Peptides

1. The following peptides (Fig. 1) are synthesized by Bracco's chemists using standard solid phase peptide synthesis [4, 7].
2. 250 μM stock solutions of peptides in 50 % DMSO are prepared and stored at −20 °C.

2.7 Neutravidin HRP Assay

1. TMB 2-component microwell peroxidase substrate (KPL).
2. Approximately 1 M phosphoric acid is prepared by diluting concentrated phosphoric acid (EM science) 1:8 with distilled water.
3. Neutravidin HRP (Pierce).

3 Methods

3.1 Transfection of 292H Cells

1. 293H medium with 10 % fetal bovine serum (FBS) ± 1 mL/L Penicillin–Streptomycin is prepared, filter sterilized using 0.2 μm filters, and stored at 4 °C to be used later for tissue culture.
2. 293H cells are grown in 293H medium supplemented with 10 % FBS plus 1 mL/L penicillin–streptomycin.
3. The cells are plated in a poly-D-lysine-coated 96-well plate at 50,000 cells per well (*see* **Note 4**) in 293H medium supplemented with 10 % FBS. The cells are grown in the presence of antibiotic for everyday culture but are split into antibiotic free medium 16–20 h prior to transfection (*see* **Note 5**).
4. Prepare DNA-Lipofectamine 2000 complexes (*see* **Note 6**). Dilute DNA (320 ng per well) in 50 μL of Opti-MEM® I medium without serum. Mix gently. Mix Lipofectamine 2000 gently before use, then dilute the appropriate amount (0.8 μL per well) in 50 μL of Opti-MEM® I medium. Mix gently and incubate for 5 min at room temperature. After the 5 min incubation, combine the diluted DNA with the diluted Lipofectamine 2000 (*see* **Note 7**). Mix gently and incubate for 20 min at room temperature to allow the DNA-Lipofectamine 2000 complexes to form (*see* **Note 8**).
5. In the 96-well plate, half the plate is mock-transfected (no DNA, *see* **Note 9**) and the remaining half is transfected with pVEGFR-2. 100 μL of DNA-Lipofectamine 2000 complexes is added to each well containing cells followed by 100 μL of medium. Mix gently by rocking the plate back and forth. Transfection is performed in serum-containing medium. The transfection reagent mix is removed from cells after 6–8 h and replaced with regular serum-containing medium. The next day cells are used for the assay.

3.2 Preparation of Peptide Solutions for Competition Assay

1. 10 μL of 250 μM biotinylated peptide stock solution and 10 μL of 33 μM Neutravidin HRP are added to 1 mL of M199 medium. Mixture is incubated on a rotator at 4 °C for 60 min (*see* **Note 10**).
2. 50 μL of soft release avidin–sepharose (50 % slurry in ddH$_2$O) is added to remove excess peptides. Incubate for 30 min on a rotator at 4 °C (*see* **Note 10**).
3. The soft release avidin–sepharose is pelleted by centrifuging at 15,300 rcf for 5 min at 4 °C and the supernatant retained (*see* **Note 10**). Keep on ice for later use.
4. Prepare 1.0 mL solutions 2.5 μM for each competing peptide to be tested (biotinylated or non-biotinylated, both have

	1 2	3 4	5 6	7 8	9 10	11 12
A	P-X (2.5 microM)	P-1 (2.5 microM)	P-2 (2.5 microM)	P-X (2.5 microM)	P-1 (2.5 microM)	P-2 (2.5 microM)
B	P-X (1.25 microM)	P-1 (1.25 microM)	P-2 (1.25 microM)	P-X (1.25 microM)	P-1 (1.25 microM)	P-2 (1.25 microM)
C	P-X (0.63 microM)	P-1 (0.63 microM)	P-2 (0.63 microM)	P-X (0.63 microM)	P-1 (0.63 microM)	P-2 (0.63 microM)
D	P-X (0.31 microM)	P-1 (0.31 microM)	P-2 (0.31 microM)	P-X (0.31 microM)	P-1 (0.31 microM)	P-2 (0.31 microM)
E	P-X (0.16 microM)	P-1 (0.16 microM)	P-2 (0.16 microM)	P-X (0.16 microM)	P-1 (0.16 microM)	P-2 (0.16 microM)
F	P-X (0.08 microM)	P-1 (0.08 microM)	P-2 (0.08 microM)	P-X (0.08 microM)	P-1 (0.08 microM)	P-2 (0.08 microM)
G	P-X (0.04 microM)	P-1 (0.04 microM)	P-2 (0.04 microM)	P-X (0.04 microM)	P-1 (0.04 microM)	P-2 (0.04 microM)
H	P-X	P-1	P-2	P-X	P-1	P-2
		MOCK Transfected			VEGFR-2 Transfected	

Fig. 2 A 96-well plate display for the cross-competition assay. The tetravalent P-1/NA-HRP complex was competed with monovalent P-1, P-2, or P-X. The amount of competing peptide used in competition is shown in the *brackets*

worked well) in separate tubes by adding 10 μL of 250 μM stock to 1 mL of M199 medium.

5. Use 0.5 mL from each of the 2.5 μM peptide solutions to do six 1:2 serial dilutions in M199 medium (make separate serial dilutions for each peptide). This provides 0.5 mL of 2.5, 1.25, 0.625, 0.3125, 0.16, 0.08, and 0.04 μM competing peptide solutions.
6. Add 5 μL of peptide/Neutravidin HRP complex prepared in **step 3** to each tube with 500 μL of competing peptide solution prepared in **step 5**. Use these solutions in the competition assay described below.

3.3 Cross-Competition Assay

1. Wash transfected cells once with M199 medium.
2. Add 80 μL of blocking solution containing 0.5 mg/mL Neutravidin to each well.
3. Incubate at 37 °C for 1 h (*see* **Note 10**).
4. Wash cells twice with M199 medium.
5. Add peptide/Neutravidin HRP solutions with different concentrations of competing peptide prepared in Subheading 3.2, **step 6** to **96** wells as shown in Fig. 2.
6. Incubate at room temperature for 2 h (*see* **Note 10**).
7. Transfer plates to 4 °C for 30 min (*see* **Note 10**).
8. Wash cells five times with ice-cold M199 medium.
9. Wash once with ice-cold PBS.
10. Add 100 μL of TMB solution to each well.
11. Incubate for 30 min (*see* **Note 10**).
12. Add 50 μL of 1 M phosphoric acid.
13. Read absorbance at 450 nm wavelength in a plate reader. Results from one such experiment are shown in Fig. 3.

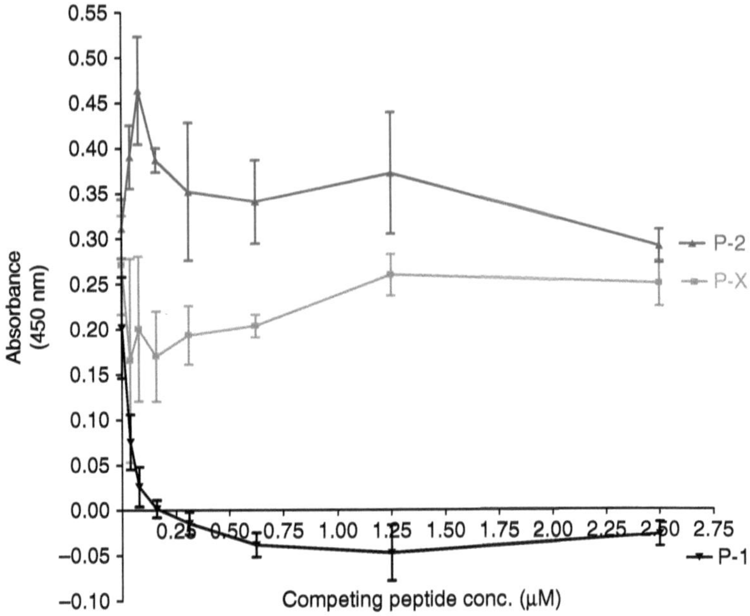

Fig. 3 Cross-competition of P-1/NA-HRP (tetravalent complex) specific binding to VEGFR-2 transfected cells with monovalent P-1, P-2, or P-X (negative control) peptides (*see* **Notes 11** and **12**). Binding was specifically competed using P-1 but no competition was observed with P-2 or P-X. Specific binding to VEGFR-2-transfected cells was obtained by subtracting binding to mock-transfected cells from VEGFR-2-transfected cells

3.4 Preparation of Peptide Mixes for Cooperativity Assays

1. To prepare peptide/Neutravidin HRP complexes for cooperativity assays, a fourfold excess of biotinylated peptide (as either a single peptide or a mixture of two peptides) is added to a 1 mL solution of 0.33 μM Neutravidin HRP (prepared by dilution of the Neutravidin HRP stock solution from Subheading 2.5 100-fold with M199 medium). Since the peptide stock solutions is prepared at a concentration of 250 μM, the total volume of biotinylated peptide added is 5.34 μL, split between the different peptide stock solutions to achieve the desired ratio (e.g., 4:0, 3:1, 2:2, and 1:3).
2. The solutions from **step 1** are diluted 100-fold by adding 10 μL to 990 μL of M199 medium. This provides solutions that are 3.33 nM with respect to NA-HRP with different ratios of the biotinylated peptides filling essentially all biotin-binding sites.

3.5 Cooperativity Assay

1. Wash transfected cells once with M199 medium.
2. Add 80 μL of blocking solution containing 0.5 mg/mL Neutravidin to each well.
3. Incubate at 37 °C for 1 h (*see* **Note 10**).

	1 2 3	4 5 6	7 8 9	10 11 12
A	P-X (4X)		P-X (4X)	
B	P-1 (4X)		P-1 (4X)	
C	P-2 (4X)		P-2 (4X)	
D	P-1 (3X) + P-2 (1X)		P-1 (3X) + P-2 (1X)	
E	P-1 (2X) + P-2 (2X)		P-1 (2X) + P-2 (2X)	
F	P-1 (1X) + P-2 (3X)		P-1 (1X) + P-2 (3X)	
G				
H				
	MOCK Transfected		VEGFR-2 Transfected	

Fig. 4 A 96-well plate display for cooperativity assay. Mixed tetravalent complexes of P-1 and P-2 with NA-HRP were compared with pure tetravalent complexes of P-1 and P-2

4. Wash cells twice with M199 medium.
5. Add peptide/Neutravidin HRP solutions from Subheading 3.4 step 2 to the appropriate wells (Fig. 4).
6. Incubate at room temperature for 2 h (*see* **Note 10**).
7. Transfer plates to 4 °C for 30 min (*see* **Note 10**).
8. Wash cells five times with ice-cold M199 medium.
9. Wash once with ice-cold PBS.
10. Add 100 μL of TMB solution to each well.
11. Incubate for 30 min (*see* **Note 10**).
12. Add 50 μL of 1 M phosphoric acid.
13. Read absorbance at 450 nm wavelength in a plate reader. Results from one such experiment are shown in Fig. 5.

4 Notes

1. Neutravidin HRP is used instead of streptavidin or avidin because it has lower nonspecific binding to molecules other than biotin [8]. Lower nonspecific binding is due to the absence of lectin-binding carbohydrate moieties [8] and also due to the absence of the cell adhesion receptor-binding RYD domain in Neutravidin [9].
2. Substitution of gelatin with BSA will not work, possibly because of biotin contamination of commercial BSA preparations.
3. Similar results are obtained when no DNA or vector DNA is used as the negative control. We use no DNA for mock transfections because of convenience.
4. Transfections are done as described in the Lipofectamine 2000 protocol from Invitrogen.

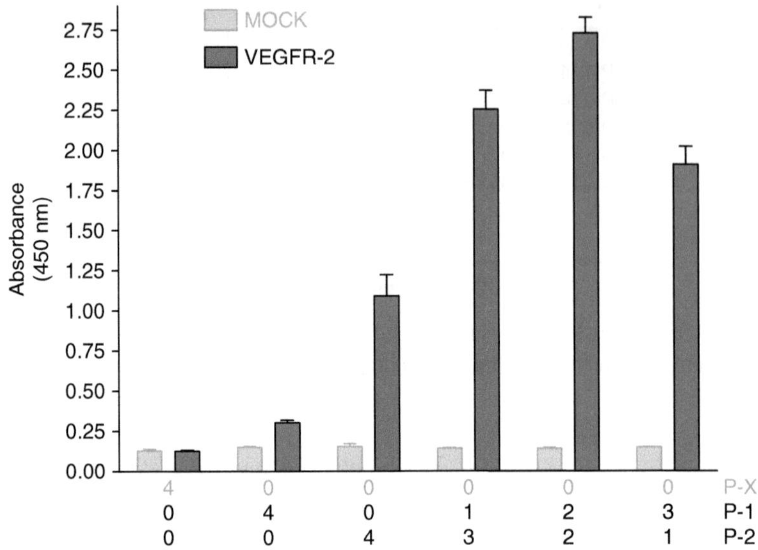

Fig. 5 Cooperativity of P-1 and P-2 in the Neutravidin HRP assay. Mixed tetravalent complexes of P-1 and P-2 on NA-HRP showed much better binding than pure tetravalent complexes of either P-1 or P-2

5. Combine the diluted Lipofectamine 2000 with the diluted DNA within 30 min. Longer incubation times may decrease activity.
6. The solution may appear cloudy, but this will not inhibit the transfection. DNA-Lipofectamine 2000 complexes are stable for 6 h at room temperature.
7. Neutravidin HRP solution should be thawed at 4 °C when ready to use for the assay.
8. The 293H cells should be ~90 % confluent at the time of transfection.
9. The presence of antibiotics reduces transfection efficiency as it causes cell death.
10. All the incubations times should be exactly followed for reproducible results.
11. Since excess of biotinylated P-1 peptide is used for making tetravalent P-1/Neutravidin HRP complex and biotin–Neutravidin HRP affinity is very high with extremely negligible off-rate. Cross-competition with non-biotinylated and biotinylated peptide gives similar results. We use biotinylated peptide for competition because of convenience.
12. Cross-competition of tetravalent complex of P-2 on Neutravidin HRP with P-1 and P-2 gives similar results.

References

1. Ragupathi G, Gathuru J, Livingston P (2005) Antibody inducing polyvalent cancer vaccines. Cancer Treat Res 123:157–180
2. Dorsam RT, Gutkind JS (2007) G-protein-coupled receptors and cancer. Nat Rev 7:79–94
3. Reubi JC (2003) Peptide receptors as molecular targets for cancer diagnosis and therapy. Endocr Rev 24:389–427
4. Shrivastava A, von Wronski MA, Sato AK, Dransfield DT, Sexton D, Bogdan N et al (2005) A distinct strategy to generate high-affinity peptide binders to receptor tyrosine kinases. Protein Eng Design Selection 18: 417–424
5. Mammen M, Choi SK, Whitesides GM (1998) Polyvalent interactions in biological systems: implications for design and use of multivalent ligands and inhibitors. Angew Chem Int Ed 37: 2754–2794
6. Bratthauer GL (1999) The avidin-biotin complex (ABC) method and other avidin-biotin binding methods. Methods Mol Biol 115:203–214
7. Pillai R, Marinelli ER, Swenson RE (2006) A flexible method for preparation of peptide homo- and heterodimers functionalized with affinity probes, chelating ligands, and latent conjugating groups. Biopolymers 84: 576–585
8. Hiller Y, Gershoni JM, Bayer EA, Wilcheck M (1987) Biotin binding to avidin. Oligosaccharide side not required for ligand association. Biochem J 248:167–171
9. Alon R, Bayer EA, Wilchek M (1990) Streptavidin contains an RYD sequence which mimics the RGD receptor domains of fibronectin. Biochem Biophys Res Commun 170: 1236–1241

Chapter 8

Substrate Phage Display for Protease Substrate Sequence Characterization: Bovine Factor Xa as a Model System

Hung-Ju Hsu and An-Suei Yang

Abstract

Regulatory proteases modulate proteomic dynamics with a spectrum of specificities against substrate proteins. Substrate phage display is one of the key methodologies in producing substrate sequence information in vitro. Factor Xa, a key regulatory protease in the blood coagulation system, is used as a model system to demonstrate a high-throughput procedure to quantitatively characterize substrate sequences and their susceptibilities for enzymatic cleavage. This methodology can be generalized to proteases for which the active forms (not necessarily purified forms) are available for the in vitro experiments.

Key words Substrate phage display, Factor Xa, Quantitative enzymatic kinetics ELISA, High-throughput platforms, Protein library, Directed evolution, Machine learning

1 Introduction

Complete characterization of a protease's specificity requires extensive in vitro experiments. High-throughput experiments have been developed to quantify protease specificity against a large number of substrates [1–4]. Nevertheless, enumerating all the possibilities in the substrate sequence space remains experimentally intractable. Moreover, the working hypothesis that the substrate-protease binding modes are unchanged regardless of the variation of the substrate sequences runs counter to an increasing body of experimental evidence [5–9].

Substrate phage display [5, 10–13] used as a platform to characterize substrate-protease specificity is not subject to the assumptions pertinent to the argument regarding subsite independency and substrate binding modes. Moreover, the phage display approach is advantageous over other high-throughput methods in that this method can effectively focus on substrate sequences with high specificities and the enzymatic susceptibility of each of the phage-displayed substrate sequences can be investigated in a high-throughput format. Although the scope of the specificity

Andrew E. Nixon (ed.), *Therapeutic Peptides: Methods and Protocols*, Methods in Molecular Biology, vol. 1088,
DOI 10.1007/978-1-62703-673-3_8, © Springer Science+Business Media, LLC 2014

measurement is still subject to the limitation of experimental resources, machine-learning algorithms, when adequately designed, can effectively make use of the substrate phage display dataset derived from the methodologies described herein to elucidate optimal rules governing the substrate specificity [14].

Specifically designed substrate phage display procedures are required for optimal machine learning [14]: Not only the most effective substrate sequences (informative positive examples) are required, the substrate sequences with a seemingly cleavable sequence pattern but that cannot be cleaved by the enzyme (informative negative examples) are also necessary in the computational learning. In addition, for each of the positive examples, the specificity needs to be quantified in high-throughput format. The dataset containing quantitative positive and negative cases provides a novel opportunity, enabling machine-learning algorithms to predict regression substrate specificity quantitatively [14]. Herein, we describe in detail the methodologies in substrate phage display and in quantitative enzyme-linked immunosorbent assay (ELISA) [5, 15] needed for generating a large quantitative dataset, which is suitable for machine-learning algorithms to model substrate specificity rules and, eventually, to automatically scan for substrate sequences in protein databases.

2 Materials

2.1 Construction of Phage-Displayed Substrate Libraries

2.1.1 Helper Phage Preparation

1. *E. coli* ER2738 strain, tetracycline resistant (New England BioLabs) (*see* **Note 1**).
2. 2YT medium (Fluka): 16 g/L tryptone, 10 g/L yeast extract, and 10 g/L NaCl. Sterilize with autoclave. Store at room temperature.
3. 2YT plate: 2YT medium with 15 g/L bacto-agar. Sterilize with autoclave. Cool to 55–65 °C. Add proper antibiotics to the sterilized medium and mix thoroughly. Pour into petri dishes and allow to solidify. Store the prepared plates at 4 °C and use within 1 month.
4. Tetracycline (Sigma): 20 mg/mL in ethanol (70 %). Store at −20 °C in the dark.
5. M13KO7 helper phage (NEB): The phage genome confers kanamycin resistance on the host *E. coli*.
6. Kanamycin (Sigma): 20 mg/mL in ddH_2O, filter-sterilized. Store at −20 °C.
7. Sterilized glycerol: glycerol (Merck) is sterilized with autoclave. Store at room temperature.

2.1.2 Construction of the Library Template Phagemid (pC-TAA)

1. 10× Duplex buffer (Integrated DNA Technologies, Inc., IDT): 100 mM potassium acetate, 30 mM HEPES, pH 7.5 (tenfold concentration).
2. pCANTAB5E phagemid (GE Healthcare) (*see* **Note 2**): *Sfi*I–*Eag*I digested, 50 ng/μL. Store at −20 °C.
3. Oligonucleotides (synthesized by IDT): 100 μM in sterilized ddH_2O. Store at −20 °C.

YV1-F: 5′-CTT TCT ATG CGG CCC AGC CGG CCC ATC ATC ATC ATC ATC ACG GCG GCT CTA GAG CTC CGC GGT ACC CGG GAG ATC TGC GGC CGC AGG TGC GCC-3′.

YV1-R: 5′-GGC GCA CCT GCG GCC GCA GAT CTC CCG GGT ACC GCG GAG CTC TAG AGC CGC CGT GAT GAT GAT GAT GAT GGG CCG GCT GGG CCG CAT AGA AAG-3′.

TAA-F: 5′-GGT GGT ACT AGT GGA TCC GGC TAA TAA TAA TCA GGT GGA GCA TCG GCC GCA GGA GCG-3′.

TAA-R: 5′-CGC TCC TGC GGC CGA TGC TCC ACC TGA TTA TTA TTA GCC GGA TCC ACT AGT ACC ACC-3′.

XA-F: 5′-GGT GGT ACT AGT GGA TCC GGC ATT GAA GGA AGA TCA GGT GGA GCA TCG GCC GCA GGA GCG-3′.

XA-R: 5′-CGC TCC TGC GGC CGA TGC TCC ACC TGA TCT TCC TTC AAT GCC GGA TCC ACT AGT ACC ACC-3′.

4. Ampicillin (Sigma): 50 mg/mL stock solution in ddH_2O, filter-sterilized. Store at −20 °C.
5. T4 DNA ligase (NEB): 400 U/μL.
6. Restriction enzymes *Sfi*I, *Eag*I, *Xba*I, *Spe*I (NEB): 10 U/μL.
7. QIAquick gel extraction kit (Qiagen).

2.1.3 Preparation of Single-Stranded Template DNA (dU-ssDNA)

1. *E. coli* dut^-/ung^- strain CJ236, chloramphenicol-resistant (NEB).
2. Chloramphenicol (Sigma): 34 μg/mL in 75 % (v/v) ethanol. Store at −20 °C.
3. Glucose (Merck): 2 M glucose in sterilized ddH_2O, filter-sterilized. Store at room temperature.
4. Uridine (Sigma): 0.25 mg/mL in ddH_2O, filter-sterilized. Store at −20 °C.
5. PEG/NaCl solution: 20 % (w/v) polyethylene glycol 8000 (J.T. Baker) and 2.5 M sodium chloride (Sigma) in ddH_2O, sterilized with autoclave. Store at room temperature.

6. PBS: 137 mM NaCl, 2.7 mM KCl, 7.7 mM Na_2HPO_4, 1.47 mM KH_2PO_4, pH 7.5 with HCl.
7. Qiaprep M13 spin kit (Qiagen).

2.1.4 Construction of Phage-Displayed Substrate Libraries

1. Oligonucleotides (synthesized by IDT): 100 μM in sterilized ddH_2O. Store at −20 °C (*see* **Note 3**).

   4X-R: 5′-CCT GCG GCC GAT GCT CCA CCT GAM NNM NNM NNM NNG CCG GAT CCA CTA GAG CCG CC-3′.

   XXGRXX-R: 5′-CCT GCG GCC GAT GCT CCA CCT GAM NNM NNG CGG CCM NNM NNG CCG GAT CCA CTA GAG CCG CC-3′.

   XRXRXX-R: 5′-CCT GCG GCC GAT GCT CCA CCT GAM NNM NNG CGM NNG CGM NNG CCG GAT CCA CTA GAG CCG CC-3′.

   where M = A, C, 50 % each; N = A, G, T, C, 25 % each.
2. T4 polynucleotide kinase (NEB): 10 U/μL.
3. ATP (adenosin 5′-triphosphate) (Sigma): 10 mM in ddH_2O, filter-sterilized. Store at −20 °C.
4. DTT (dithiothreitol) (Sigma): 100 mM in ddH_2O, filter-sterilized. Store at −20 °C.
5. 10× TM buffer: 0.5 M Tris and 0.1 M $MgCl_2$, pH7.5.
6. 10 mM dNTP (deoxyribonucleotide triphosphate mixture) (Promega): 10 mM of dATP, dCTP, dTTP, and dGTP.
7. T7 DNA polymerase (NEB): 10 U/μL.
8. T4 DNA ligase (NEB): 400 U/μL.

2.1.5 Electroporation and Phage Rescue

1. LB medium (Sigma): 10 g/L tryptone, 5 g/L yeast extract, and 5 g/L sodium chloride in ddH_2O, sterilized with autoclave. Store at room temperature.
2. 10% (v/v) glycerol (Merck): in ddH_2O, sterilized with autoclave.
3. pUC18 DNA (Stratagene): 1 ng/μL in ddH_2O. Store at −80 °C.
4. SOC medium (Fluka): 20 g/L tryptone, 5 g/L yeast extract, 0.5 g/L sodium chloride, 0.2 g/L potassium chloride. Adjust pH to 7.0 in 1 L of ddH_2O, sterilized with autoclave. Add 5 mL autoclaved 2 M magnesium chloride and 20 mL of filter-sterilized 1 M glucose. Store at 4 °C.

2.2 Substrate Phage Selection and Screening

2.2.1 Substrate Phage Selection

1. Ni-NTA agarose (Qiagen), stored at 4 °C.
2. Binding buffer: 50 mM $NaHPO_4$ pH 8.0, 300 mM NaCl in ddH_2O, sterilized with autoclave. Store at room temperature.
3. Micro-filter column, 0.5 μm pore size (Millipore).

4. Wash buffer: 50 mM $NaHPO_4$ pH 8.0, 300 mM NaCl, and 10 mM immidazole (*see* **Note 4**), sterilized with autoclave. Store at room temperature.
5. Bovine factor Xa (fXa) reaction buffer: 20 mM Tris–HCl pH 6.8, 50 mM NaCl, 1 mM $CaCl_2$ in ddH_2O, filter-sterilized. Store at room temperature.
6. Bovine factor Xa protease 2 U/μL, stored at −20 °C (QIAGEN).
7. Elution buffer: 50 mM $NaHPO_4$ pH 8.0, 300 mM NaCl, and 200 mM imidazole in ddH_2O, filter-sterilized. Store at room temperature.

2.2.2 Single Colony Analysis

1. Deep 96-well polypropylene plate, 2.2 ml/well with lid.
2. Colony picker (Qpix2, Genetix).
3. Micro-plate shaker.
4. Ni-NTA His-sorb plate (Qiagen).
5. Plate sealer (NUNC).
6. PBST: PBS with 0.05 % (v/v) Tween 20.
7. Anti-M13-HRP: anti-M13-HRP conjugate (GE Healthcare) 1:5,000 (v/v) in PBS with 2.5 % (w/v) skim milk.
8. ABTS: 2, 2′-azino-bis (3-ethylbenz-thiazoline-6-sulfonic acid) one component reagent (ABTS, BioFX).
9. Stop solution: 1 % (w/v) SDS in ddH_2O. Store at room temperature.
10. ELISA reader at 405 nm (BioTek).

2.2.3 Phage ssDNA Sequencing

1. Iodide buffer (NaI buffer): 10 mM Tris–HCl pH 8.0, 1 mM EDTA, 4 M NaI (Merck) in deionized water. Store at room temperature in the dark.
2. Absolute ethanol (Merck).
3. 75% (v/v) ethanol: 75 mL of absolute ethanol. Add ddH_2O to final volume 100 mL.
4. Primer PhD-96, 5′-CCC TCA TAG TTA GCG TAA CG-3′, 5 μM, Store at −20 °C (NEB).

2.2.4 Quantitative ELISA for Enzymatic Kinetics

1. Quench solution: 1.5 M of guanidine pH 8.0 in 1× fXa reaction buffer. Store at room temperature.

3 Methods

All the experimental procedures need to be carefully carried out with fully sterilized equipments and working space. Filtered pipette tips are used during transferring phage solutions to prevent cross contamination. The methodology for phage library construction below is modified based on the protocol of Sidhu and Weiss [16].

3.1 Construction of Phage-Displayed Substrate Libraries

3.1.1 Helper Phage Preparation

1. Inoculate 2 mL 2YT medium containing 20 μg/mL tetracycline with a single colony of *E. coli* strain ER2738 from a plate with colonies less than 1 week old. Shake at 200 rpm for 6 h at 37 °C.
2. Add ~10^8 cfu of M13KO7 helper phage to 1 mL of the ER2738 culture and incubate at 37 °C for 30 min.
3. Streak the infected ER2738 cells onto a 2YT plate containing 20 μg/mL kanamycin. Incubate the plate at 37 °C overnight.
4. Inoculate 50 mL 2YT medium containing 50 μg/mL kanamycin with a single colony from the plate. Incubate the culture at 37 °C with shaking at 200 rpm overnight.
5. Centrifuge the culture at 10,000 × *g* for 15 min at 4 °C and discard the pellet. Repeat this step again. Store the supernatant phage solution at 4 °C in conical tubes. The phage solution should be viable for 6 months. For longer storage, add sterilized glycerol to final concentration of 50 % and store at −80 °C (*see* **Note 5**).
6. Titer the phage concentration with freshly prepared *E. coli* ER2738 at early-log phase (OD600 ~ 0.4). The phage titer should be around 10^{10} cfu/mL (*see* **Note 6**).

3.1.2 Construct Library Template Phagemid (pC-TAA)

1. pC-YV1 construction (Fig. 1 and *see* **Note 7**): mix 2 μL of each of the complementary primers YV1-F and YV1-R and 5 μL 10× duplex buffer in a 200-μL PCR tube. Adjust to final volume of 50 μL with sterilized ddH_2O. Put the tube in a PCR machine for annealing process: heat to 94 °C for 3 min, followed by cooling to 20 °C (0.5 °C per 30-s step). Annealed duplex DNA is precipitated by alcohol, followed by *Sfi*I–*Eag*I restriction enzyme digestion and gel-purification with QIAquick gel extraction kit in 50 μL final volume. Mix 1 μL of the restriction enzyme-digested duplex DNA, 5 μL *Sfi*I–*Eag*I digested pCANTAB5E, 5 μL 10× ligase buffer, and ddH_2O to final volume of 49 μL. Add 1 μL of T4 DNA ligase to the mixture and incubate the reaction mixture at 16 °C overnight. Electroporate 1 μL of the ligated phagemid into 80 μL electro-competent *E. coli* strain ER2738 (details of the electro-competent cell preparation and electroporation procedures are shown in Subheading 3.1.5). Double-stranded phagemid DNA is purified from transformed bacteria. pC-YV1 phagemid construct is verified with restriction enzyme digestion and DNA sequencing.

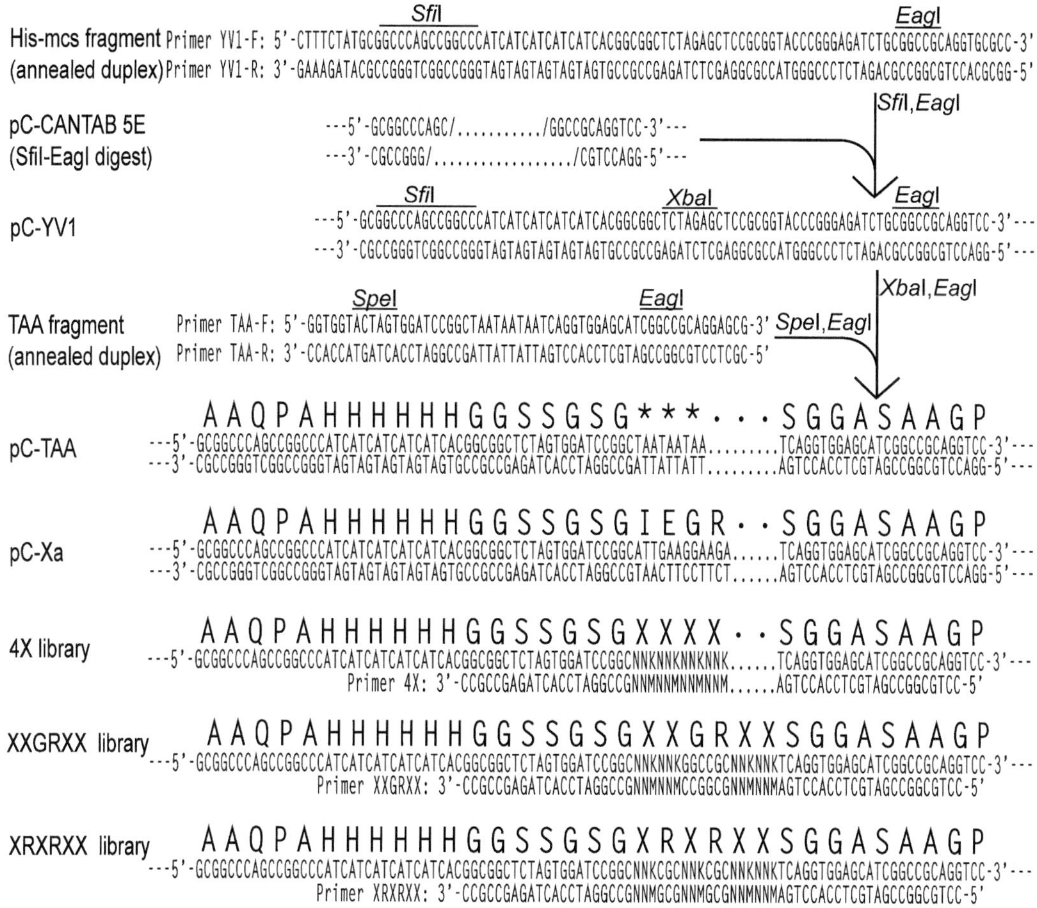

Fig. 1 Constructs of phagemids for template, control, and phage-displayed substrate libraries

2. pC-TAA construction (Fig. 1 and *see* **Note 8**): mix 2 μL of each of the complementary primers TAA-F and TAA-R and 5 μL 10× duplex buffer in a 200-μL PCR tube. Adjust to final volume of 50 μL with sterilized ddH_2O. Put the tube in a PCR machine for annealing process as described above. Annealed duplex DNA is precipitated by alcohol, followed by *Spe*I–*Eag*I restriction enzyme digestion and gel-purification with QIAquick gel extraction kit in 50 μL final volume. Digest 2.5 μg pC-YV1 dsDNA with *Xba*I–*Eag*I restriction enzymes and then gel-purify with QIAquick gel extraction kit in 50 μL final volume. Mix 3 μL of the restriction enzyme-digested duplex DNA, 1 μL *Spe*I–*Eag*I digested pC-YV1, 2 μL 10× ligase buffer, and ddH_2O to final volume of 19 μL. Add 1 μL of T4 DNA ligase to the mixture and incubate the reaction mixture at 16 °C overnight. Electroporate 1 μL of the ligated phagemid into 80 μL electro-competent *E. coli* strain ER2738

(details of the electro-competent cell preparation and electroporation procedures are shown in Subheading 3.1.5). Double-stranded phagemid DNA is purified from transformed *E. coli* ER2738. pC-TAA phagemid construct is verified with restriction enzyme digestion and DNA sequencing.

3. Inoculate 2 mL 2YT medium containing 100 μg/mL ampicillin and 10^8 cfu of M13KO7 helper phage (Subheading 3.1.1) with a single colony of *E. coli* ER2738 harboring pC-TAA phagemid. Incubate the culture at 37 °C with shaking at 200 rpm for 30 min.
4. Inoculate 50 mL 2YT containing 100 μg/mL ampicillin and 50 μg/mL kanamycin with the culture above. Incubate at 37 °C with shaking at 200 rpm overnight.
5. Centrifuge the overnight culture at 10,000 × *g* for 15 min at 4 °C and discard the pellet. Repeat the process again. Collect the phage supernatant to a clean Falcon tube and store at 4 °C. The pC-TAA phage solution is used for the following dU-ssDNA preparation.
6. Construction of the control phage pC-Xa (Fig. 1) follows the same procedure described in **step 2** of this section, using the oligonucleotides XA-F and XA-R.

3.1.3 Preparation of Single-Stranded Template DNA (dU-ssDNA)

1. Inoculate 2 mL 2YT medium containing 17 μg/mL chloramphenicol with a single CJ236 colony and shake the culture at 200 rpm for 5–6 h at 37 °C (OD_{600} ~ 0.5).
2. Add 10^8 cfu of pC-TAA phages to 1 mL of the CJ236 culture and incubate the mixture at 37 °C for 30 min.
3. Streak the infected CJ236 cell onto 2YT plate containing 100 μg/mL ampicillin and incubate the plate at 37 °C overnight.
4. Inoculate 2 mL 2YT medium containing 100 μg/mL ampicillin and 5×10^8 cfu of M13KO7 helper phage with a single colony from the plate above. Shake the culture at 200 rpm for 1 h at 37 °C.
5. Inoculate with the above culture to 50 mL 2YT medium containing 100 μg/mL ampicillin, 50 μg/mL kanamycin, 20 mM glucose (*see* **Note 9**), and 0.25 μg/mL uridine (*see* **Note 10**). Incubate at 37 °C with shaking at 200 rpm overnight.
6. Centrifuge the culture at 10,000 × *g* for 15 min at 4 °C and discard the cell pellet.
7. Mix 35 mL of phage supernatant with 14 mL PEG/NaCl in a 50-mL conical tube. Incubate the mixture on ice for 30 min. Centrifuge at 12,000 × *g* for 25 min at 4 °C. Decant the supernatant carefully and invert the tube on a paper towel to drain the remaining liquid. Resuspend the pellet in 600 μL PBS with vigorous vortex.

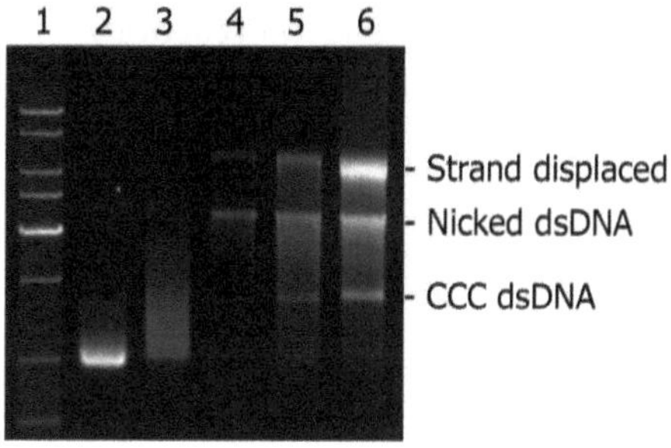

Fig. 2 Gel analysis of the library dsDNA synthesis. Lane *1*:1 kb dsDNA marker. Lane *2*: 200 ng of purified dU-ssDNA. Lane *3*: A mock reaction without T4 polymerase. Lanes *4–6*: Molar ratio (primer:template) increases from 1:1, 3:1, to 15:1

8. dU-ssDNA from the phage solution is extracted with Qiaprep M13 spin kit following the manufacturer's instruction. The yield of dU-ssDNA extracted from 35 mL phage supernatant is about 45–60 μg. Determine the dU-ssDNA concentration by spectrophotometery and adjust to the concentration to 0.2 μg/μL with EB buffer (from the Qiaprep M13 spin kit). Apply 1 μL of the dU-ssDNA solution on 1.2 % agarose gel: the dU-ssDNA should appear at about 3 kb to the dsDNA markers (Fig. 2).

3.1.4 Construction of Phage-Displayed Substrate Libraries

1. Polynucleotide kinase reaction:

   0.6 μL oligo DNA (100 μM) (4X-R, XXGRXX-R, or XRXRXX-R, see Subheading 2.1.4 and *see* **Note 3**).

   2 μL 10× T4 polynucleotide kinase buffer.

   1 μL 10 mM ATP.

   Adjust total volume to 18 μL with ddH_2O.

   Add 2 μL T4 polynucleotide kinase.

2. Incubate at 37 °C for 2 h.
3. Template-Primer Annealing Reaction:

   20–*X*μL autoclaved ddH_2O.

   20 μL template dU-ssDNA (from **step 8** of Subheading 3.1.3).

   *X*μL kinased oligo (from **step 1** of this section).

   5 μL 10× TM buffer.

   Adjust total volume to 50 μL with ddH_2O.

   X is determined by the molar ratio of primer:template: (1:1, 3:1, 15:1, *see* **Note 11**). Heat the mixture to 90 °C and cool slowly (in the step of 0.5 °C/30 s) to 20 °C in a thermal cycler. The annealing efficiency could vary depending on the primer design.

4. Extension reaction:

   50 μL annealing mixture (from **step 3** of this section).

   5 μL 10 mM dNTP.

   2 μL 10 mM ATP.

   3 μL 100 mM DTT.

   0.8 μL T4 DNA ligase.

   0.8 μL T7 DNA polymerase.

   Total 61.6 μL.

5. Incubate the reaction mixture at 20 °C overnight. Analyze the reaction outcome by running 2 μL of sample on a 1.2 % TAE agarose gel (Fig. 2).
6. Purify dsDNA with QIAGEN gel extraction kit following the manufacturer's instruction with only one modification: To elute DNA, add 35 μL ddH_2O to the center of the column membrane and incubate at 50 °C for 10 min (*see* **Note 12**). Centrifuge at 12,000 × *g* for 1 min to collect DNA. Repeat this elution step with 25 μL ddH_2O. Combine the two elutions.

3.1.5 Electroporation and Phage Rescue

1. Inoculate 50 mL 2YT medium containing 20 μg/mL tetracycline with a single ER2738 colony from a 2YT plate. Incubate at 37 °C with shaking at 200 rpm overnight.
2. Inoculate each of four 2-L flasks, containing 250 mL LB, 20 μg/mL tetracycline, with 2 mL of overnight culture from the previous step. Incubate the cultures at 37 °C with shaking (200 rpm) until the OD_{600} ~ 0.5 (approximately 2–3 h).
3. Chill the cells on ice for 1 h. Centrifuge at 3,000 × *g* for 15 min at 4 °C.
4. Wash the cell pellet with 400 mL pre-chilled 10 % glycerol. Centrifuge at 3,000 × *g* for 15 min at 4 °C. Discard the supernatant. Repeat the wash step four times.
5. Suspend the cells in 0.5 mL 10 % glycerol. Dispense 80 μL cells to 1.5 mL eppendorf tube on ice. This protocol provides 12 × 80 μL tubes of electro-competent cells with concentration around 10^{11} cfu/tube.
6. The electro-competency of the cells is evaluated by transforming 10 ng of pUC18 to 80 μL of the electro-competent cells. This test should generate ~10^9 ampicillin-resistant colonies. Blank test should also be carried out to control for contamination in the electro-competent cells.
7. For a single electroporation, mix 6 μL dsDNA from the **step 6** of Subheading 3.1.4 with 80 μL competent cells in a pre-chilled 2 mm cuvette on ice. Electroporate the electro-competent cells with the following setting: 2.5 kV, 200 Ω resistance, and 25 μF capacitance. Immediately add 1 mL of

SOC to each of the cuvettes after electroporation. The time constant of high-voltage discharge should be ~5.0 ms (*see* **Note 13**).

8. Repeat **step** 7 ten times. Pool the transformed cells to a 250-mL flask, and incubate at 37 °C for 30 min for recovery.
9. Spread serially diluted recovered cells on 2YT plates containing 100 μg/mL ampicillin and 20 mM glucose to evaluate the transformation efficiency. The total number of variants obtained should reach $\sim 10^9$ for the combined transformed cells. Pick around 20 colonies from the plate, rescue the phage, and extract ssDNA from the phage particles for sequencing (Subheading 3.2.3). At least 60 % of the single colonies should harbor the transformed phagemid.
10. Add ampicillin to the transformed cells to a final concentration of 100 μg/mL. Incubate at 37 °C with shaking at 200 rpm for 30 min.
11. Add 20 mL of helper phages (stock in 10^{10-11} cfu/mL; moi 10–100) to the transformed cells. Incubate at 37 °C with shaking at 200 rpm for 1 h.
12. Adjust the transformed cell to a volume of 100 mL with 2YT medium and add ampicillin and kanamycin to final concentration of 100 μg/mL and 50 μg/mL, respectively. Incubate at 37 °C with shaking at 200 rpm overnight.
13. Centrifuge the culture at 10,000 × *g* for 30 min at 4 °C and discard the cell pellet. The supernatant contains phage-displayed substrate library. Determine the titer of the phage stock using freshly cultured *E. coli*, strain ER2738, in early phase ($OD_{600} \sim 0.5$). The phage titer should be $\sim 10^{11}$ cfu/mL. Once the substrate library is prepared, it should be used within 2 weeks (*see* **Note 14**).

3.2 Substrate Phage Selection and Screening

3.2.1 Substrate Phage Selection (Fig. 2 Shows the Flowchart)

1. Substrate phage library (4X, XXGRXX, or XRXRXX library, Fig. 1) is precipitated by mixing 35 mL of phage supernatant with 14 mL PEG/NaCl (Subheading 3.1.3). Centrifuge at 12,000 × *g* for 30 min at 4 °C. Decant the supernatant and resuspend the pellet in 1 mL binding buffer.
2. Wash 200 μL of Ni-NTA in a 2-mL eppendorf tube with 1 mL binding buffer. Centrifuge at 1,000 × *g* for 30 s. Remove buffer by aspiration.
3. Add the phage library from **step 1** (1 mL) to the Ni-NTA beads (**step 2**) at room temperature for 2 h with rotating at 25 rpm (*see* **Note 15**).
4. Transfer the phage-Ni-NTA mixture to a micro-filter. Spin the micro-filter at 50 × g for 30 s. Retain the flow-through for phage titer.

5. Wash the Ni-NTA beads 21 times, each with 500 μL wash buffer and centrifugation at 50 × *g* for 30 s. This step intends to eliminate the phages with pC-TAA phagemid.
6. Add 500 μL of fXa reaction buffer to the Ni-NTA beads and centrifuge at 50 × *g* for 30 s. Retain the flow-through for phage titer.
7. The phage remaining on the Ni-NTA beads are treated with 5 U fXa protease in 500 μL fXa reaction buffer. Incubate the micro-filter column sealed with parafilm at room temperature for 2 h with rotating at 25 rpm (*see* **Note 16**).
8. Centrifuge the column at 50 × *g* for 30 s and retain the flow-through for phage titer.
9. After the enzymatic treatment, the Ni-NTA beads are washed nine times with wash buffer (*see* **Note 4**). Amplify the washed phage in 2 mL early-log *E. coli* ER2738. Incubate at 37 °C for 30 min. Add 5 mL M13KO7 helper phage (10^{10} cfu/mL) for 1 h. Adjust the volume to 100 mL with 2YT medium and add ampicillin and kanamycin to final concentration of 100 μg/mL and 50 μg/mL, respectively. Incubate at 37 °C with shaking at 200 rpm overnight. Centrifuge the culture at 10,000 × *g* for 30 min at 4 °C and discard the cell pellet. The supernatant contains phage-displayed substrate library. Titer the phage concentration with freshly cultured *E. coli* ER2738 strain at early phase (OD_{600} ~ 0.5), and the phage titer should be around ~10^{11} cfu/mL. Use this amplified phage library as input library in **step 1** to start the next round of selection (*see* Fig. 3 for the complete process).
10. Elute the Ni-NTA beads with elution buffer six times. Save the elution buffer for analysis. These are potential informative negative cases. At the end of the last cycle, single colonies of the selected phages are cultured for further analysis (Subheading 3.2.2). These are potential informative positive cases. All the positive and negative phage variants need to be confirmed with the kinetic measurements described below.

3.2.2 Single Colony Analysis

1. Infect host *E. coli* ER2738 cells with selected phages. Plate about 200–500 cfu of infected cells on a 10-cm diameter 2YT/ampicillin (100 μg/mL)/Glucose (20 mM) plate, or about 1,000–4,000 cfu on a 22-cm^2 bioassay plate. Incubate the plate at 37 °C overnight.
2. Set up Qpix2 colony picker to transfer overnight single colonies from the plate to a deep 96-well plate. Each well contains 1 mL 2YT with 100 μg/mL ampicillin and 5 × 10^8 cfu M13KO7 helper phage. Seal and incubate the plate at 37 °C for 2 h. Shake with micro-plate shaker set at low speed.

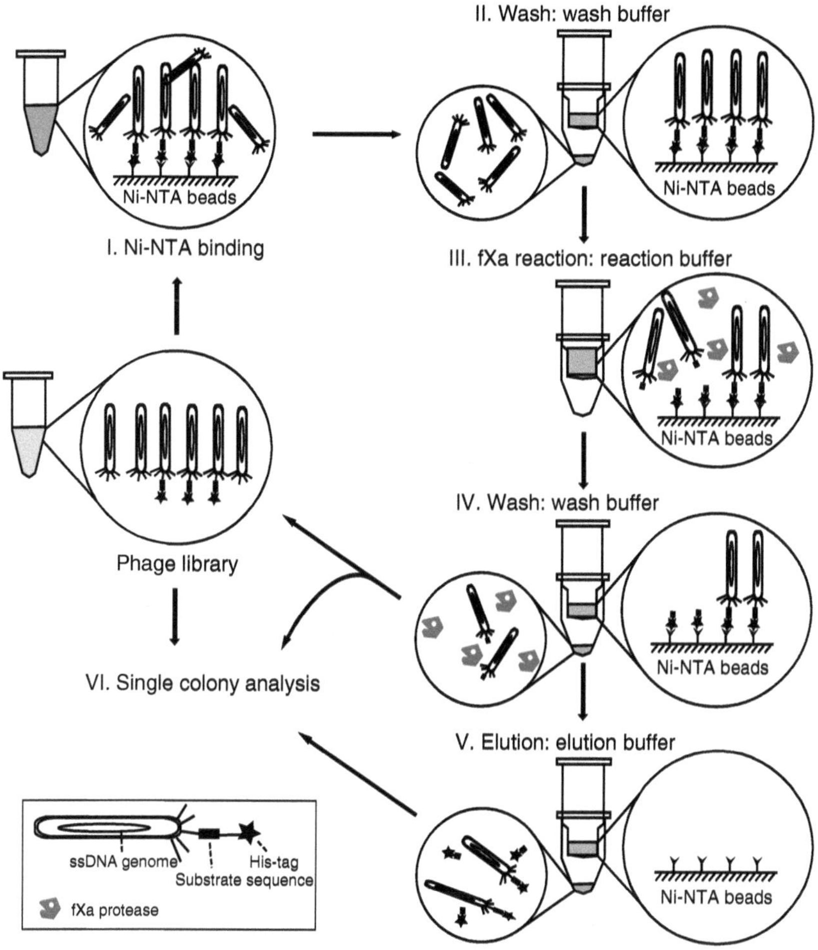

Fig. 3 Flowchart for the library selection procedure of substrate phage. Library phages (4X, XXGRXX, and XRXRXX library) displaying substrate peptide and His tag N′-terminal to the pIII capsid protein are captured on Ni-NTA beads (I). After washing to remove the non-bound phages (II), the captured phages are treated with fXa protease in situ (III–IV). Cleaved substrate phages are amplified and rescued for next round of selection (I). Selected phage clones are picked for single colony analysis (VI)

3. Add to each well 200 μL 2YT containing 300 μg/mL ampicillin and 300 μg/mL kanamycin. Incubate the plate at 37 °C with vigorous shaking overnight. Each well should contain ~10^{11} cfu/mL of cultured phage.
4. Centrifuge at 3,000 × *g* at 4 °C for 30 min. Transfer 10 μL supernatant from each well of the culture plate to the corresponding well of a 96-well His-sorb plate. Add 90 μL fXa reaction buffer with 0.2 U fXa protease. Repeat the process to fill a control 96-well His-sorb plate, where only fXa reaction buffer (without fXa enzyme) is added to each corresponding well. Incubate the pair of plates at room temperature for 1 h with agitation.

5. After 1 h of reaction, the His-sorb plates are washed three times with 250 μL PBST each and rinsed once with 250 μL PBS.
6. Add 100 μL anti-M13-HRP to each well and incubate for 1 h with low-speed shaking.
7. Wash the His-sorb plates three times with 250 μL PBST each and rinsed once with 250 μL PBS.
8. Add 100 μL of ABTS to each well for 10 min. Stop the chromogenic reaction by adding 100 μL of 1 % SDS to each well.
9. Read OD of each well at 405 nm. Phage in the wells with $OD_{control} > 1.0$ and $OD_{experiment}/OD_{control} < 0.9$ are selected for sequencing (Subheading 3.2.3) and enzymatic kinetics measurement with quantitative ELISA (Subheading 3.2.4).

3.2.3 Phage ssDNA Sequencing

1. Transfer 500 μL of the phage supernatant in the selected wells of the phage culture plate (from Subheading 3.2.2) to an eppendorf tube. Add 200 μL PEG/NaCl and mix by inverting several times. Incubate on ice for 10 min.
2. Following the incubation on ice centrifuge at 12,000 × *g* for 30 min at 4 °C. Discard the supernatant.
3. Dissolve the phage pellet thoroughly in 100 μL NaI buffer. Add 300 μL of chilled absolute ethanol. Incubate 5 min at room temperature.
4. Centrifuge at 12,000 × *g* for 30 min and discard supernatant. Rinse the pellet with 200 μL of chilled 75 % ethanol. Dry under vacuum for 10 min.
5. Dissolve the pellet in 30 μL deionized water.
6. Run 2 μL of sample on 1.2 % agarose gel (Fig. 4, also *see* **Note 17**).
7. Add 4 μL ssDNA (~0.25 μg) and 1 μL (5 pmol) primer PhD-96 to make up 10 μL mixture with deionized water. Send for sequencing.

3.2.4 Quantitative ELISA for Enzymatic Kinetics

1. For each phage, prepare six identical reaction mixtures with 20 μL phage solution and 80 μL fXa reaction buffer containing 0.2 U fXa protease in each well of a low-absorbent plate (polypropylene plate). Stop the reaction at 0, 10, 20, 30, 40, 50 min, respectively, by adding 50 μL of quench solution.
2. Transfer 100 μL of each quenched reaction mixture to wells of a His-sorb plate and incubate at room temperature for 1 h with agitation. Follow the same process from **step 5** to **9** in Subheading 3.2.2 to obtain the OD values at 405 nm. The quantitative measurement of the phage binding to the His-sorb surface, $[S]_t$ (t is the enzymatic reaction time), is calibrated with a standard curve derived from the same plate—where a

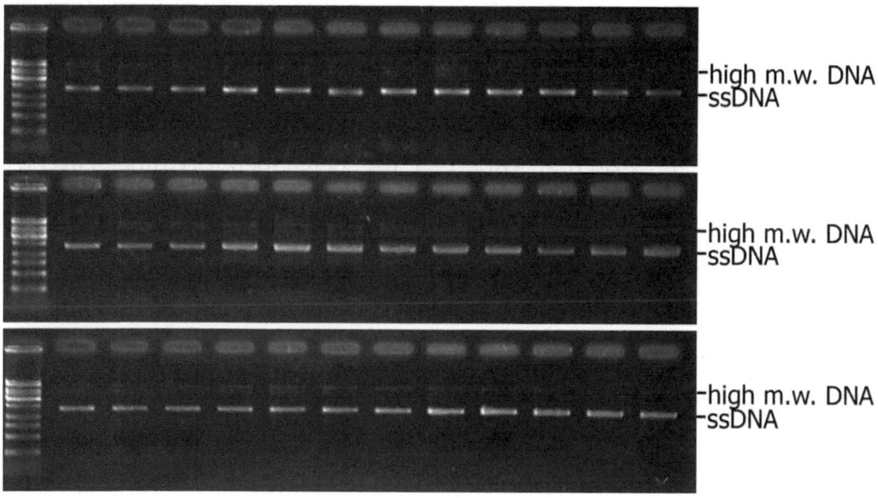

Fig. 4 ssDNA extracted from selected phage and analyzed on agarose gel. Major band of the phagemid ssDNA appears at about 3 kb of dsDNA marker. Minor bands of high-molecule weight DNA are host genome fragments or helper phage genome. In this gel, ssDNA of phage clones are selected from XRXRXX library

standard pC-Xa phage solution is diluted linearly in a series of six wells to generate a standard curve of OD_{405nm} versus known phage concentration (*see* **Note 18**).

3. The concentration of the substrate phage in the reaction mixture is on the order of 10^{-13} M, which is much lower than the lower limit of K_m of fXa (10^{-6} M as estimated by Bianchini et al. [17]). For cases where $[S] \ll K_m$, the reaction rate is governed by the approximated Michaelis–Menten equation:

$$\ln\{[S]_t / [S]_{t=0}\} = -t[E](k_{cat} / K_m)$$

where the hydrolysis kinetics is a pseudo-first-order reaction with the reaction constant linearly proportional to k_{cat}/K_m, provided that the enzyme concentration $[E]$ is a constant. Moreover, the pseudo-first-order reaction constant is not affected by the concentration of the substrate when $[S] \ll K_m$. Based on the Michaelis–Menten mechanism, we plot $\ln\{[S]_t/[S]_{t=0}\}$ versus t (in minute) to derive the slope $-k_{obs}$,

$$k_{obs}(\text{min}^{-1}) = [E](k_{cat} / K_m)$$

The concentrations of the substrate phage $[S]_t$ are measured with the quantitative ELISA described in **steps 1** and **2**. k_{obs} is used as a quantitative measurement for the phage-displayed substrate sequence specificity.

4 Notes

1. *E. coli* glycerol stocks are prepared by adding sterilized glycerol to an overnight culture to final concentration of 50 %, and are stored at −80 °C 1 mL per eppendorf tube. Glycerol stocks stored at −80 °C are viable for 5 years.
2. The pCANTAB5E phagemid/helper phage system is a monovalent form of peptide display on the phage pIII protein. This monovalent system is advantageous in quantitative phage ELISA for measurement of enzyme kinetics.
3. The following oligonucleotide primer design principle has been empirically successful for primers shorter than ~90 bases. The Tm (melting temperature) of the flanking sequences (complementary to the template DNA sequence, in contrast to the region of degenerate codon NNK) should be in the temperature range of 60–70 °C (in 50 mM salt). The length of the complementary stretch flanking the degenerate codons should be around 25 ± 5 residues to match this Tm range. Another consideration is the intramolecular secondary structure propensities of the designed primers. Tendency for secondary structure formation in the primer would reduce the overall yield of dsDNA synthesis. Both the Tm and intramolecular secondary structure propensity for the designed primer sequences can be predicted using http://www.basic.northwestern.edu/biotools/oligocalc.html.
4. Wash buffer containing 10 mM imidazole is essential for minimizing nonspecific phage binding to the Ni-NTA beads.
5. For concentrated helper phage stocks stored at −80 °C, PEG/NaCl precipitation (Subheading 3.1.3) is carried out before adding glycerol to 50 %.
6. It is important to titer phage (in particular, for M13KO7 helper phage) with the *E. coli* host in early-log phase ($OD_{600} \sim 0.4$); older hosts may underestimate the phage concentration by 10–100 folds. If phage titers are to be compared quantitatively, the host growth phase needs to be controlled at the same stage for phage titer throughout the experiments.
7. pC-YV1 is constructed to encode hexa-HIS tag N-terminal to the multiple cloning site, where substrate sequences and peptide libraries can be conveniently inserted.
8. The pC-TAA phagemid does not encode an expressible pIII fusion protein, making this phagemid an ideal template for library construction. Otherwise, library construction with an expressible phagemid as the template would generate a library heavily biased (>40 %) toward the polypeptide encoded by the template.

9. 20 m*M* glucose in culture medium inhibits pIII coat protein expression, which is driven by *lac* promoter in pCANTAB5E. Adding glucose increases the yield of dU-ssDNA from 45 to 60 μg/35 mL phage supernatant.
10. Addition of uridine to the culture of dut^-/ung^- *E. coli* CJ236 incorporates uracil into dU-ssDNA, which can be efficiently removed in dut^+/ung^+ *E. coli* strains, leaving only the nascent strand replicated.
11. Molar ratio between primer and template phagemid affects the yield of dsDNA synthesis. Molar ratio of 15:1 (primer:template) is used for constructing the 4X, XXGRXX, and XRXRXX libraries.
12. ddH_2O is used in place of EB buffer to reduce the salt concentration in the eluted dsDNA solution. The low salt concentration decreases the possibility of arcing in the following step of electroporation of the dsDNA into the host cells.
13. Electroporation fails when an arc appears through the metal electrode plates. Several reasons could account for the failure: (1) The electro-competent cells are not washed extensively, leaving relatively high salt concentration. (2) DNA solution contains high concentration of salt. (3) The cuvette is defective.
14. Substrate phage display libraries rescued directly from transformed host cells are less biased in complexity distribution than those from library amplification due to differences in phage infectivity, displayed peptide toxicity, and phage reproduction. Thus it is recommended to always start the panning process from substrate phage display libraries rescued from transformed host cells.
15. Phosphate buffer has been evaluated and found to be a better buffer for Ni-NTA-HIS tag binding.
16. fXa incubation times determine the selection pressure. We select the 4X library with a 2 h reaction time. Selections with XXGRXX and XRXRXX library start with 120 min reaction time in the first round selection, followed by 60 min, 45 min, and finally 30 min reaction time.
17. Phagemid ssDNA should appear as a single band (Fig. 4). If more than one band appears, care must be taken since the helper phage's genome could mix with the phagemid DNA. This is due to adding too much helper phage for rescuing selected phage. Should this happen sequencing of the phagemid will likely fail.
18. The linear range of the standard curves was found between $OD_{405\ nm} = 0.1–2$. Hence the supernatant phage concentration is adjusted such that the $OD_{405\ nm}$ for the 0 min enzymatic reaction is in the range of 0.4–1.8, which ensures that the substrate phage concentration remains in the linear range of the standard curves throughout the quantitative ELISA measurements.

Acknowledgment

This work is supported by grants from NHRI and NSC of Taiwan.

References

1. Gosalia DN, Salisbury CM, Maly DJ, Ellman JA, Diamond SL (2005) Profiling serine protease substrate specificity with solution phase fluorogenic peptide microarrays. Proteomics 5:1292–1298
2. Salisbury CM, Maly DJ, Ellman JA (2002) Peptide microarrays for the determination of protease substrate specificity. J Am Chem Soc 124:14868–14870
3. Harris JL, Backes BJ, Leonetti F, Mahrus S, Ellman JA, Craik CS (2000) Rapid and general profiling of protease specificity by using combinatorial fluorogenic substrate libraries. Proc Natl Acad Sci U S A 97:7754–7759
4. Marnett AB, Craik CS (2005) Papa's got a brand new tag: advances in identification of proteases and their substrates. Trends Biotechnol 23:59–64
5. Hsu HJ, Tsai KC, Sun YK, Chang HJ, Huang YJ, Yu HM, Lin CH, Mao SS, Yang AS (2008) Factor Xa active site substrate specificity with substrate phage display and computational molecular modeling. J Biol Chem 283: 12343–12353
6. Ding X, Rasmussen BF, Petsko GA, Ringe D (2006) Direct crystallographic observation of an acyl-enzyme intermediate in the elastase-catalyzed hydrolysis of a peptidyl ester substrate: Exploiting the "glass transition" in protein dynamics. Bioorg Chem 34:410–423
7. Laskowski M, Qasim MA (2000) What can the structures of enzyme-inhibitor complexes tell us about the structures of enzyme substrate complexes? Biochim Biophys Acta 1477:324–337
8. Coombs GS, Rao MS, Olson AJ, Dawson PE, Madison EL (1999) Revisiting catalysis by chymotrypsin family serine proteases using peptide substrates and inhibitors with unnatural main chains. J Biol Chem 274: 24074–24079
9. Brandstetter H, Kuhne A, Bode W, Huber R, von der Saal W, Wirthensohn K, Engh RA (1996) X-ray structure of active site-inhibited clotting factor Xa implications for drug design and substrate recognition. J Biol Chem 271:29988–29992
10. Deperthes D (2002) Phage display substrate: a blind method for determining protease specificity. Biol Chem 383:1107–1112
11. Ohkubo S, Miyadera K, Sugimoto Y, Matsuo K, Wierzba K, Yamada Y (2001) Substrate phage as a tool to identify novel substrate sequences of proteases. Comb Chem High Throughput Screening 4:573–583
12. Smith MM, Shi L, Navre M (1995) Rapid identification of highly active and selective substrates for stromelysin and matrilysin using bacteriophage peptide display libraries. J Biol Chem 270:6440–6449
13. Matthews DJ, Wells JA (1993) Substrate phage: selection of protease substrates by monovalent phage display. Science 260:1113–1117
14. Chen CT, Yang EW, Hsu HJ, Sun YK, Hsu WL, Yang AS (2008) Protease substrate site predictors derived from machine learning on multilevel substrate phage display data. Bioinformatics 24:2691–2697
15. Sharkov NA, Davis RM, Reidhaar-Olson JF, Navre M, Cai D (2001) Reaction kinetics of protease with substrate phage. Kinetic model developed using stromelysin. J Biol Chem 276:10788–10793
16. Sidhu SS, Weiss GA (2004) Constructing phage display libraries by oligonucleotide-directed mutagenesis: Phage display (Clackson T, Lowman HB eds.), Oxford University Press, NY, pp 27–41
17. Bianchini EP, Louvain VB, Marque PE, Juliano MA, Juliano L, Le Bonniec BF (2002) Mapping of the catalytic groove preferences of factor Xa reveals an inadequate selectivity for its macromolecule substrates. J Biol Chem 277:20527–20534

Chapter 9

Engineering Peptide Therapeutics Using MIMETIBODY™ Technology

Kristen Picha, Chichi Huang, Peter Bugelski, and Karyn O'Neil

Abstract

The MIMETIBODY™ platform was developed to expand the opportunities for application of biotherapeutics. While the utility of antibodies as antagonists has been well demonstrated, their application as agonists has been more challenging. For steric reasons, antibodies may be less well suited to perform as agonists or as inhibitors of GPCRs. In contrast, many bioactive peptides function as agonists or by interaction with GPCRs but their development as therapeutics has been challenging due to their small size and metabolic lability. The MIMETIBODY™ platform has been used to develop a variety of stable, long-lived molecules with intrinsic activities similar to that of their parent peptides. This chapter describes methods for construction of expression plasmids, expression and purification strategies, and methods for characterizing the activity of these novel proteins.

Key words MIMETIBODY™, Glucagon-like 1 peptide, Erythropoietin, Transient expression, Pharamocokinetics, Cell proliferation, Hemoglobin, cAMP, Blood glucose

1 Introduction

Protein therapeutics represent the fastest growing segment in the pharmaceutical industry. The increased success of antibody therapeutics has fueled an interest in expanding the utility of protein therapeutics to address other unmet medical needs. The biophysical properties of antibodies limit their utility for some applications, especially those involving receptor agonism. In contrast, peptides represent a rich source of therapeutic candidates, despite the fact that unmodified peptides have not traditionally made good drugs. Transformation of a metabolically labile peptide into a drug remains challenging, yet many important biological processes are regulated by peptides. In contrast, considerable advances have been made in the development of antibody therapeutics [1, 2]. Protein engineering tools have developed to a point where antibody libraries are routinely generated and screened to identify optimal therapeutic candidates [3, 4]. A technology that could

Andrew E. Nixon (ed.), *Therapeutic Peptides: Methods and Protocols*, Methods in Molecular Biology, vol. 1088,
DOI 10.1007/978-1-62703-673-3_9, © Springer Science+Business Media, LLC 2014

couple the desirable pharmacological properties of antibodies with the bioactivity of small peptides would be a valuable extension of antibody technology. To address this need, we have developed the MIMETIBODY™ platform as a novel technology for the display and delivery of bioactive peptides [5, 6]. A variety of bioactive peptides have been engineered into the platform to demonstrate the utility and versatility of the MIMETIBODY™ platform.

The MIMETIBODY™ platform involves linking a bioactive peptide sequence to an Fc segment of an antibody. Development of protein therapeutics in the MIMETIBODY™ platform often requires iterative protein engineering steps to address potency, stability, solubility, or aggregation issues. Preliminary constructs often have reduced potency compared to the unmodified peptide and manipulation of linker length and composition can frequently be used to improve activity, solubility, and stability. Alternatively, site-specific modifications in the peptide moiety or the linker can be used to improve the properties of therapeutic candidates. Finally, it is possible to change the Fc region to a different isotype or an engineered variant to modify characteristics in the molecule that are Fc mediated including FcRn binding, Fc receptor binding, and complement activation. In this chapter, we provide an overview of the strategies and methods used to construct, express, and purify gene candidates and to characterize the activity and biophysical properties of molecules engineered using the MIMETIBODY™ platform.

2 Materials

2.1 Expression Vector Construction

1. Oligonucleotides are synthesized at Integrated DNA Technologies (IDT, Coralville, IA), resuspended in water to 1 mg/mL and stored at −20 °C (*see* **Note 1**).
2. Platinum Pfx polymerase kit, including $MgSO_4$, enhancer, reaction buffer and pfx polymerase and 10 mM dNTP stock solutions (Invitrogen, Carlsbad, CA) are stored at −20 °C (*see* **Note 2**).
3. All restriction enzymes, their reaction buffers (New England Biolabs, Ipswich, MA) are stored at −20 °C.
4. Alkaline Phosphatase (Roche Applied Science, Indianapolis, IN) is stored at −20 °C.
5. Rapid DNA ligation kit (Roche) is stored at −20 °C (*see* **Note 3**).
6. QIAquick Gel extraction kit and QIAquick PCR purification kit (Qiagen, Valencia, CA) (*see* **Note 4**).
7. Agarose (Sigma Aldrich, St. Louis, MO) is stored at room temperature.
8. Ethidium Bromide (Bio-Rad, Hercules, CA) is stored at room temperature.

9. TBE buffer (Invitrogen) is stored at room temperature.
10. Agarose Ready Gels (Bio-Rad) are stored at room temperature.
11. MAXEfficiency DH10B chemically competent *E. coli* (Invitrogen) is stored at −80 °C (*see* **Note 5**).
12. Luria Broth (Invitrogen).
13. Luria Broth Agar + ampicillin plates (TEKNova, Hollister, CA).
14. QIAprep spin miniprep kit and HiSpeed Maxiprep kit (Qiagen).
15. Big Dye Terminator v3.1 Cycle Sequencing Kit (Applied Biosystems, Foster City, CA) is stored at −20 °C.

2.2 Transient Expression

1. Sterilized plasmid DNA is stored at −20 °C and ready for use (*see* **Note 6**). Dulbecco's Modified Eagle's Medium (DMEM) (Gibco/Invitrogen, Carlsbad, CA) supplemented with 10 % fetal bovine serum (FBS) (Gibco) and 0.25 mg/mL of geneticin (G418) (Gibco) is stored at 4 °C.
2. Opti-Mem I media (Invitrogen).
3. 10× Dulbecco's phosphate-buffered saline (PBS) (Invitrogen).
4. Lipofectamine 2000 (Gibco) (*see* **Note 7**).
5. 293 SFM II media (Gibco) supplemented with 4 mM L-glutamine (20 mL/L) (Gibco) + 5 mM sodium butyrate (Alfa Aesar, Ward Hill, MA) is stored at 4 °C.
6. HEK 293E cells (Invitrogen).

2.3 Enzyme-Linked Immunosorbent Assay (ELISA)

1. Carbonate buffer: Add 65 mL of 0.2 M sodium carbonate, 185 mL sodium bicarbonate and make up to 1 L with H_2O.
2. Antibody coating buffer: Dilute AffiniPure goat anti-human IgG, Fcγ fragment specific antibody (Jackson ImmunoResearch) in carbonate buffer to a final concentration of 10 mg/mL.
3. High-absorbance ELISA plates (Corning Costar #3369) (*see* **Note 8**).
4. TBST: Dilute 10× TBS (TEKNova) to 1× in water and add TWEEN-20 to a final concentration of 0.05 %.
5. Blocking buffer: Use TBST supplemented with 5 % powdered milk (*see* **Note 9**).
6. Secondary antibody solution: Dilute alkaline phosphatase conjugated goat anti-Hu IgG Fc (Jackson ImmunoResearch) in TBST (1:2,000). 5 mL of the secondary antibody solution is used per 96-well ELISA plate (*see* **Note 10**).
7. Phosphatase substrate solution: Add 20 μL of alkaline buffer (Sigma Aldrich) into 10 mL of NaCl (0.9 %). Dissolve two phosphatase substrate tablets 5 mg each (Sigma Aldrich) in the salt solution. 10 mL of the phosphatase substrate solution is required per 96-well ELISA plate)

2.4 Sodium Dodecyl Sulfate Polyacrylamide Gel Electrophoresis (SDS-PAGE)

1. NuPAGE 10× Sample Reducing Agent, NuPAGE LDS Sample Buffer, NuPAGE MOPS running buffer and Tris-Acetate SDS Running buffer (Invitrogen). Store at room temperature store others at 4 °C.
2. 4–12 % NuPAGE Bis-Tris gels and 3–8 % NuPAGE Tris-Acetate gels (Invitrogen). Store at 4 °C.
3. SimplyBlue™ Safestain (Invitrogen) or other protein stain for SDS-PAGE.

2.5 Western Blotting

1. Immun-Blot PVDF membranes (Bio-Rad, Hercules, CA) (*see* **Note 11**).
2. Trans-Blot SD Semi-Dry Electrophoretic Transfer Cell (Bio-Rad, Hercules, CA) (*see* **Note 12**).
3. NuPAGE Transfer Buffer (Invitrogen) (*see* **Note 13**).
4. TBST is prepared as described in Subheading 2.3, **item 4**
5. Blocking buffer is prepared as described in Subheading 2.3, **item 5**.
6. Anti-peptide region antibody (CNTO 3443) was generated in house, and is stored in PBS at 4 °C (*see* **Note 14**).
7. HRP-conjugated anti-mouse IgG (Amersham Biosciences/GE, Pittsburgh, PA) is diluted 1:40,000 in TBST (*see* **Note 15**).
8. The ECL kit, Hypercassette and Hyperfilm ECL (Amersham Biosciences) (*see* **Note 16**).

2.6 MIMETIBODY™ Protein Purification

1. Pierce ImmunoPure Immobilized Protein A (Pierce, Rockford, IL) (*see* **Note 17**).
2. Immunopure Gentle Ag/Ab binding buffer (Pierce) and Immunopure Gentle Ag/Ab elution buffer (Pierce Chemical Co, St. Louis, MO). Store at 4 °C.
3. Tris buffer solution (MP Biomedicals, Solon, OH).
4. Superdex 200 column (1.5 L volume; Amersham Biosciences).
5. Amicon-15 unit (Millipore)
6. TSKgel 3000SWxl (Toshoh Biosciences, 8541)

2.7 UT-7 Cell Proliferation Assay

1. Dulbecco's PBS (Gibco)
2. I10Q media: Prepare by supplementing Iscove's Modified Dulbecco's Medium (IMDM) with 10 % FBS and 2 mM L-GLUTAMINE. Stored at 4 °C.
3. I5Q media: Prepare by supplementing IMDM with 5 % FBS and 2 mM L-glutamine.
4. MTS solution (Promega, Madison, WI) is stored at −20 °C and is thawed before use (maximum two freeze–thaw cycles).
5. UT-7_{EPO} cells are grown and expanded in I10Q media supplemented with 1 U/mL of human recombinant EPO. Recombinant human EPO stock solution is stored at −80 °C and used for a maximum of two freeze–thaw cycles [7] (*see* **Note 18**).

2.8 Rat Pharmacokinetic and Pharmacodynamic Analysis

1. Assay buffer: Starting Block TBS (Pierce).
2. Capture antibody: Anti-EMP1 mAb, CNTO3443 (Centocor).
3. Detection antibody: Biotinylated anti-human IgG1 or IgG4 (*see* **Note 19**).
4. Detection reagent: Streptavidin–europium solution (Perkin Elmer).
5. Enhancement solution (Perkin Elmer).
6. Hematology analyzer: Advia 120 (Siemens Diagnostics, Tarrytown, NY).

2.9 cAMP Assay

1. INS-1E cells were kindly provided by C.B. Wollheim (*see* **Note 20**).
2. Supplemented RPMI 1640 media: RPMI 1640 media (Invitrogen) plus FBS (10 %), L-glutamine (1 %), sodium pyruvate (1 %), nonessential amino acids (1 %), and β-mercaptoethanol (50 μM).
3. Lance cAMP kit (Perkin Elmer) (*see* **Note 21**) .

2.10 Mouse Pharmacokinetic Study

1. C57Bl/6J mice (Jackson Laboratory; Bar Harbor, ME).
2. 96-well Maxisorp plates (Nunc).
3. Protease inhibitor cocktail: Roche complete EDTA free (Roche Applied Science) (*see* **Note 22**).
4. Linco intact GLP-1 ELISA plates (Linco) (*see* **Note 23**).
5. Detection antibody: Goat anti-human H + L alkaline phosphatase conjugate (Jackson Immunoresearch). For ELISA, the antibody was diluted 1:1,000 in PBS (*see* **Note 24**).
6. Alkaline phosphatase substrate: 4 methylumbelliferyl phosphate (Millipore) (*see* **Note 25**).
7. Stop solution: EP-AP (Millipore).

2.11 ip Glucose Tolerance Test

1. Glucometer (Lifescan).
2. D-glucose (Sigma, St. Louis, MO).
3. C57Bl/6J mice (Jackson Laboratory; Bar Harbor, ME).
4. Purina Test Diet (58126) (*see* **Note 26**).

3 Methods

MIMETIBODY™ constructs are expressed in mammalian cell lines, such as HEK or CHO cell lines. Figure 1 shows a typical expression plasmid map for the CNTO736 MIMETIBODY™. The plasmid is based on pCEP4 plasmid (Invitrogen). The MIMETIBODY™ coding region consists of an IgG VH signal peptide, a bioactive peptide, linker/VH/Hinge, and a human Fc.

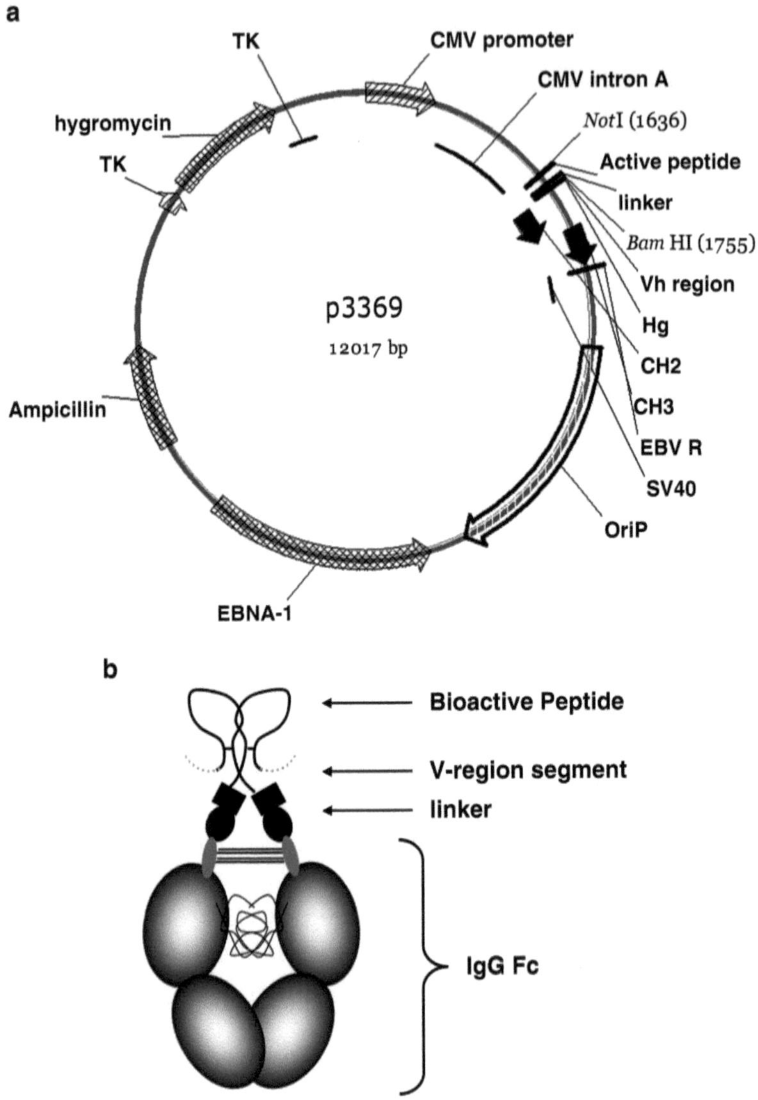

Fig. 1 (**a**) Schematic illustration of a plasmid for expression of MIMETIBODY™ constructs in mammalian cells. The vector is derived from pCEP4 to include a signal peptide together with the mimetic peptide, linker, VH, hinge, and human Fc sequences that encode the MIMETIBODY™ gene. (**b**) Schematic illustration of the structure of CNTO736 including the CH2 and CH3 domains of the Fc, the hinge including the disulfide bonds, a linker segment, and two peptide segments

Several restriction enzyme digestion sites (such as *Nhe*I, *Not*I, and *Bam*HI) are included in the plasmid to enable convenient cloning of different Mimetibody™ constructs. Standard recombinant DNA methods are used to construct the plasmids.

To generate sufficient quantities of protein to quickly evaluate MIMETIBODY™ constructs for biological activity and biophysical properties, the various proteins are transiently expressed in HEK 293E cells. ELISA and Western Blot are used to monitor the amount and quality of the protein transiently expressed in cell supernatants. Protein A affinity column and size exclusion chromatography are used to purify MIMETIBODY™ proteins.

Functional characterization of the MIMETIBODY™ constructs is evaluated both in vitro and in vivo. CNTO 530, an erythropoietin (EPO) receptor agonist, is evaluated in a UT-7 assay and the pharmacokinetics and pharmacodynamics of the molecule are monitored in normal rats [6]. CNTO736 and CNTO2628 are both glucagon-like peptide-1 (GLP-1) receptor agonists designed to interrogate different linker segments [5]. Both constructs are evaluated in an in vitro cAMP assay, and the pharmacokinetics and pharmacodynamics of the molecule are evaluated in normal and diabetic mice, respectively.

3.1 Expression Vector Construction

1. To assemble the peptide coding region, two long oligonucleotides covering the peptide coding regions are designed for overlap PCR so that 17–21 bases of the 3′ end of the first oligonucleotide complement 17–21 bases of the 3′ end the second oligonucleotide. Two additional shorter primers are designed as 5′ and 3′ PCR amplification primers (*see* **Note 27**).
2. The two long oligonucleotides are annealed by mixing 100 ng of each primer in a PCR tube with PCR reaction buffer in a total volume of 100 μL. Heat the tube at 95 °C for 30 s and then cool to 4 °C for at the rate of 1 °C per second. The annealing reaction can be conveniently done in a PCR machine.
3. 10 mM dNTP, 50 mM $MgSO_4$, 10× enhancer, 100 ng of each amplification primer,10× Pfx polymerase buffer and Pfx polymerase are added to the tube containing the annealed primers according to the manufacture's protocol and five cycles of PCR performed (15 s at 94 °C, 30 s at 45 °C, and 30 s at 68 °C).
4. Add 10 ng each of the 5′ and 3′ short primers and perform another 15–20 cycles of PCR (94 °C for 15 s, 55 °C for 30 s, and 68 °C for 30 s). Store the tube at 4 °C after PCR is completed.
5. Purify PCR products with the QIAquick PCR Purification Kit following manufacturer's protocol.
6. Digest the purified PCR product and the vector plasmid DNA (Fig. 1) with the proper restriction enzymes (e.g., *Bam*HI and *Not*I). Purify the digested PCR product and expression vector DNA by gel electrophoresis. The correct size bands are visualized by UV light and cut from the gel. The digested PCR product and vector DNA are recovered from the gel using QIAquick Gel Extraction kit (*see* **Note 28**).

7. Mix the restriction enzyme digest, gel purified PCR product and vector DNA in a ligation reaction using the Rapid DNA ligation kit at an approximate molar ratio of 3:1 in a total reaction volume of 20 μL. Incubate the reaction at room temperature for 10–20 min.
8. Add 5 μL of ligation mixture to 50 μL of chemically competent DH10B *E. coli* following the manufacturer's recommended protocol (30 min on ice, 45 s at 42 °C, 2 min on ice, and 1 h recovery shaking at 37 °C). Store the remainder of the ligation mixture at −20 °C in case a repeat transformation is required.
9. Plate transformed bacteria onto LB-agar plates supplemented with 100 μg/mL of ampicillin and incubate the plates at 37 °C overnight to allow bacterial growth.
10. For each transformation, ten colonies are picked to screen for peptide insertion. Colonies are grown in 3 mL of LB broth supplemented with 100 μg/mL of ampicillin. Cultures are incubated overnight (approximately 16 h) at 37 °C in a shaking incubator.
11. Pellet the bacteria using a microcentrifuge for 3 min. Isolate plasmid DNA from the bacteria pellets using the QIAprep Spin Miniprep kit following manufacturer's protocol.
12. Restriction enzyme digestion is used to identify clones with the appropriate size insert (*see* **Note 29**).
13. Sequence the plasmid DNA. Positive clones are confirmed using sequencing primers specific to vector sequence immediately 5′ or 3′ to the coding sequence.
14. Larger amounts of plasmid DNA for expression experiments are isolated from 250 mL of bacterial cultures using a Maxiprep kit. Plasmids must be kept sterile and stored at −20 °C.

3.2 Transient Expression

1. HEK293E cells are expanded and maintained in DMEM supplemented with FBS and G418. Cells are routinely split every 3–4 days.
2. Seed 8×10^6 cells into a T-150 flask with 30 mL DMEM 24 h before the transfection procedure.
3. Warm Opti-MEM I to room temperature and combine 15 μg of sterilized plasmid DNA with 1 mL of prewarmed Opti-MEM I. Incubate at room temperature for 5 min.
4. In a separate tube, combine 90 μL of Lipofectamine 2000 with 1 mL of prewarmed Opti-MEM I and incubate at room temperature for 5 min.
5. Combine plasmid DNA tube with the Lipofectamine 2000 tube and incubate at room temperature for 20 min.

6. Transfer the mixture to the T-150 flask pre-seeded with HEK 293E cells. Incubate at 37 °C in a tissue culture incubator with 5 % CO_2 for 24 h.
7. Exchange the media with serum-free medium 293 SFM II supplemented with L-glutamine and sodium butyrate.
8. Collect cell supernatants from tissue culture flasks 3–5 days after the media exchange and centrifuge at 850 × *g* for 10 min at 4 °C.
9. Collect clear supernatant and filter it through a 0.22 μm filter.
10. Concentrate samples with Millipore Centricon Plus-70 by centrifugation at 2,800 × *g*, 4 °C. Final volume should be less than 20 mL. Store concentrated samples at 4 °C for purification.

3.3 ELISA

1. A sandwich ELISA is used to monitor protein concentration in cell supernatants. Add 50 μL of goat anti-Fc antibody diluted in carbonate buffer to each well of the high-absorbance plate and incubate overnight at 4 °C (*see* **Note 30**).
2. Wash plate three times with TBST on a plate washer.
3. Add 200 μL of blocking buffer to each well and incubate the plate at room temperature for 1 h.
4. Standard curve preparation: dilute purified human IgG in TBST to 1 μg/mL. In a 96-well round-bottomed plate, add 100 μL of the IgG-TBST solution to the first well of the column and continue with 1:2 serial dilutions down the plate (*see* **Note 31**).
5. Test sample preparation: dilute cell supernatant 1:20 in the TBST. In a 96-well round-bottomed plate, add 100 μL of the diluted sample solution to the first well of the column and continue with 1:10 serial dilutions down the plate.
6. Transfer 50 μL from each well of the prep plate to the ELISA plate. Standards and test samples are routinely run in duplicate or triplicate rows.
7. Incubate the ELISA plates for 1 h at room temperature.
8. Wash plate three times with TBST.
9. Add 50 μL of secondary antibody solution to each well. Incubate at room temperature for 30 min.
10. Wash plate three times with TBST on the plate washer.
11. Add 100 μL alkaline phosphatase substrate per well. Monitor color development. Stop reactions by adding 50 μL 3 M NaOH per well if necessary.
12. Read plate using an ELISA plate reader at 405 nm.

3.4 SDS-PAGE

1. Mix appropriate amount of protein (0.25 μg for silver staining or Western Blotting; 2.5 μg for SimplyBlue staining) with NuPAGE SDS Sample Buffer and NuPAGE Sample Reducing

Agent (for reducing gel only) and heat the samples at 70 °C for 10 min.

2. Load the reducing gel sample to a pre-made 4–12 % NuPAGE Bis-Tris gel or nonreducing gel samples onto a 3–8 % NuPAGE Tris-Acetate gel.
3. Run the gel electrophoresis at 200 V for 60 min, or until the dye front migrates to the bottom edge of the gel. Use NuPAGE MOPS SDS Running Buffer for reduced samples (with NuPage Antioxidant) or Tris-Acetate SDS Running Buffer for nonreducing samples.
4. Remove the gel from the cassette and rinse the gel three times for 5 min with 100 mL deionized water to remove SDS and buffer salts.
5. Stain the gel with SimplyBlue™ SafeStain to cover the gel and shake gently for 1 h at room temperature. Destain by washing three times with 100 mL deionized water for 5 min.

3.5 Western Blotting

1. After running the protein samples on SDS-PAGE (described above), wash the gel in NuPAGE Transfer Buffer containing 10 % methanol for 20 min with gentle agitation.
2. Transfer the proteins to an Immobilon-P PVDF membrane using the Trans-Blot SD Semi-Dry Electrophoretic Transfer Cell, following manufacture's protocol.
3. Soak the membrane in blocking buffer overnight at 4 °C with gentle agitation.
4. Incubate the membrane with the primary antibody CNTO 3443 (1 μg/mL) in TBST at room temperature for 1 h (*see* **Note 32**).
5. Wash the membrane four times for 5 min in TBST with gentle shaking.
6. Incubate the membrane with an HRP-conjugated anti-mouse IgG (1:40,000 dilution in TBST) at room temperature for an hour.
7. Wash the membrane six times for 5 min in TBST with gentle shaking.
8. Using a Kimwipe, gently blot the membrane to remove excess liquid. Place the membrane on a piece of parafilm, protein side up; Detection Reagent 1 and Detection Reagent 2 included in the ECL kit is mixed in equal volumes (3 mL of each per membrane) and added directly to the membrane surface. Wait for 1 min before draining the blot.
9. Using a Hypercassette, expose the Hyperfilm to the membrane for a series of time points to provide the optimal exposure.

3.6 MIMETIBODY™ Protein Purification

1. To purify MIMETIBODY™ constructs which contain the human IgG constant region, Protein A affinity column is used as the first purification step followed by size exclusion Chromatography. Equilibrate 1 mL protein A column with 5 mL Immunopure Gentle Ag/Ab binding buffer.
2. Dilute the sample at least 1:1 with Immunopure Gentle Ag/Ab binding buffer before applying to the Protein A column to maintain the proper ionic strength and pH for optimal binding.
3. Load the diluted sample to the column and allow it to flow completely into the gel.
4. Wash the Protein A column with 15 mL of the Binding Buffer.
5. Elute proteins with 5 mL of Elution Buffer into 500 μL neutralization buffer (1 M Tris–HCl, pH 9.0). Alternatively, proteins are directly eluted using Gentle Elution buffer followed by buffer exchange into PBS using a Superdex 200 gel filtration column at a flowrate of 4.5 mL/min. MIMETIBODY™ constructs that are sensitive to low pH are best eluted using the Gentle Elution buffer.
6. Check the protein of individual fractions by monitoring OD_{280} and confirm via SDS-PAGE.
7. Pool the selected fractions and dialyze the protein into 20 mM Tris, pH 7.4 at 4 °C overnight.
8. Collect the dialyzed protein and concentrate it with an Amicon-15 unit, 2,800 × *g* for 10–20 min, at 4 °C.
9. Sample is purified further via size exclusion chromatography on a Superdex 200 column using PBS as the running buffer at a flow rate of 4.5 mL/min (*see* **Note 33**).
10. Pool the selected fractions, measure the protein concentration using a appropriate extinction coefficient from the absorbance measurement at 280 nm. Sterilize purified proteins by filtration using a sterile filter with a 0.22 μm cutoff (*see* **Note 34**).
11. If necessary, concentrate protein using an Amicon-15 unit, 2,800 × *g* for 10–20′, 4 °C. Store protein samples at 4 °C.
12. To determine if the purified MIMETIBODY™ protein form aggregates, analytical size exclusion chromatography is used. 15 μg of each protein dialyzed in PBS is loaded onto a TOSOH TSKfel G3000SWXL column (Tosoh).
13. The samples are separated at a flow rate of 1 mL/min with PBS as the running buffer.

3.7 UT-7 Cell Proliferation Assay

1. Wash UT-7 cells to remove residual EPO by adding 50 mL DPBS and pelleting the cells for 10 min at 100 × *g*. Repeat three times.

2. Resuspend the cells in I5Q media without EPO and incubate cells at 37 °C with 5 % CO_2 for 24 h to starve the cells.
3. Wash starved cells once in 50 mL DPBS. Count cells and resuspend them to the 6×10^5 cells/mL in I5Q media (*see* **Note 35**).
4. EPO standard preparation: Dilute EPO stock solution in the I5Q media to10 ng/mL (3.28×10^{-10} M). In a 96-well round-bottomed plate, add diluted EPO solution to the first well of the row and continue with 1:2 serial dilutions across the plate, omitting the last well of the row (media only).
5. Sample preparation: Dilute MIMETIBODY™ or other EPO-R agonists to a starting concentration of 5.04×10^{-8} M. On a 96-well round-bottomed plate, add diluted EPO solution to first well of the row and continue with 1:3 serial dilutions across the plate, omitting the last well of the row (media only).
6. Transfer 50 μL of the UT-7 cell suspension to the test wells on a 96-well flat-bottomed plate. Final cell density is 30,000 cells/well.
7. Transfer 50 μL of the EPO standards or the sample dilutions to each well. Negative controls containing media only or PBS serve as the negative control wells. The standards and test samples are routinely run in duplicate or triplicate.
8. Incubate test plates at 37 °C with 5 % CO_2 for 48 h.
9. Add 20 μL of MTS solution to each well. Incubate test plates at 37 °C with 5 % CO_2.
10. Read the plate and record the OD_{490}–OD_{650} value on an ELISA plate reader at 1, 2, and 3 h after adding the MTS solution. The 3-h data point is typically used. An example of the proliferation data generated is shown in Fig. 2a.

3.8 Rat Pharmacokinetic and Pharmcodynamic Analysis

1. Sprague–Dawley rats (4/group) are dosed intravenously with CNTO528 or CNTO530 (1 mg/kg for PK experiments or 0.3 mg/kg for PD experiments) (*see* **Note 36**).
2. At various times after dosing, while under light anesthesia, collect blood from the retro-orbital sinus using uncoated hematocrit tubes with a CO_2–air mixture, followed by clotting for PK experiments or EDTA-coated tubes for PD experiments.
3. For PK experiments, centrifuge the blood (10–15 min at 2,500 × *g*) and recover the serum. Store the serum at −80 °C.
4. Hemoglobin is measured in whole blood with an Advia 120 Hematology Analyzer using multi-species software according to the manufacturer's instructions. Representative hemoglobin data is shown in Fig. 2b.
5. The concentration of intact MIMETIBODY™ is measured using a time-resolved immunofluorometric assay (DELFIA) as described [6].

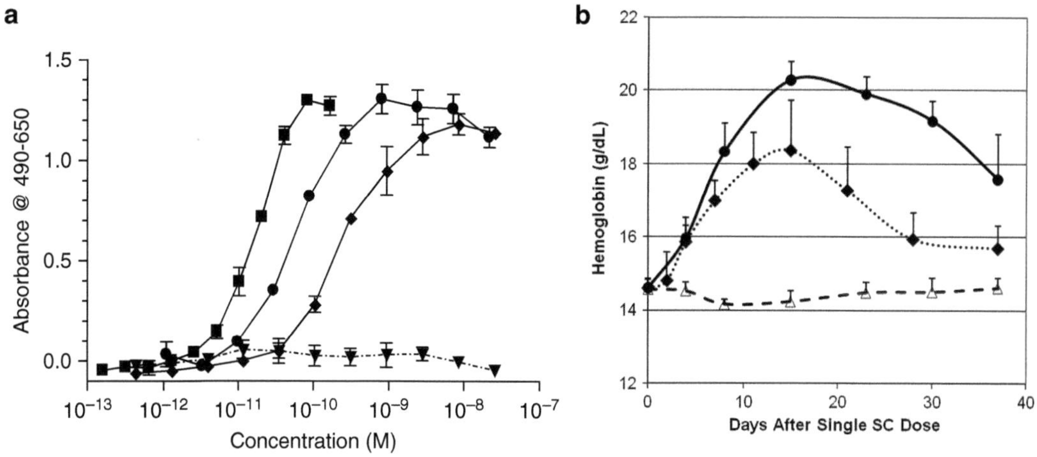

c

Drug	ID	Cmax (ng/mL)	AUC ng.day/mL	t½ day	CL mL/day/kg	Vz mL/kg
CNTO528	Mean	29979.29	17744.90	1.54	56.81	126.18
CNTO528	SD	4917.82	1849.85	1.46–1.61	5.89	17.06
CNTO530	Mean	27319.26	53790.72	3.03	19.00	93.58
CNTO530	SD	3987.52	9023.21	2.59–4.73	3.23	38.00

Fig. 2 (**a**) Supernatants obtained by transient transfection for CNTO528 (*diamonds*, EC50 = 2.28×10^{-10}M) or CNTO530 (*circles*, EC50 = 5.00×10^{-11} M) were tested for proliferative activity in UT-7 cells. Viable cells, monitored by MTS absorbance at 490 nm is plotted as a function EPO receptor agonist concentration. Human recombinant EPO (*squares*, EC50 = 2.14×10^{-11} M) was used as positive control and cell supernatant from a mock transfection was used as negative control (*triangles*). Data are presented as the mean ± SD (n = 2). (**b**) Hemoglobin monitored in normal rats as a function of time following subcutaneous dosing with PBS (*triangles, dashed line*), CNTO528 (*diamonds, dotted line*), or CNTO530 (*circles, solid line*). The data are presented as the mean ± SD (n = 4). (**c**) Pharmacokinetic properties for CNTO528 and CNTO530 dosed subcutaneously in normal rats at 1 mg/kg. Mean data values and standard deviations are shown (n = 4)

6. White Fluoro Nunc C-96-well maxisorp plates are coated with murine anti-EMP1 monoclonal antibody (CNTO3443, a mouse monoclonal antibody generated by immunization with a conjugate protein fused to EMP1.).
7. Plates were blocked with 200 μL of assay buffer for 1 h at room temperature.
8. Samples, standards, and controls are diluted 1:10, 1:100, and 1:1,000, respectively, for assay. Plates are washed with PBS and 50 μL of diluted samples or standards are added to the appropriate wells with incubation for at room temperature for 1 h.
9. Plates are washed three times and 50 μL biotinylated mouse anti-human IgG1 or IgG4 is added and incubated for 1 h at room temperature.

10. Plates are washed three times and 50 μL of streptavidin–europium solution is added to each well and incubated for 1 h at room temperature. Plates are washed, 50 μL of enhancement solution is added, plates are incubated for 5 min and read on EnVision Multilabel Counter.
11. The concentration of MIMETIBODY™ was calculated from the standard curve. Data are analyzed by WinNonlin 4.0.1 (Pharsight, Mountain View, CA). Representative pharmacokinetic parameters are shown in Fig. 2c.

3.9 cAMP Assay

1. INS-1E cells (1×10^5 cells/well) are plated in 96-well plates in supplemented RPMI 1640 media and cultured for 4 days at 37 °C with 5 % CO_2.
2. Media is aspirated from the wells and 24 μL of Alexa Fluor 647 anti-cAMP antibody is added followed by 24 μL of the CNTO736 or CNTO2628 MIMETIBODY™ (in PBS/0.5 % BSA/0.5 mM IBMX).
3. The cells are stimulated at room temperature for 7 min.
4. The cells are lysed according to the manufacturer's protocol (Perkin Elmer).
5. The plates are incubated at room temperature for 1 h and the fluorescence intensity is measured using an excitation wavelength of 340 nm and emission wavelength of 665 nm.
6. cAMP concentrations are determined using a standard curve prepared according to the manufacturer's instructions (Perkin Elmer). An example of typical cAMP data is shown in Fig. 3a.

3.10 Pharmacokinetic Analysis in Mice

1. C57BL/6J mice are dosed intravenously with CNTO736 or CNTO2628 (1 mg/kg) (*see* **Note 37**).
2. At various times after dosing, three animals are sacrificed and blood is collected via cardiac puncture in 3.8 % sodium citrate containing protease inhibitors.
3. Blood is centrifuged (10–15 min at 2,500 × *g*) and plasma is recovered and stored at −80 °C.
4. Plasma samples are prepared at dilutions of 1:10, 1:50, 1:100, 1:200, 1:400, 1:800, 1:1,200. Standards are prepared for each MIMETIBODY™ starting a 0.1 μg/mL with 1:2 serial dilutions.
5. The concentration of intact MIMETIBODY™ is measured using a modified form of a purchased ELISA designed to detect intact GLP-1 (Linco). The pre-coated, blocked plates received from Linco are washed and 100 μL assay buffer and 100 μL standard or plasma samples are added to the appropriate wells. Plates are incubated overnight at 4 °C.

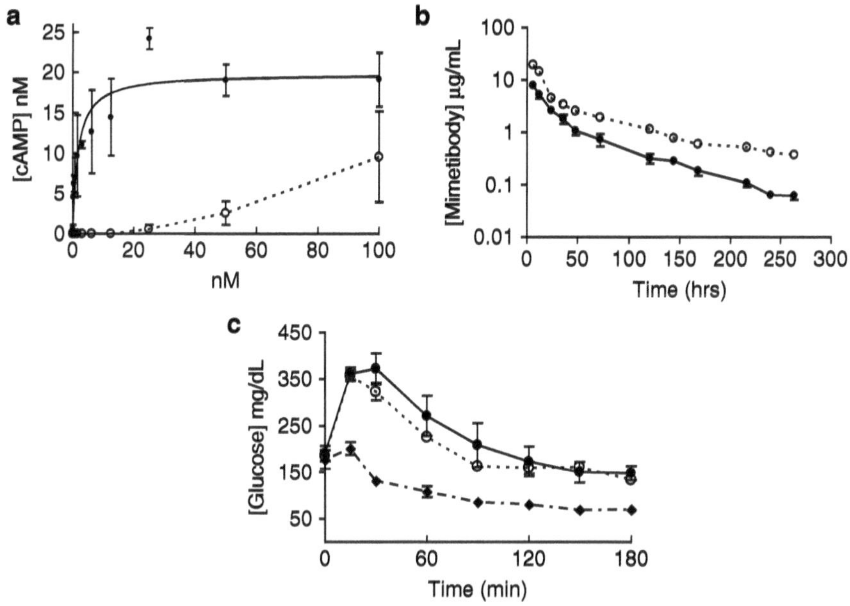

Fig. 3 (**a**) The concentration of cAMP was measured in INS-1E cells after addition of increasing concentrations of CNTO736 (*solid line*) or CNTO2628 (*dotted line*). The fit to the CNTO736 data provided an EC_{50} of 1.8 nM (**b**) The concentration of CNTO736 or CNTO2628 was determined at various times following an intravenous dose of 1 mg/kg in normal mice. Data for CNTO736 (*solid circles* and *line*) or CNTO2628 (*circles, dashed line*) are presented as the mean ± SEM ($n=3$). (**c**) Fasted mice (DIO ($n=5$)) were dosed intravenously with vehicle (*black circles*), CNTO736 (0.5 mg/kg (*black diamonds*)), or CNTO2628 ((*white circles*) 0.5 mg/kg) 10 min prior to an ipGTT. The results are presented as the mean ± SEM ($n=5$)

6. Plates are washed, 100 µL of diluted goat anti-human H+L alk-phos conjugate is added to each well and plates are incubated for 2 h at 37°C.
7. The plates are washed and bound-conjugated antibody is detected by addition of substrate (4-Methylumbelliferyl Phosphate, Millipore) and incubation at room temperature. The reaction is stopped by addition of stop solution after 20 min.
8. The plates are read on a fluorescent plate reader (Molecular Devices) by exciting at 355 nm and monitoring emission at 460 nm.
9. The concentration of MIMETIBODY™ is calculated from the standard curve. Examples of the data for CNTO736 and CNOT2628 are shown in Fig. 3b.

3.11 ip Glucose Tolerance Test (GTT) in Diet-Induced Obese Mice

1. C57Bl/6J mice are fed a high fat diet (Purina Test Diets) beginning at 4 weeks of age. All animals achieve 3 consecutive weeks of fasting blood glucose greater than 120 mg/dL prior to inclusion in the study.

2. Male DIO mice are randomized ($n = 5$) based on fasting blood glucose.
3. Mice are dosed intravenously with the MIMETIBODY™.
4. Five minutes later ($T = -5$ min relative to GTT), fasting glucose is measured from tail vein blood using a hand-held glucometer (Lifescan) (*see* **Note 38**).
5. Five minutes later ($T = 0$), mice are given a bolus dose of D-GLUCOSE (1 g/kg) intraperitoneally.
6. Blood glucose is measured after 15, 30, 60, 90, 120, 150, and 180 min using tail vein blood and a hand-held glucometer (Lifescan). Representative data for CNTO736 and CNOT2628 are shown in Fig. 3c.

4 Notes

1. Oligonucleotides can be ordered from many suppliers. We generally use oligonucleotides as received from the vendor without further purification. For long oligonucleotides it may be necessary to request purification from the vendor or to gel purify using standard methods.
2. It is possible to use other polymerases. Pfx polymerase has good fidelity and high processivity and works well in our hands. If another polymerase is substituted, optimization according to the manufacturer's instructions may be required.
3. A variety of other DNA ligation kits are available including ones from Agilent, New England Biolabs and Promega. The Rapid DNA ligation kit from Roche is rapid and robust and works well for many ligation applications.
4. Other kits for extraction of nucleic acids from agarose gels and clean up of PCR reactions are available from a variety of vendors. Alternatively, nucleic acids can be recovered from agarose gels by performing a phenol extraction and ethanol precipitation.
5. We routinely use MAXEfficiency competent cells from Invitrogen due to their high transformation efficiency and ease of use. Most other competent *E. coli* strains could be substituted and used according to the manufacturer's directions.
6. Plasmid is sterilized by addition of 0.1 volume of 3 M Sodium Acetate and 2.5 volumes of 100 % ethanol. Mix the tube well and incubate at −80 °C for 30 min. Spin in a microcentrifuge at 16,000 × *g* for 15 min at 4 °C and carefully decant the supernatant being careful not to disturb the pellet. Wash once with 1 mL of 70 % ethanol and spin at 4 °C at 16,000 × *g*. Aspirate supernatant and air dry DNA. Add sterile TE buffer to final DNA concentration of 1 mg/mL.

7. Lipofectamine 2000 (Invitrogen) is a reliable method for rapid transfection of mammalian cells including HEK 293E cells. Many alternatives for performing transient transfections exist including other cationic lipid reagents such as Transfectam (Promega), Superfect (Qiagen), Fugene (Roche). In addition to lipid-based reagents, other methods such as calcium phosphate transfections, Polyethylenimine (PEI)-assisted transfection and electroporation are all common and effective methods of transfection.
8. For optimal ELISA signals, it is important to use a high protein binding ELISA plate. We have used high protein binding plates from Corning Costar or Maxisorb plates from Nunc routinely although it may be possible to use high protein binding plates from other vendors. Plates treated for tissue culture use are not appropriate. Maximal coating efficiency should be determined empirically.
9. Alternative blocking strategies could be used. We generally use 5 % dry powdered milk however this may not be appropriate if a streptavidin–biotin detection system will be used due to the presence of biotinylated proteins in milk. Alternatives include blocking buffers from Pierce and Millipore.
10. The choice of detection reagent is based on the choice of Fc chosen for preparation of the MIMETIBODY™. For simplicity and convenience we have described the use of an alkaline phosphatase anti-hu IgG Fc detection reagent. If a murine Fc containing MIMETIBODY™ is prepared, an appropriate detection reagent must be substituted. It is also possible to use many other detection systems including horse radish peroxidase or biotin–Streptavidin systems. Appropriate substrates must be coordinated with the conjugated anti-Fc reagents.
11. We have had the most success with reproducible Western blots using PVDF membranes. Other membranes may be substituted and used according to the manufacturer's directions.
12. The Trans-blot SD Semi-dry Electrophoretic Transfer cell from Bio-Rad enable fast, high efficiency transfer of proteins for Western blots. Other electrophoretic transfer apparatus are available and could be substituted and used according to the manufacturer's instructions. If a different apparatus is substituted, it may be necessary to adjust the protein loading to account for differences in transfer efficiency. Colored molecular weight markers, available from various vendors, can be used to assess the protein transfer efficiency.
13. If an SDS-PAGE system different from the NuPAGE system is used, transfer buffers will need to be modified according to the manufacturer's instructions.

14. To enable accurate assessment of the presence of the bioactive peptide on the MIMETIBODY™ construct, it is essential to have a peptide that recognizes the intact N-terminus of the peptide. We routinely prepare anti-peptide antibodies in-house by a variety of standard methods and several commercial vendors supply services to prepare appropriate antibodies. As one of the primary routes for metabolism of MIMETIBODY™ constructs is cleavage at sites in the bioactive peptide, it is important that this detection reagent bind the amino terminus of the peptide and not bind if proteolysis has generated incomplete peptide fragments.
15. In combination with the anti-peptide antibody described in *see* **Note 14**, the anti-Fc reagent chosen for detection of the intact MIMETIBODY™ is determined by the Fc species and isotype for the construct. It is important to use a high specificity anti-Fc reagent, although other conjugates, including alkaline phosphatase, could be substituted.
16. The detection reagent must be paired with the conjugation chemistry used for the detection antibody. ECL is a high sensitivity method that we have used successfully, although other detection systems could be used according to the manufacturer's instructions.
17. A variety of Protein A affinity chromatography media can be used to capture the MIMETIBODY™ from the cell supernatant. We have had good success with Immobilized Protein A columns from Pierce and MabSelect columns from GE Healthcare Life Sciences.
18. It is essential to use UT-7/$_{EPO}$ cells for optimal signal/noise in proliferation assays. Other UT-7 cells that are not EPO dependent cannot be substituted.
19. For pharmacokinetic assays, it is important to measure the intact MIMETIBODY™ (*see* **Note 14**). The detection antibody can be either directly conjugated with HRP or alkaline phosphatase, although biotinylation of the detection antibody followed by treatment with streptavidin-HRP or streptavidin-EU is significantly more sensitive.
20. Other GLP-1 receptor expressing cells can be used including Rin-m or min-6 cells, although we have found INS-1 cells to be the easiest to maintain and most reproducible.
21. Other assays to quantify cAMP can be used. One alternative the homogeneous AlphaScreen assay from Perkin Elmer. However, the AlphaScreen assay requires an instrument that is Alpha-enabled. The Lance cAMP assay is a homogeneous time-resolved fluorescence resonance (TRF) energy transfer assay that is very sensitive and robust. This assay requires an instrument capable of measuring TRF (i.e., Perkin Elmer EnVision™).

22. An accurate assessment of the amount of intact MIMETIO-BODY™ is essential. Protease inhibitors are added to eliminate the potential for post-collection proteolysis. We routinely use the Roche complete EDTA-free protease inhibitor cocktail however other protease inhibitor cocktails may also work.
23. It is important to measure the intact molecule and not simply the Fc portion of the molecule. One mechanism for clearance of the intact MIMETIBODY™ is proteolytic degradation somewhere between the N-terminus of the bioactive peptide and the intact Fc. Thus, an antibody that recognizes the intact N-terminus of the bioactive moiety is essential. The Linco kit provides an antibody specific to the N-terminus of GLP-1 and proteolysis of even 2 N-terminal residues renders the antibody incapable of binding.
24. The specificity of the detection antibody is important to minimize background signal and improve S/N ratios. We have found the Jackson ImmunoResearch goat anti-human (H+L) antibody (cat no. 109-055-003) to be well behaved. Other alternatives may be available. It is also possible to use a detection antibody with a different label (i.e., horse radish peroxidase, HRP), however in this case, it is important to match the substrate appropriately.
25. Other alkaline phosphatase substrates could be used. Some alternatives include PhosphoGLO (KPL), Attophos (Promega), or pNPP (*p*-nitrophenylphosphate, Sigma). If an alternatively conjugated antibody is used, the detection substrate must be paired appropriately.
26. It is essential that mice have maintained 3 consecutive weeks of high blood glucose. In our experience, mice maintained on Purina Test Diet #58126 where 60 % of the energy is derived from fat become diabetic at approximately 10–12 weeks of age however, most we use mice maintained on this diet for 18–20 weeks routinely. Other rodent diets are available with a lower percentage of the energy from fat, however mice maintained on lower fat diets take longer to become diabetic.
27. It is important to ensure that the annealing temperature of the overlap regions in the oligonucleotides (or primers) is between 62 and 68 °C to maximize extension efficiency. We routinely use synthetic oligonucleotides up to 70–80 bp; for longer peptide segments it may be necessary to prepare multiple overlapping oligonucleotides.
28. PCR products and vector DNA can be purified by agarose gel electrophoresis. Depending on the size of the amplified peptide segment, it may be necessary to use a higher percentage of agarose. An appropriate DNA fragment size marker should be included to enable recovery of the correct fragment.

29. It is convenient if a unique restriction site is encoded in the bioactive peptide sequence. In this case digestion with a single restriction enzyme followed by comparison of the digested and undigested vector by agarose gel electrophoresis can confirm insertion of the fragment. If no unique sites are introduced, digestion using enzymes with sites near the insertion site (i.e., *Not*I and/or *Bam*HI) can be used.
30. It is important to optimize the coating conditions for capture antibody used to coat the well. If a different capture antibody is used, a screen to evaluate coating efficiency under various buffer and concentration conditions will yield improved signal/noise ratios for the ELISA. Alternatively, it is possible to use IgG MIMETIBODY™ concentration.
31. An appropriate IgG standard should be used to enable accurate quantification of the MIMETIBODY™. The IgG standard should match the species and isotype for the MIMETIBODY™ construct. Quantified IgG standards are available from a variety of vendors including Jackson Immunoresearch Labs and Millipore.
32. For Western blotting, it may be necessary to optimize the concentration of primary antibody used to bind to the target MIMETIBODY™. One easy way to do this is to run several lanes of the protein sample for detection on the same gel. After transfer, the individual lanes can be cut apart and treated with different concentrations of the primary antibody. The concentration of antibody required for good visualization will depend on the affinity and specificity of the primary antibody.
33. For purification of greater than 20 mg of protein, we routinely use the 1.5 L Superdex 200 column with a flow rate of 4.5 mL/min. For small amounts of protein, a 200 mL Superdex 200 column can be used with a flow rate of 1–2 mL/min.
34. The extinction coefficient can be calculated from the amino acid sequence by a variety of software packages. One available package can be found at www.expasy.ch/tools/protparam.htmL.
35. Cells are counted by trypan blue exclusion using a hemocytometer according to the manufacturer's instructions. Alternatively, cells could be counted using a cell counting instrument from Coulter or Cedex.
36. We routinely use Sprague–Dawley rats since it is possible to obtain multiple bleeds from one rat. In contrast, it is only possible to take one blood sample from a mouse. For PK studies, we generally use 3–4 animals for each time point and rats are bled 4–5 times over the course of the study. Time points range from 5 min to 2 weeks and blood samples are 300 μL.

37. Assessment of pharmacokinetic properties should be performed in the same strain of mice where the pharmacodynamic assessment is done. Since C57Bl/6J mice are used for the blood glucose studies, they are also used for the PK analyses.
38. A hand-held glucometer is the easiest and fastest to use. If a hand-held glucometer is not available, blood glucose can also be measured using a glucose oxidase assay (Invitrogen).

References

1. Reichert JM, Rosensweig CJ, Faden LB, Dewitz MC (2005) Monoclonal antibody successes in the clinic. Nat Biotechnol 23:1073–1078
2. Reichert JM (2009) Global antibody development trends. mAbs 1:86–87
3. Kretzschmar T, von Ruden T (2002) Antibody discovery: phage display. Curr Opin Biotechnol 13:598–602
4. Lerner RA (2006) Manufacturing immunity to disease in a test tube: the magic bullet realized. Angew Chem Int Ed Engl 45:8106–8125
5. Picha KM, Cunningham MR, Drucker DJ, Mathur A, Ort T, Scully M, Soderman A, Spinka-Doms T, Stojanovic-Susulic V, Thomas BA, O'Neil KT (2008) Protein engineering strategies for sustained glucagon-like peptide receptor-dependent control of glucose homeostasis. Diabetes 57:1926–1934
6. Bugelski PJ, Capocasale RJ, Makropoulos D, Marshall D, Fisher PW, Lu J, Acthuthanandam R, Spinka-Doms T, Kwok D, Graden D, Volk A, Nesspor T, James IE, Huang C (2008) CNTO530: molecular pharmacology in human UT-7_{EPO} cells and pharmacokinetics and pharmacodynamics in mice. J Biotechnol 134:171–180
7. Komatsu N, Yamamoto M, Fujita H, Miwa A, Hatake K, Endo T, Okano H, Katsube T, Fukumaki Y, Sassa S, Miura Y (1993) Establishment and characterization of an erythropoietin-dependent subline, UT-7/Epo, derived from human leukemia cell line, UT-7. Blood 82:456–464

Chapter 10

Evaluation of Peptides as Protease Inhibitors and Stimulators

Hannu Koistinen, Can Hekim, Ping Wu, Ale Närvänen, and Ulf-Håkan Stenman

Abstract

Changes in proteolytic activity are associated with several diseases, including cancer. Proteases are potential drug targets and targeting of proteases is used for treatment of various conditions/diseases, like high blood pressure and HIV. We present here detailed protocols for basic evaluation of the effects of peptides on the activity of proteases, using kallikrein-related peptidases KLK2 and KLK3 (also known as hK2 and PSA), and trypsin as examples. KLK2 and KLK3 are major prostatic proteases, and they are potential targets for prostate cancer treatment. KLK2 has trypsin-like activity and KLK3 chymotrypsin-like activity. By phage display technology, we have developed peptides that specifically stimulate KLK3-activity and other peptides that inhibit KLK2 or trypsin. The effect of the peptides on the proteolytic activity of proteases can be studied using substrates, the cleavage of which generates detectable signal, allowing rapid evaluation of protease activity. The cleavage of protein substrates can be detected by SDS-PAGE, followed by staining of the proteins. We also describe graphical analysis of the IC50-value, the effect of a peptide on Michaelis-Menten constant (K_m) and the maximal reaction rate (V_{max}).

Key words Protease, Peptide, Inhibition, Kinetics, SDS-PAGE, Kallikrein, hK2, PSA, Phage display

1 Introduction

Increased proteolytic activity is associated with several diseases, including cancer [1]. On the other hand, some proteases also have tumor suppressive effects [2]. Proteases have been estimated to represent 5–10 % of the potential drug targets [1] and targeting of proteases is used for treatment of various conditions/diseases, like high blood pressure and HIV [3]. We present here detailed protocols for the basic evaluation of the effects of peptides on the proteolytic activity of proteases, using kallikrein-related peptidases KLK2 (kallikrein-related peptidase 2, also known as human kallikrein 2, hK2) and KLK3 (kallikrein-related peptidase 3, prostate-specific antigen, PSA), and trypsin as examples. KLK2 and KLK3 are major

Andrew E. Nixon (ed.), *Therapeutic Peptides: Methods and Protocols*, Methods in Molecular Biology, vol. 1088, DOI 10.1007/978-1-62703-673-3_10, proteases in the prostate, and they are potential targets for prostate cancer treatment [4]. KLK2 has trypsin-like activity and KLK3 chymotrypsin-like activity. According to our hypothesis, it may be possible to control prostate cancer growth by modulating the proteolytic activity of these proteases. To this end we have developed, by phage display technology, peptides that specifically stimulate KLK3-activity and other peptides that inhibit KLK2 [5, 6]. We have also developed peptides that inhibit trypsin [7]. Noteworthy, the panning of phage libraries was much more efficient when the proteases were captured to a solid phase by an antibody that exposes the active site of the proteases as compared to the proteases that were directly allowed to adsorb to the solid phase.

The effect of peptides on the proteolytic activity of proteases can be studied using substrates, the cleavage of which generates detectable signal. This allows rapid evaluation of protease activity. Here we describe the use of chemically synthesized chromogenic peptide substrates. However, as the properties of such substrates may differ from the physiological substrates or are not readily available, we present also a protocol using a protein substrate. The cleavage of protein substrates can be detected by several methods, including separation of the fragments by SDS-PAGE, followed by protein staining. While the detailed kinetic studies are beyond this chapter, we describe, for reversible inhibitors, graphical analysis of the inhibitor concentration causing 50 % inhibition of a given concentration of the enzyme (IC50-value), the effect of a peptide on Michaelis-Menten constant (K_m) and the maximal reaction rate (V_{max}). More accurate analysis of kinetic parameters by nonlinear regression can be performed using one of the several graphics programs available.

2 Materials

All reagents should be of analytical grade. H_2O refers to laboratory grade water (e.g., water purified using Milli-Q® Ultrapure Water Purification System, Millipore), although for some protocols deionized water would be sufficient.

2.1 Enzyme Activity Assay

1. Make all dilutions in 50 mM Tris–HCl buffer, pH 7.7, containing 9 g/l (0.15 M) NaCl (TBS) and 1 g/l bovine serum albumin (BSA) (0.1 % BSA/TBS) (*see* **Note 1**).
2. Proteases: We use KLK3 (isoform B) purified from seminal plasma by immunoaffinity chromatography, followed by separation of the different isoforms by anion-exchange chromatography as described [8] and recombinant KLK2 [9]. These enzymes can also be obtained commercially. Working solution

of KLK3 is 3.5 μM (i.e., 100 μg/ml) and KLK2 1.7 μM (50 μg/ml) (*see* **Notes 2** and **3**). Final concentrations in the activity assay are one-tenth of these. Useful concentrations for some other proteases and their substrates can be found in *see* **Note 13**.

3. Chromogenic substrates: 1 mM S-2302 (H-D-Pro-Phe-Arg-pNA·HCl, plasma kallikrein substrate) for KLK2 and 1 mM S-2586 (MeO-Suc-Arg-Pro-Tyr-pNA·2HCl, chymotrypsin substrate) for KLK3 (both in H_2O) (both from Chromogenix Instrumentation Laboratory, Milan, Italy) (*see* **Note 4**). Final concentrations in the activity assay are 0.2 mM.
4. Peptides: Peptides have been produced in house or as custom peptides (Anaspec, San Jose, CA). For stock solutions, peptides are dissolved in H_2O at 2 mg/ml concentration (corresponds to about 1.3–1.9 mM, depending on the peptide, final concentrations in activity assay are one-tenth of these) (*see* **Notes 5** and **6**). The peptides used here are KLK3 stimulating peptides C4 (CVAYCIEHHCWTC, disulfide bridges between C5 and C10, and C1 and C13) and B2 (CVFAHNYDYLVC, disulfide bridge between C1 and C12), and KLK2-inhibiting peptide ARFKVWWG [5, 6].
5. Clear 96-well microtiter plates and a plate reader capable of measuring absorbance at 405 nm (e.g., Victor 1420 Multilabel fluorometer, Wallac-Perkin-Elmer, Turku, Finland).

2.2 Degradation of Protein Substrates

1. Low protein binding tubes (Protein LoBind Tube 0.5 ml, Eppendorf, Hamburg, Germany) (*see* **Note 7**).
2. Recombinant trypsin-1 (produced in house, [7]), 0.1 μM (2.6 μg/ml, final concentration in reaction 0.033 μM) in TBS. Trypsin-inhibiting peptide (GIPCAWFRTGCA, cyclic, [7]), 0.6 mg/ml in TBS (470 μM, final concentration 157 μM).
3. α-Casein (Sigma), 1 mg/ml in TBS (final concentration 0.33 mg/ml) (*see* **Note 8**).

2.3 Detection of Protein Fragmentation

2.3.1 SDS-PAGE

Unless otherwise stated all materials are from Invitrogen.

1. Sample buffer: Per-sample, 5 μl NuPAGE® LDS Sample Buffer (allow to reach room temperature before use) and 2 μl NuPAGE® Reducing Agent. Make a fresh mix, enough for all samples, just before the use.
2. Molecular weight marker: Precision Plus Kaleidoscope Standards (Bio-Rad, Hercules, CA).
3. Running buffer: Mix 50 ml 20× NuPAGE® MES SDS Running Buffer with 950 ml H_2O (*see* **Note 9**). Separate 200 ml to which 500 μl NuPAGE® Antioxidant is added just before the

electrophoresis for use in the upper buffer chamber of the XCell *SureLock*™ Mini-Cell. Mix thoroughly.

4. NuPAGE® 4–12 % Bis-Tris 10-well gels (*see* **Note 9**).
5. XCell *SureLock*™ Mini-Cell electrophoresis unit, gel knife, 20 ml syringe and 22 gauge needle.

2.3.2 Silver Staining

1. Fixation solution: 30 % ethanol, 0.5 % acetic acid.
2. Rinse solution: 20 % ethanol.
3. 0.02 % sodium thiosulfate in H_2O.
4. 0.2 % $AgNO_3$ in H_2O.
5. Development solution: Mix 70 μl 37 % formaldehyde and 5 ml 0.02 % sodium thiosulfate in 60 ml H_2O. Add 24 ml 12.5 % Na_2CO_3 and fill to 100 ml with H_2O. Prepare fresh solution immediately before use.
6. Stop solution: 5 % Tris-base, 2.5 % acetic acid.

2.4 Estimation of IC50, K_m and V_{max}

2.4.1 IC50

1. As in Subheading 2.1, but several dilutions of peptides are used. For KLK2-inhibiting peptide 1:2 dilution series ranging from 31 to 4,000 μM (0.033–4.2 mg/ml) is suitable.

2.4.2 K_m and V_{max}

1. As in Subheading 2.1, but three different dilutions of peptide (the one in Subheading 2.1., and 1:10 and 1:100 dilutions, i.e., 2, 0.2, 0.02 mg/ml) and several dilutions of substrate covering the range 0.1–3.4 mM are used.

3 Methods

3.1 Enzyme Activity Assay

1. All steps are performed at room temperature and all samples should be analyzed at least as duplicates.
2. Include negative controls, i.e., substrate without protease, and protease and peptides without substrate. Also the activity of enzyme without peptides should be evaluated. All samples should be in the same buffer (i.e., if proteases or the solubilized peptides are in other buffer than BSA/TBS, that buffer should be present at the same amount in all wells).
3. Add 60 μl 0.1 % BSA/TBS into the wells of 96-well microtiter plate (*see* **Note 10**). Add 10 μl protease and 10 μl peptide, and preincubate for 30 min (*see* **Note 11**). In controls, the protease or the peptides may be omitted (*see* **step 2**), instead the same amount of corresponding buffer should be added.
4. Add 20 μl substrate and measure absorbance at 405 nm immediately and at 5 min intervals for 30 min (*see* **Note 12**).
5. Plot the observed absorbance against the reaction time and read the reaction velocity (absorbance) from the area where the absorbance increases in linear fashion (see Fig. 1a, b).

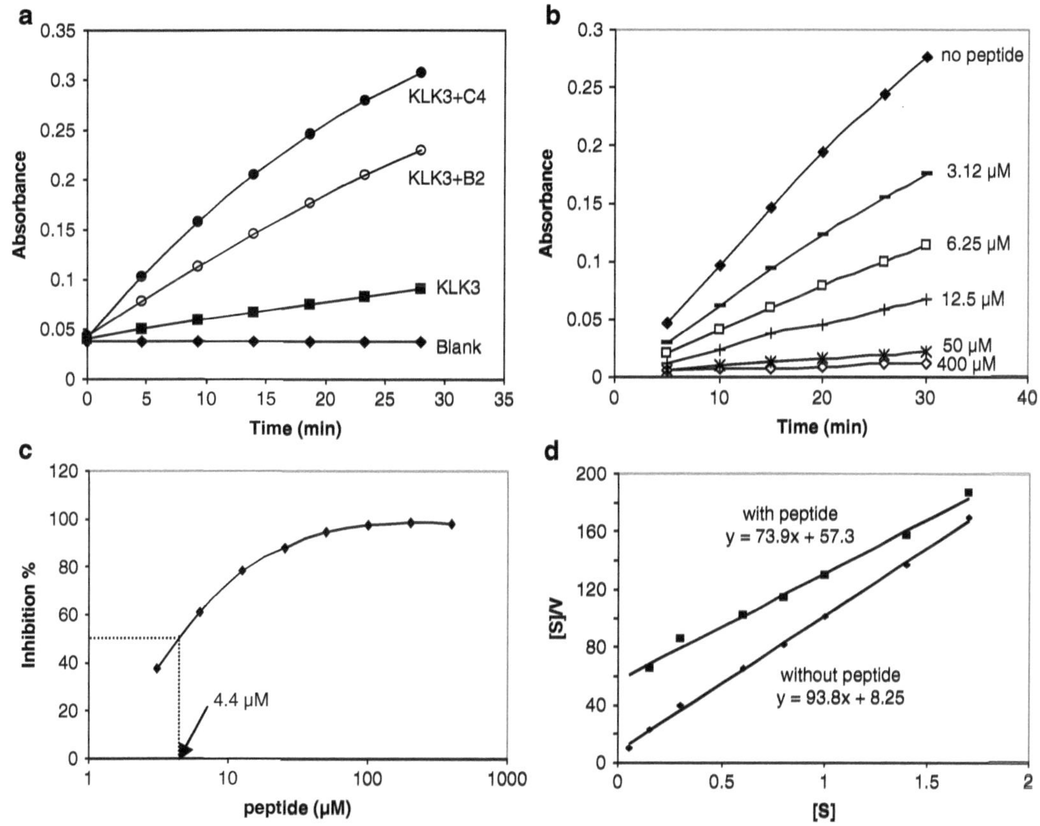

Fig. 1 The effect of the peptides on KLK3 (**a**) and KLK2 (**b**) activity, determined as in Subheading 3.1. Note that in (**a**), especially in the case of KLK3 + C4, the reaction starts to reach plateau and the values should be read from earlier time points. The inhibition/stimulation % can be calculated according to formula in Subheading 3.1, **step 6**. For example, C4 peptide stimulates KLK3 activity by 470 % [100 % − 100 × (0.120/0.021)% = −470%. The data is from 10 min time point. Note that background, i.e., blank, has been deducted from the absorbances]. (**c**) Determination of IC50-value (Subheading 3.4.1). Inhibition % is calculated from the data in (**b**), in which, due to clarity, only some of the tested peptide concentrations are shown. The concentration of peptide inhibiting KLK2 by 50 % is 4.4 μM (note that this is the final concentration of the peptide in total volume of the reaction mixture). (**d**) Determination of K_m and V_{max} using Hanes-Woolf plot (Subheading 3.4.2). Reaction velocity (V) is expressed as change of absorbance/min (ΔA/min) and initial substrate concentration ([S]) as mM (final concentration in the reaction). For the uninhibited KLK2, K_m = (8.25/93.8) mM = 0.088 mM and V_{max} = (1/93.8) ΔA/min = 0.011 ΔA/min (*see* **Note 20**), and for KLK2 inhibited by peptide K_m = (57.3/73.9) mM = 0.78 mM and V_{max} = (1/73.9) ΔA/min = 0.014 ΔA/min. The relatively similar V_{max}-values and increase in K_m-value, by the peptide, suggest that peptide acts as a competitive inhibitor. In this figure only 4 μM peptide concentration is shown, but optimally the effect of the peptide on V_{max}- and K_m-values should be determined using at least three different enzyme concentrations (*see* **Note 19**)

6. Calculate inhibition/stimulation percentage. First subtract background caused by the substrate alone (use corresponding time point). If the enzyme or peptides give a high signal without substrate, that should be also subtracted from the absorbance of the enzyme and peptide containing wells, respectively. Then use these values for calculation of the inhibition percentage according to the formula: 100 − 100 × (absorbance with peptide/absorbance without peptide) %. Negative values indicate stimulation (*see* **Note 13**).

3.2 Degradation of Protein Substrates

1. Incubate 5 μl trypsin-1 with 5 μl peptide or TBS for 15 min at room temperature.
2. Add 5 μl α-casein and incubate for 3 h at 37 °C to allow degradation of casein to take place. In addition set up a tube containing 5 μl casein and 10 μl TBS.
3. Proceed to SDS-PAGE (Subheading 3.3.1).

3.3 Detection of Protein Fragmentation (See Note 14)

3.3.1 SDS-PAGE

Instructions regarding electrophoresis using NuPAGE® Gels in the XCell *SureLock*™ Mini-Cell electrophoresis unit and are slightly modified from manufacturer's instructions (IM-1002 and IM-9003, available at www.invitrogen.com). If further details of the assembly of the electrophoresis unit are needed the readers are referred to these instructions. Other electrophoresis systems can be used with equal results, but buffers and setup of the system might be different.

1. Mix 10 μl sample with 5 μl sample buffer containing reducing agent. Heat at 70 °C for 10 min. Mix and centrifuge to pellet any precipitate.
2. Remove the NuPAGE® Gel from the pouch and rinse the gel cassette with H_2O. Remove the tape from the bottom of the cassette.
3. Orient two gel cassettes in the Mini-Cell the cassette side with the comb facing inwards towards the upper buffer chamber (slot on the bottom of the gel towards the outer buffer chamber). Push the cassettes to the bottom of the Mini-Cell and lock into place with the gel tension wedge. If only one gel is used, the plastic buffer dam replaces the second gel cassette.
4. Completely fill the upper buffer chamber with the 1× running buffer, to which NuPAGE® antioxidant has been added. Gently pull the comb out of the cassette (be careful that the wells are not disrupted). Rinse the wells carefully with the buffer in the chamber using a needle attached to syringe.
5. Fill the lower (outer) buffer chamber with 600 ml of running buffer. It does not matter if some buffer from the upper chamber has over flown to the lower chamber, but if upper chamber is leaking the gels must be locked in place again.
6. Carefully load 10 μl samples and molecular weight marker into the wells using gel-loading pipette tips (or normal 100 μl or smaller pipette tips). Be careful not to push pipette too deep as it might result to detachment of the gel from the plates and the leakage of the samples from the wells.
7. Connect the electrophoresis unit to power supply and run gels for approximately 40 min (until the blue front almost reaches the bottom slot of the gel) at 200 V.
8. After electrophoresis is complete, shut off the power, disconnect electrodes, and remove gel cassette(s).

9. Lay the cassette on table the side of the cassette with a slot in the bottom facing down, and separate all sides, except the one with the wells, by inserting the gel knife into the gap between the cassette's two plates and pulling up the handle of the knife. Remove carefully and discard the top plate, allowing the gel to remain on the bottom plate. Remove the top (wells) and bottom (bottom slot below the blue line) of the gel by cutting with the gel knife. Proceed to silver staining of the proteins (Subheading 3.3.2).

3.3.2 Silver Staining (See **Note 15**)

Use gloves throughout the staining procedure.

1. Remove the gel from the plate by holding the plate and gel, with the gel facing downward, over a container containing fixation solution, use the gel knife to carefully loosen the gel and allow it to drop into the fixation solution. Incubate for 1 h.
2. Wash the gel for 10 min in 20 % ethanol and 10 min in H_2O.
3. Incubate in 0.02 % sodium thiosulfate for exactly 1 min.
4. Rinse two times for 20 s with H_2O.
5. Rinse briefly with 0.2 % $AgNO_3$, followed by 30 min incubation in the same solution.
6. Rinse with H_2O for 10 s.
7. Briefly rinse with fresh development solution and add the same solution for development until the desired protein bands are visible. Typically this takes few minutes, but might be faster or slower.
8. Stop the reaction with stop solution (*see* **Note 16**).
9. Analyze fragmentation (*see* Fig. 2). Large differences are easy to see by eye, but, especially if kinetic parameters will be analyzed, the degree of fragmentation should be analyzed by densitometry (*see* **Note 17**).

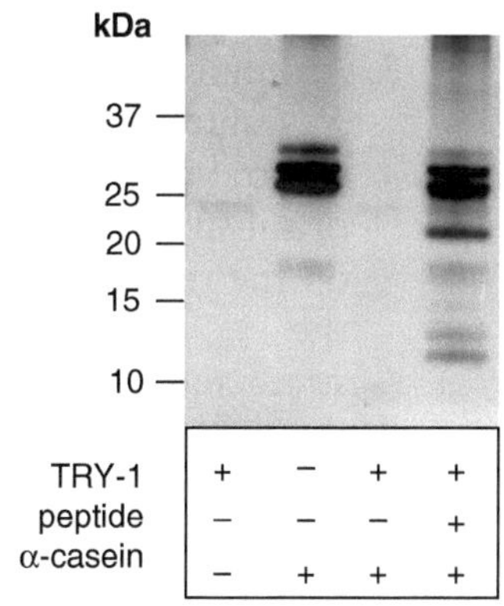

Fig. 2 Detection of degradation of protein substrate by SDS-PAGE, followed by silver staining. Lane *1*, i.e., trypsin-1 (TRY1) alone, contains fivefold more trypsin-1 than the other lanes. Trypsin-1 completely degrades α-casein, which can be inhibited by trypsin-1-inhibiting peptide

3.4 Estimation of IC50, K_m and V_{max}

K_m and V_{max} can be used to distinguish between competitive, non-competitive and uncompetitive mechanism of inhibition, and, thus, the binding site of the peptides, i.e., competitive inhibitors bind to the active site, while the non-competitive ones bind somewhere else, affecting the activity of the enzyme by changing its structure (*see* **Note 18**). The graphical methods presented here give approximate results. For more accurate detection of kinetic parameters nonlinear regression should be performed using one of the several graphics programs available.

3.4.1 IC50

1. Estimate the activity of protease with several different peptide concentrations as above in Subheading 3.1. In the case of KLK2 we have used 0–4,000 μM peptide (0–400 μM final concentration).
2. Plot inhibition % against peptide concentration (Fig. 1c) and estimate the concentration of peptide that inhibits the activity by 50 %, i.e., IC50.

3.4.2 K_m and V_{max}

1. Determine the reaction velocity (absorbance), as in Subheading 3.1, but using several different concentrations of substrate (0.1–3.4 mM, final concentration half of that) and, to reach high enough substrate concentrations, 50 μl substrate (reduce the amount of 0.1 % BSA/TBS accordingly, i.e., to 30 μl). With all substrate concentrations determine reaction velocity with three different peptide concentrations (0.02, 0.2, and 2 mg/ml, final concentrations in the activity assay are one-tenth of these) and without the peptide (*see* **Note 19**).
2. Plot the ratio of initial substrate concentration (i.e., the concentration of substrate, assuming no degradation has taken place) to reaction velocity (absorbance) against the initial substrate concentration (this is called Hanes-Woolf plot, Fig. 1d). From this plot the K_m- and V_{max}-values can be read from the slope of the fitted line ($1/V_{max}$) and *x*-intercept ($-K_m$) (*see* **Note 20**).
3. An inhibitor is considered as competitive if it increases the K_m-value without changing the V_{max}, i.e., the inhibitor and substrate compete for access to the active site and, thus, excess of substrate can be used to overcome the effect of inhibitor on the reaction velocity. If the K_m-value is not changed, but the V_{max} is decreased, inhibition is non-competitive, i.e., substrate and inhibitor does not compete with each other, and the reaction velocity without inhibitor can not be reached with the inhibitor even at high substrate concentrations. There are also special cases of inhibition, like product inhibition and uncompetitive inhibition, in which both V_{max}- and K_m-values are decreased.

4 Notes

1. Different proteases have different buffer requirements, e.g., the activity of some enzymes is greatly dependent on pH, salt concentration, and other factors, such as divalent cations, that are needed especially for the activity of metalloproteases. Dilute enzymes and substrates into a buffer in which the enzyme is active. BSA acts as a carrier protein preventing adsorption of the proteins and peptides onto the wells. However, some peptides may bind to BSA or BSA may be a substrate for some proteases. NaN_3, 8 mM, can be used as bacteriostatic preservative in TBS.
2. Some proteases, like trypsin, are unstable due to self-digestion. When applicable, it is recommended that proenzymes are activated shortly prior the assay, although active enzymes are generally stable for several months at −20 °C, and even more at −80 °C. Repeated freeze/thaw cycles should be avoided.
3. Optimal enzyme and substrate concentrations have to be determined for each enzyme/substrate separately, so that the reaction does not reach completion (plateau) too quickly or slowly. We use usually concentrations that allow the detection of reaction velocity within several minutes to several hours. However, excess of substrate should be avoided (*see* **Note 6**).
4. We have also used fluorescent KLK3 substrate 4-Morpholinecarbonyl-HSSKLQ-AMC [11] (custom peptide from JPT, Berlin, Germany) as 40 mM stock solution in DMSO. This increases sensitivity of the assay significantly, reducing the consumption of protease and peptides. In this assay 5 μl of 60 nM PSA, 5 μl 5.6 μM B2-peptide, and 10 μl of 600 μM substrate has been used in a 384-well plate [12]. The fluorescence (355/460 nm) was followed for several hours.
5. Generally, the peptides we have used have been stable as powder at +4 °C for several years. During storage moisture should be avoided. We usually keep the peptides in small 1.5–2.5 ml tube that is stored inside 50 ml Falcon tube containing silica gel. Peptides dissolved in H_2O (usually 2–5 mg/ml) have been generally stored at −20 °C, and they have been stable for over a year. Some peptides can be kept even at 37 °C for several weeks without losing the activity. However, not all peptides are as stable. Moreover, the solubility of the peptides varies considerably and some are not readily dissolved in H_2O.
6. The peptide concentration as compared to the substrate concentration should be high enough so that, in the case of competitive inhibition, excess of substrate does not mask the effect.

7. To prevent adsorption of proteases to tube walls, low protein binding tubes are used instead of carrier proteins, which may co-migrate with the substrate proteins, and, thus, complicate the interpretation of the results in silver staining.
8. For silver staining the purity of the proteins has to be high enough to allow observation of degradation.
9. Choice of running buffer and gel type is dependent on the size of the separated proteins and their fragments (for migration patterns of protein standards on NuPAGE® Novex Gels see www.invitrogen.com).
10. We normally carry out the experiment in 96-well microtiter plates in 100 μl final volume. However, for fluorometric assay, scaling down to 20 μl reaction volume in 384-well plate works well (*see* also **Note 4**).
11. We generally use 15–30 min preincubation, but this may be omitted. Some weakly inhibiting peptides might be competitive substrates. With these, long enough preincubation times should abolish the inhibitory activity.
12. For highly active proteases, like trypsin, shorter measurement intervals might have to be used, and vice versa for proteases with low activity.
13. Specificity of the peptides should be analyzed by studying their effect on the activity of other proteases with similar structure or activity. In our studies [5, 6], we have used chymotrypsin (10 nM final concentration), bovine trypsin (4 nM), human trypsin-1, -2, -3, and -C (4 nM), plasmin (20 nM), and plasma kallikrein (20 nM). The assays have been performed as described for KLK2 and KLK3 (Subheading 3.1), but the substrate for chymotrypsin was S-2586 (MeO-Suc-Arg-Pro-Tyr-pNA), for plasmin S-2403 (pyroGlu-Phe-Lys-pNA·HCL), for plasma kallikrein S-2302 (H-D-Pro-Phe-Arg-pNA·2HCl), and for the various trypsin isoenzymes S-2222 (Bz-Ile-Glu-Gly-Arg-pNA) (all from Chromogenix Instrumentation Laboratory). All substrates were used at 0.2 mM final concentration, except S-2403 which was 0.4 mM. The activities were measured in 0.1 % BSA/TBS, except for plasmin, trypsin, and kallikrein, where NaCl was omitted and pHs used were 7.4, 8.0, and 9.0, respectively The peptides were used as in Subheading 2.1 (2 mg/ml).
14. For identification/quantification of the cleavage of protein substrates we generally use SDS-PAGE, followed by silver staining or Western immunoblotting, and mass spectrometry. Other staining methods or radioactive substrates may also be used. For higher throughput and easier quantification of the cleavage of IGFBP-3 (a KLK2, KLK3, and trypsin substrate

[13]), we have used two immunoassays, one measuring only intact IGFBP-3, while the other one recognizes also fragmented IGFBP-3 [10, 13]. For proKLK3 activation by KLK2, and its inhibition by KLK2-inhibiting peptides, we have used KLK3 activity assay [6].

15. This protocol is suitable for the staining of proteins that will be subsequently identified by mass spectrometry.
16. The gel can be stored for several days in H_2O.
17. Staining signal should not be too strong (saturated) as it does not anymore correlate well with the amount of protein. Several programs, like the freely available ImageJ (Rasband, W.S., ImageJ, US National Institutes of Health, Bethesda, MD, http://rsb.info.nih.gov/ij/, 1997–2008), can be used for quantification of scanned images.
18. For peptides that bind irreversibly to the target protease the effect on K_m and V_{max} are not relevant.
19. K_m and V_{max} are dependent on the substrate used, and may vary significantly between different substrates. Reaction velocity should be determined with and without inhibitory peptide using several different concentrations of substrate, ranging from several times higher than K_m to lower than K_m. For very high and low concentrations of substrate/peptide the data might not fit well to the plot. For reliable results, at least three different concentrations of the peptide, ranging from efficient inhibition to slight inhibition, should be used. If the K_m for the given enzyme and substrate is not known, it can also be estimated, accurately enough for selection of proper concentration range for further analyses, by measuring reaction velocity (absorbance), using wide range of substrate concentrations. Absorbance is plotted against substrate concentration and the concentration of substrate which gives the reaction velocity half of the maximal velocity (V_{max}), corresponds to K_m.
20. For the purpose of estimating the effect of peptides on V_{max} the reaction velocity can be expressed as absorbance (or change of absorbance/min, ΔA/min). However, absorbance can be converted to moles (of reacted substrate) by using *p*-nitroaniline standard (this is toxic). Alternatively substrate reaction may be allowed to reach completion with sufficiently high concentration of protease. It is then assumed that the amount of substrate added in the reaction, when completely cleaved, yields the given absorbance. This assumes that the reaction is completed (i.e., absorbance does not increase anymore) due to cleavage of all substrate molecules, not because the possible inactivation of the enzyme. The latter can be tested by adding more substrate to the reaction.

Acknowledgments

The authors wish to thank Dr Leena Valmu for the silver staining protocol and Ms. Helena Taskinen for technical assistance. Our original studies referred to have been supported by National Technology Agency of Finland, University of Helsinki, Helsinki University Central Hospital, the Finnish Cancer Foundation, the Academy of Finland, Juselius Foundation and Finska Läkaresällskapet.

References

1. Overall CM, Kleifeld O (2006) Tumour microenvironment—opinion: validating matrix metalloproteinases as drug targets and anti-targets for cancer therapy. Nat Rev Cancer 6(3):227–239
2. Lopez-Otin C, Matrisian LM (2007) Emerging roles of proteases in tumour suppression. Nat Rev Cancer 7(10):800–808
3. Turk B (2006) Targeting proteases: successes, failures and future prospects. Nat Rev Drug Discov 5(9):785–799
4. Koistinen H, Närvänen A, Pakkala M, Hekim C, Mattsson JM, Zhu L, Laakkonen P, Stenman UH (2008) Development of peptides specifically modulating the activity of KLK2 and KLK3. Biol Chem 389:633–642
5. Wu P, Leinonen J, Koivunen E, Lankinen H, Stenman UH (2000) Identification of novel prostate-specific antigen-binding peptides modulating its enzyme activity. Eur J Biochem 267(20):6212–6220
6. Hekim C, Leinonen J, Närvänen A, Koistinen H, Zhu L, Koivunen E, Väisänen V, Stenman UH (2006) Novel peptide inhibitors of human kallikrein 2. J Biol Chem 281(18): 12555–12560
7. Wu P, Weisell J, Pakkala M, Peräkylä M, Zhu L, Koistinen R, Koivunen E, Stenman UH, Närvänen A, Koistinen H (2010) Identification of novel peptide inhibitors for human trypsins. Biol Chem 391:283–293
8. Wu P, Stenman UH, Pakkala M, Närvänen A, Leinonen J (2004) Separation of enzymatically active and inactive prostate-specific antigen (PSA) by peptide affinity chromatography. Prostate 58(4):345–353
9. Lövgren J, Tian S, Lundwall A, Karp M, Lilja H (1999) Production and activation of recombinant hK2 with propeptide mutations resulting in high expression levels. Eur J Biochem 266(3):1050–1055
10. Koistinen H, Seppala M, Koistinen R (1994) Different forms of insulin-like growth factor-binding protein-3 detected in serum and seminal plasma by immunofluorometric assay with monoclonal antibodies. Clin Chem 40(4): 531–536
11. Denmeade SR, Lou W, Lovgren J, Malm J, Lilja H, Isaacs JT (1997) Specific and efficient peptide substrates for assaying the proteolytic activity of prostate-specific antigen. Cancer Res 57(21):4924–4930
12. Koistinen H, Wohlfahrt G, Mattsson JM, Wu P, Lahdenpera J, Stenman UH (2008) Novel small molecule inhibitors for prostate-specific antigen. Prostate 68:1143–1151
13. Koistinen H, Paju A, Koistinen R, Finne P, Lövgren J, Wu P, Seppälä M, Stenman UH (2002) Prostate-specific antigen and other prostate-derived proteases cleave IGFBP-3, but prostate cancer is not associated with proteolytically cleaved circulating IGFBP-3. Prostate 50(2):112–118

Chapter 11

Assessment of Antimicrobial (Host Defense) Peptides as Anti-Cancer Agents

Susan Douglas, David W. Hoskin, and Ashley L. Hilchie

Abstract

Cationic antimicrobial (host defense) peptides (CAPs) are able to kill microorganisms and cancer cells, leading to their consideration as novel candidate therapeutic agents in human medicine. CAPs can physically associate with anionic membrane structures, such as those found on cancer cells, causing pore formation, intracellular disturbances, and leakage of cell contents. In contrast, normal cells are less negatively-charged and are typically not susceptible to CAP-mediated cell death. Because the interaction of CAPs with cells is based on charge properties rather than cell proliferation, both rapidly dividing and quiescent cancer cells, as well as multidrug-resistant cancer cells, are targeted by CAPs, making CAPS potentially valuable as anti-cancer agents.

CAPs often exist as families of peptides with slightly different amino acid sequences. In addition, libraries of synthetic peptide variants based on naturally occurring CAP templates can be generated in order to improve upon their action. High-throughput screens are needed to quickly and efficiently assess the suitability of each CAP variant. Here we present the methods for assessing CAP-mediated cytotoxicity against cancer cells (suspension and adherent) and untransformed cells (measured using the tritiated thymidine-release or MTT assay), and for discriminating between cell death caused by necrosis (measured using lactate dehydrogenase- or ^{51}Cr-release assays), or apoptosis and necrosis (single-stranded DNA content measured by flow cytometry). In addition the clonogenic assay, which assesses the ability of single transformed cells to multiply and produce colonies, is described.

Key words Antimicrobial (host defense) peptides, Anti-cancer, Therapeutic, Cytotoxicity, MTT, Tritiated thymidine-release, Lactate dehydrogenase, Clonogenic, Single-stranded DNA, ^{51}Cr-release

1 Introduction

There are numerous methods for assessing cancer cell death by apoptosis or necrosis in response to cationic antimicrobial peptides (CAPs) [1–3], including synthetic variants [4]. Here we present several commonly used methods that we find appropriate to assay the cytotoxic effect of CAPs on suspension and adherent cells. Cytotoxicity of CAPs against erythrocytes is measured using a

Andrew E. Nixon (ed.), *Therapeutic Peptides: Methods and Protocols*, Methods in Molecular Biology, vol. 1088,
DOI 10.1007/978-1-62703-673-3_11, © Springer Science+Business Media, LLC 2014

modification of the standard hemolysis assay developed by Vogel [5] using fresh blood from human volunteers. CAP-mediated killing of suspension and adherent cells is measured using a variety of assays, including the standard MTT test for the metabolic activity (mitochondrial succinate dehydrogenase) of viable cells and other assays based on the selective release of labeled cellular components. It is noteworthy that the MTT and lactate dehydrogenase (LDH)-release assays cannot be conducted under conditions in which reducing agents (e.g., reduced glutathione) are present as these agents generate formazan crystals that lead to false-negative results with respect to cytotoxicity [6]. Furthermore, MTT binds albumin, which prevents the entry of MTT into cells, leading to false-positive results (i.e., albumin appears toxic; [6]; therefore, we recommend using another assay if the possible interaction of CAP with albumin is being investigated. In addition, the MTT assay should not be conducted on cancer cells that express multidrug-efflux pumps such as *P*-glycoprotein since the expression of these proteins also interferes with the MTT assay [7]. Finally, cholesterol interferes with the MTT assay [8]; therefore, the effect of agents that alter the cholesterol content of the cells and thereby affect their sensitivity to CAPs should not be assessed by the MTT assay. Under these circumstances other assays, such as the tritiated thymidine-release assay, can be used to assess CAP-mediated cytotoxicity.

LDH is a cytoplasmic enzyme that is released following damage to the cell membrane. The enzyme catalyzes the oxidation of lactate to pyruvate, reducing NAD+ to NADH/H+, which can be measured using a colorimetric assay. Like the LDH-release assay, the ^{51}Cr-release assay assesses membrane disruption in late apoptotic or necrotic cells. Cells take up radioactive sodium chromate ($Na_2{}^{51}CrO_4$) into the cytosol where it binds to proteins that are then released when the cell membrane is damaged. Release of ^{51}Cr into culture medium can be measured by liquid scintillation counting (Compton-electron energy absorption) or gamma counting. The tritiated thymidine-release assay (sometimes called the JAM assay) uses liquid scintillation counting to measure DNA fragmentation, which occurs in apoptotic and necrotic cells. The tritiated thymidine-release assay can be superior to the ^{51}Cr-release assay because of lower background counts. However, because cells may differ in their ability to be labeled with tritiated thymidine or radioactive sodium chromate, pilot experiments may be necessary to select the appropriate assay.

DNA denaturation (using formamide, a mild denaturing agent, and heating at low temperatures) that yields single-stranded DNA (ssDNA) can be achieved in apoptotic but not necrotic cells [9]. Apoptosis in response to CAPs can then be analyzed by flow cytometric measurement of ssDNA recognition by fluorochrome-labeled antibody.

Finally, the clonogenic assay is the gold standard for the evaluation of the cytotoxic activity of CAPs on cancer cells. The protocol that we describe must be conducted on adherent cells that have the capacity to form colonies in tissue culture plates (e.g., MCF7 breast cancer cells). Although the clonogenic assay is very sensitive, it requires approximately 2 weeks to complete rather than the 2 days that are needed to complete the other assays described in this chapter.

2 Materials

2.1 Cell Culture

1. RPMI 1640 medium (Invitrogen, Carlsbad, CA).
2. Dulbecco's Modified Eagle's Medium (DMEM; Invitrogen).
3. 0.25 % Trypsin in 1 mM EDTA (Invitrogen).
4. TrypLE™ (Invitrogen).
5. FBS, heat-inactivated (Invitrogen).
6. 10,000 U penicillin, 10 mg/mL streptomycin (Penicillin/streptomycin 10× solution; Sigma-Aldrich, Oakville, ON, Canada).
7. T75 Cell culture flasks (Corning Inc., Tewksbury, MA).
8. Polystyrene flat-bottomed 6-well, 24-well, or 96-well tissue culture plates (Sarstedt Inc., Montreal, QC, Canada).
9. Polypropylene round-bottomed microtiter plates (Sarstedt).

2.2 Peptides

Peptides (>80 % purity) were synthesized by commercial suppliers using F-moc chemistry and purified by HPLC. Peptides were dissolved in deionized distilled water or serum-free medium to make stock solutions and aliquots were frozen at −20 °C. Prior to use, peptides were thawed and diluted into the appropriate medium for use in assays.

2.3 Erythrocyte Lysis (Hemolysis) Assay

1. Dulbecco's phosphate-buffered saline (D-PBS; Invitrogen).
2. 2 % Triton X-100 solution.
3. 150 mM Sodium chloride solution.
4. 100 mM, pH 7.4 Sodium phosphate buffer.

2.4 MTT Assay for Apoptosis/Necrosis in Suspension and Adherent Cells

1. MTT(3-(4,5-dimethyl-2-thiazolyl)-2,5-diphenyl-2H-tetrazolium bromide) (Sigma-Aldrich).
2. Anhydrous isopropanol/0.1 N HCl (freshly prepared) or dimethyl sulfoxide (DMSO).

2.5 Tritiated Thymidine-Release (JAM) Assay for Early Apoptosis/Necrosis in Suspension and Adherent Cancer Cells

1. 72 Ci/mmol [^{3}H]-methyl-thymidine (MP Biomedicals, Irving, CA).
2. Glass fiber filter mats (Skatron Instruments, Sterling, VA).
3. Liquid scintillation tubes (Beckman-Coulter Canada, Inc., Mississauga, ON, Canada).

2.6 ^{51}Cr-Release for Late Apoptosis/Necrosis in Suspension and Adherent Cells

1. 5 mCi/ml, pH 8 Sodium chromate ($Na_2{}^{51}CrO_4$) in normal saline (MP Biomedicals).
2. 2 % Triton X-100 solution.
3. Liquid scintillation tubes (Beckman-Coulter Canada Inc.).

2.7 LDH-Release Assay for Necrosis in Adherent Cells (MDA-MB-231)

1. Cytotoxicity Detection Kit (LDH) (Roche Applied Science, Laval, QC, Canada).
2. 2 % Triton X-100 solution.

2.8 ssDNA Detection by Flow Cytometry for Measuring Apoptosis in Suspension and Adherent Cells

1. Anti-ssDNA IgM monoclonal antibody (clone MAb-F7-26, 10 μg/ml in PBS; Millipore, Temecula, CA).
2. Mouse IgM isotype control (clone GC323, 10 μg/ml in PBS; Millipore).
3. Methanol.
4. Formamide.
5. 1 % (w/v) skim milk in PBS.
6. Phycoerythrin (PE)-conjugated anti-mouse IgM (1:50 in PBS; Jackson ImmunoResearch, West Grove, PA).

2.9 Clonogenic Assay for Proliferating Adherent Cancer Cells

1. 4 % (w/v in methanol) crystal violet (Sigma-Aldrich).

3 Methods

3.1 Cell Culture

1. Cell lines are obtained from ATCC (Manassas, VA) and cultured according to ATCC recommendations. In general, suspension and adherent cells are cultured at 37 °C in RPMI 1640 or DMEM medium containing 0.5 % (v/v) penicillin/streptomycin antibiotic preparation, 5 or 10 % fetal bovine serum (FBS) in a 5 or 10 % CO_2 humidified atmosphere, respectively. Suspension cells are plated in flat-bottomed microtiter plates such that the cells are in the exponential growth phase at the end of the experiment, which can be estimated by the color change of phenol red-containing medium (i.e., orange color of the medium indicates exponential cell growth without medium depletion). Adherent cells are plated in flat-bottomed microtiter plates such that the cells reach

75–90 % confluency by the end of the assay. Flat-bottomed polystyrene microtiter plates are used for suspension and adherent cells to maximize culture area and to allow for the measurement of light absorbance using a microplate reader. Cells are routinely tested for mycoplasma contamination using the MycoSensor PCR Assay Kit (Agilent Technologies Canada, Inc., Mississauga, ON) or the MycoAlert® mycoplasma detection kit (Lonza Walkersville, Inc., Walkersville, MD). Detection of mycoplasma contamination using the latter system is rapid (i.e., 15 min), but requires the use of a luminometer.

2. Adherent cells (e.g., MDA-MB-231) are passaged in T75 cell culture flasks. Spent medium is aspirated and discarded and 3 mL trypsin–EDTA is added to disrupt monolayers and lift adherent cells (flasks are left for several minutes and then rapped sharply to detach cells). Trypsin–EDTA is neutralized by the addition of 3.5 mL of DMEM containing 10 % FBS. Aliquots of cell suspensions (amounts vary according to cell line) are transferred to 12 mL of fresh DMEM supplemented with 10 % FBS in a T75 flask for propagation. Cells are allowed to grow until 70–80 % confluency is reached.
3. Cells are counted by adding 10 μL of 0.1 % Trypan blue to 10 μL of cell suspension. An aliquot of 10 μL of the cell: Trypan blue mixture is counted using a hemocytometer or automated cell counter (Countess instrument, Invitrogen) and resuspended in medium containing little (2.5 %) or no FBS to minimize subsequent CAP adsorption to anionic serum components. Dead cells stain blue, while viable cells remain colorless. Only cultures that have >90 % viable cells are used.
4. Cells are diluted to the desired concentration (usually 1×10^5 cells/mL) and 100 μL aliquots are distributed into 96-well microtiter plates for use in assays.

3.2 Peptides

1. Peptide dilutions can be set up several ways but we find the following to work well: wells of the 96-well microtiter plate are filled with 100 μL of appropriate medium. To the top wells, add a further 90 μL of medium and 10 μL of stock peptide. Mix by pipetting up and down several times. Take 100 μL from the top well and add to the next well. Mix well. Repeat dilutions with 100 μL to generate a dilution series. Discard the final 100 μL. Using a multichannel pipettor and starting from the bottom well add 100 μL cells to each well. This results in a peptide concentration of 1/40 stock concentration in the top well, then twofold dilutions in each successive well. Alternatively, CAP stocks can be diluted in culture tubes as desired and added directly to the plated cells (1:2 dilution). This is particularly useful for adherent cells that require 24 h to adhere to the tissue culture plate prior to exposure to the CAP, or if multiple

compounds will be added to the same well (in which 1:4 dilutions may be necessary).

2. Peptides are diluted from stocks into serum-free medium and are added to cell cultures to achieve the desired final concentration of CAPs and FBS. Alternatively, CAPs may be diluted in PBS for the hemolysis assay.

3.3 Erythrocyte Lysis (Hemolysis) Assay

Freshly prepared erythrocytes are exposed to twofold dilutions of CAP for the desired length of time, lysed erythrocytes are pelleted by centrifugation and the absorbance of the hemoglobin in the supernatant is measured using a spectrophotometer. The negative control contains erythrocytes but no CAP, the positive control contains erythrocytes plus 1 % Triton X-100, and the blank does not contain any erythrocytes.

1. Collect blood in 5 mL EDTA-containing vacutainers and centrifuge at 500 × *g* in a clinical centrifuge.
2. Remove serum and replace with the same volume of 150 mM NaCl.
3. Wash blood cells three times in 150 mM NaCl.
4. Resuspend erythrocytes in the same volume of 100 mM sodium phosphate buffer (pH 7.4).
5. Dilute 1 mL of erythrocytes tenfold in 100 mM sodium phosphate buffer, yielding ~5×10^8 cells/mL.
6. Incubate 100 μL of erythrocytes with 100 μL of twofold dilutions of CAP in polypropylene round-bottomed 96-well plates for 1–8 h at 37 °C (*see* Subheading 3.2, **step 1** for plate setup). Samples are mixed after 30 min by placing the plates on a shaking platform.
7. Centrifuge plates at 500 × *g* for 10 min to sediment cell debris, and transfer supernatants to polystyrene flat-bottomed 96-well plates.
8. Measure the absorbance at 541 nm using a spectrophotometer. Percent hemolysis is calculated as follows:

$$\% \text{ hemolysis} = 100 \times [[(E_{Abs} - B_{Abs}) - (N_{Abs} - B_{Abs})]/[P_{Abs} - N_{Abs} - B_{Abs}]],$$

where *E*, *B*, *N*, and *P* denote the absorbance of CAP (experimental)-treated, blank, medium (negative control)-treated, and Triton-X (positive control)-treated samples, respectively.

3.4 MTT Assay for Apoptosis/Necrosis in Suspension and Adherent Cells

Cells are incubated with twofold dilutions of CAP (*see* Subheading 3.2, **step 1** for plate setup) and MTT for 2–6 h at 37 °C in a 5 % CO_2 humidified atmosphere in flat-bottomed 96-well tissue culture plates. Mitochondrial dehydrogenases of viable cells cleave the tetrazolium ring of MTT, yielding purple formazan crystals that are insoluble in aqueous solutions but soluble in acidified isopropanol or DMSO. The resulting purple solution is measured at 490 nm using a spectrophotometer. The negative control contains cells but no CAP and the blank does not contain cells.

1. Make MTT stock (5 mg/mL in PBS); be sure to make enough for all plates (~2 mL/plate). Sometimes it takes thorough mixing to dissolve the MTT powder (*see* **Note 1**). Protect the MTT solution from exposure to light.
2. Add 100 μL of CAP dilutions to 100 μL of cells at desired concentration and mix by pipetting.
3. Using a multichannel pipettor, add 20 μL of MTT stock to each well 2–6 h before the end of the assay, mix after adding to each row.
4. Incubate the plate for 2 h at 37 °C, centrifuge at 1,400 × *g* for 5 min, quickly remove medium by expelling the contents of the plate into a sink in one fluid motion (*see* **Note 2**).
5. Add 100 μL of 0.1 N HCl/anhydrous isopropanol or DMSO to each well, agitate the plate on a rotating platform for ~10 min to dissolve formazan residue or if necessary pipette up and down several times to resuspend the dye.
6. Transfer the resuspended dye solution to a 96-well polystyrene flat-bottomed plate if cytotoxicity toward suspension cells is being assessed in a round-bottomed plate.
7. Measure the absorbance at 490 nm using a spectrophotometer. Percent cytotoxicity is calculated as follows:

   % cytotoxicity $= 100 \times [(E_{Abs} - B_{Abs})/(N_{Abs} - B_{Abs})]$, where E, B, and N denote the absorbance from CAP (experimental)-treated, blank, and medium (negative control)-treated samples, respectively.

3.5 Tritiated Thymidine-Release (JAM) Assay for Early Apoptosis/Necrosis in Suspension and Adherent Cancer Cells

Cells are prelabeled with [^{3}H]-methyl thymidine, which incorporates into newly synthesized DNA. After exposure to CAPs, the cells are lysed and their contents passed over glass fiber filters using a multiple sample harvester. Intact radiolabeled DNA is retained on the filters, whereas DNA fragments pass through.

1. Label cells with 5 μCi/mL [^{3}H]-methyl-thymidine for 2 h (*see* **Note 3**).
2. Lift cells with trypsin solution or TrypLE (if adherent) and wash three times with DMEM or RPMI 1640 medium.
3. Seed the desired number of cells in a 100 μL volume in a 96-well flat-bottomed tissue culture plate.
   (a) Resuspend suspension cells in medium containing twice the desired concentration of FBS.

(b) Resuspend adherent cells in complete medium and allow the cells to adhere overnight at 37 °C in a 10 % CO_2 humidified atmosphere. Replace the medium with medium containing twice the desired concentration of FBS prior to exposing the cells to the CAP.

4. Treat the [^{3}H]-methyl-thymidine-labeled cells with 100 μL of serum-free medium alone or twice the desired concentration of CAP diluted in serum-free medium, thereby bringing the concentration of FBS and CAP to the desired concentrations.
5. At the endpoint of the assay, harvest intact DNA from suspension cells directly onto glass fiber filter mats using a multiple sample harvester (Skatron Instruments AS, Lier, Norway). Adherent cells should be lysed by several freeze/thaw cycles to ensure cell lysis prior to harvesting since some cells are resistant to osmotic shock (*see* **Note 4**).
6. Tritiated thymidine-labeled DNA is subsequently quantified by liquid scintillation counting once the glass fiber mats dry (~12 h). Percent DNA fragmentation is calculated as follows:

$$\%\ \text{DNA fragmentation} = 100 \times [(C_{\text{CPM}} - E_{\text{CPM}}/C_{\text{CPM}}],$$

where *C* and *E* denote the amount of intact DNA from control- and CAP-treated cells, respectively.

3.6 ^{51}Cr-Release Assay for Late Apoptosis/Necrosis in Suspension and Adherent Cells

Cells are prelabeled with $Na_2{}^{51}CrO_4$, incubated with CAP, and the radioactive chromium released into the culture medium by damaged cells is measured by liquid scintillation counting or gamma counting.

1. Collect cells (suspension or adherent) by centrifugation at 500 × *g* for 5 min and incubate the pellet with an equal volume of $Na_2{}^{51}CrO_4$ for 1 h at 37 °C.
2. Wash the cells at least five times with medium containing 5–10 % FBS.
3. Resuspend and treat the cells as described in Subheading 3.5 (**steps 3** and **4**).
4. In addition to the negative control (medium plus cells), include a positive control (Triton X-100 plus cells).
5. At the end of the assay, centrifuge the plate at 500 × *g* for 5 min and transfer 100 μL of the cell-free supernatant to sample vials.
6. Measure the radioactivity of the samples by liquid scintillation counting. Percent cytotoxicity can be calculated as follows:

$$\text{Percent cytotoxicity} = 100 \times ([E_{\text{CPM}} - N_{\text{CPM}}]/[P_{\text{CPM}} - N_{\text{CPM}}]),$$

where *E*, *N*, and *P* denote chromium release from CAP (experimental)-treated, medium (negative control)-treated, and Triton-X (positive control)-treated samples, respectively.

7. Alternatively, ^{51}Cr-release can be determined by gamma counting.

3.7 LDH-Release Assay for Necrosis in Adherent Cells

The Cytotoxicity Detection Kit (LDH) from Roche Applied Science quantifies the LDH activity in cell-free culture supernatants using a diaphorase/NAD^+ mixture and tetrazolium salt. Diaphorase oxidizes the NADH/H^+ formed by LDH and in so doing the H/H^+ is transferred to a yellow tetrazolium salt, forming a red formazan salt, which is quantified using a spectrophotometer. The negative control contains cells and medium but no CAP, the blank contains medium but no cells, and the positive control contains 1 % Triton X-100 plus cells.

1. Seed a 96-well flat-bottomed plate with 100 μL of cells (at the desired concentration) per well.
2. Incubate the plate at 37 °C overnight to allow for attachment of the cells.
3. Carefully remove medium and wash the attached cells with fresh medium containing 2.5 % FBS.
4. Add 100 μL of CAP dilutions per well and 100 μL of medium or 2 % Triton X-100 to control wells.
5. Incubate the plate for 3 h at 37 °C.
6. Centrifuge the plate for 10 min at 250 × *g*.
7. Remove 100 μL of the cell-free supernatant and transfer it into a flat-bottomed 96-well plate.
8. Add 100 μL of the reaction mixture to each of the wells.
9. Incubate the plate for 30 min, protected from light, and measure absorbance at 490 and 650 nm (*see* **Note 5**). Percent cytotoxicity is calculated as follows:

   $$\% \text{ cytotoxicity} = 100 \times [(E_{\text{Abs}} - B_{\text{Abs}} - N_{\text{Abs}})/(P_{\text{Abs}} - N_{\text{Abs}} - B_{\text{Ab}})],$$

   where *E*, *B*, *N*, and *P* denote the absorbance of CAP (experimental)-treated, blank, medium (negative control)-treated, and Triton-X (positive control)-treated samples, respectively.

3.8 ssDNA Detection by Flow Cytometry for Measuring Apoptosis in Suspension and Adherent Cells

Cells that have been exposed to CAP are treated with formamide and mild heat to denature ssDNA, which is then detected by flow cytometric analysis using a fluorochrome-labeled anti-ssDNA antibody from Millipore.

1. Seed 24- or 6-well tissue culture plates with 500 μL of suspension cells or 1 mL of adherent cells, respectively, at the desired concentrations. Allow adherent cells to adhere overnight as described in **step 3**(b) of Subheading 3.5.
2. Incubate suspension or adherent cells with 500 μL (24-well) or 1 mL (6-well) of serum-free medium or CAP in serum-free medium. Include a positive control (i.e., a known inducer of apoptosis such as actinomycin D).

3. At the end of the assay, pellet the cells by centrifugation for 5 min at 500 × *g* and resuspend the cells in 1 mL of ice-cold PBS.
4. While gently vortexing the resuspended cells, add 6 mL of methanol chilled to −20 °C. Store for 1–3 day at −20 °C.
5. Pellet cells by centrifugation for 10 min at 500 × *g* and discard the supernatant.
6. Gently resuspend each cell pellet in 250 μL of formamide.
7. Incubate for 10 min at 75 °C, then move the cells to a room temperature water bath for 5 min.
8. Gently resuspend cells in 2 mL of 1 % skim milk and incubate for 10 min at room temperature.
9. Split samples into two 1.5 mL Eppendorf tubes and centrifuge for 10 min at 500 × *g*.
10. Discard the supernatant and resuspend one set of samples in 100 μL of anti-ssDNA antibody and the other set in 100 μL of isotype control antibody. Incubate the tubes at room temperature for 30 min.
11. Pellet the cells by centrifugation for 5 min at 500 × *g* and wash the pellets with PBS twice.
12. Incubate pellets in 100 μL of PE-conjugated anti-mouse IgM for 15 min at room temperature.
13. Pellet the cells by centrifugation for 5 min at 500 × *g* and wash the pellets with PBS twice.
14. Resuspend the pellets in 500 μL of PBS and transfer to flow cytometry tubes.
15. Read the samples on a flow cytometer (FACSCalibur; BD Biosciences) using the FL-2 setting. A positive result (apoptosis) is indicated by a rightward shift in fluorescence in CAP-treated cells labeled with the anti-ssDNA antibody when compared to the negative control.

3.9 Clonogenic Assay for Apoptosis/Necrosis in Adherent Cells

Adherent cells that have been exposed to the CAP are plated in complete medium and allowed to grow until individual colonies form. The colonies are then stained with crystal violet and counted.

1. Seed the desired number of cells in complete medium in a 6-well flat-bottomed tissue culture plate.
2. Allow cells to adhere overnight.
3. Remove medium and add 1 mL of medium containing twice the desired concentration of FBS.
4. Add 1 mL of serum-free medium or CAP in serum-free medium.

5. After incubating the cells for the desired amount of time, lift cells using trypsin solution or TrypLE and count cells by Trypan blue dye exclusion. At this point you can proceed by plating known cell numbers (a) or by conducting serial dilutions (b). The remainder of this experiment is conducted in complete medium.
   (a) After counting the cells, dilute each sample to a known cell number (e.g., 200, 500, 2,500 cells/mL). Add 1 mL of each diluted cell sample to a fresh 6-well plate (*see* **Note 6**).
   (b) After counting the cells, conduct serial dilutions on each sample. Plate 1 mL of each diluted sample in a 6-well plate. At the end of the assay the colony number can be multiplied by the dilution factor to calculate the number of colony-forming units in each treatment group.
6. Add an additional 1 mL of medium to each well and incubate the cells until visible colonies form (approximately 10–14 days). Change the culture medium every third day.
7. Wash colonies once with PBS.
8. Add 500 μL of crystal violet solution and incubate for 15 min.
9. Remove the crystal violet solution and wash the colonies with deionized distilled water (*see* **Note 7**).
10. Count the colonies. The data can be expressed as the number of colony-forming units or percent cytotoxicity, which is calculated as follows:

$$\%\ \text{cytotoxicity} = 100 \times [1 - (E_{\text{CFU}}/N_{\text{CFU}})],$$

where E and N denote the number of colony-forming units from the CAP (experimental)-treated and medium (negative control)-treated samples, respectively.

4 Notes

1. MTT stock is filter-sterilized to remove any particles that do not dissolve. The stock solution is stable for several months at 4 °C.
2. Try not to repeat tapping as it may loosen the dye pellet.
3. Not all cells will label well with tritiated thymidine. Certain cells label very poorly (e.g., some breast cancer cell lines) precluding the use of this assay, whereas other cells (e.g., certain ovarian cancer cell lines) simply require greater levels of tritiated thymidine and longer exposure times for sufficient labeling. Some troubleshooting may be necessary to determine whether this assay is appropriate for the cell line of interest. Control counts should be consistent and in the tens of thousands range.

4. The number of freeze/thaw cycles necessary to achieve cell lysis depends on the cell line. Certain cell lines (e.g., prostate cancer cells) only require one cycle, whereas other cell lines (e.g., ovarian cancer cells) require multiple cycles.
5. Absorbance readings of samples are obtained by subtracting the A_{650} from the A_{490}. Incubation times of between 2 and 24 h are recommended. Since CAPs act fairly quickly, 3 h was chosen in our study. However, a time-course should be conducted to determine the optimum incubation time for each cell line and CAP.
6. Knowing how many cells of each treatment group to add to each well will require a fair amount of troubleshooting. Generally, medium-treated cells are better at forming colonies and require the plating of fewer cells (e.g., 500), whereas CAP-treated cells lose their ability to form colonies and require the plating of more cells for good colony formation (e.g., 2,500). This method requires cell numbers to be as accurate as possible; therefore, averaging of multiple cell counts may be necessary.
7. A squirt bottle can be used to wash the colonies in tissue culture plates. Alternatively, each plate can be carefully rinsed in a large beaker filled with water.

References

1. Zhang L, Falla TJ (2010) Potential therapeutic application of host defense peptides. Methods Mol Biol 618:303–327
2. Hilchie AL, Hoskin DW (2010) The application of cationic antimicrobial peptides in cancer treatment: laboratory investigations and clinical potential: . John Wiley & Sons Inc, Hoboken, NJ
3. Douglas SE, Patrzykat A, Pytyck J, Gallant JW (2003) Identification, structure and differential expression of novel pleurocidins clustered on the genome of the winter flounder, *Pseudopleuronectes americanus* (Walbaum). Eur J Biochem 270:3720–3730
4. Hilpert K, Winkler DF, Hancock RE (2007) Peptide arrays on cellulose support: SPOT synthesis, a time and cost efficient method for synthesis of large numbers of peptides in a parallel and addressable fashion. Nat Protoc 2:1333–1349
5. Vogel CW, Pluckthun A, Muller-Eberhard HJ, Dennis EA (1981) Hemolytic assay for venom phospholipase A2. Anal Biochem 118: 262–268
6. Funk D, Schrenk HH, Frei E (2007) Serum albumin leads to false-positive results in the XTT and the MTT assay. Biotechniques 43:178, 180, 182 passim
7. Vellonen KS, Honkakoski P, Urtti A (2004) Substrates and inhibitors of efflux proteins interfere with the MTT assay in cells and may lead to underestimation of drug toxicity. Eur J Pharm Sci 23:181–188
8. Ahmad S, Ahmad A, Schneider KB, White CW (2006) Cholesterol interferes with the MTT assay in human epithelial-like (A549) and endothelial (HLMVE and HCAE) cells. Int J Toxicol 25:17–23
9. Frankfurt OS, Krishan A (2001) Identification of apoptotic cells by formamide-induced DNA denaturation in condensed chromatin. J Histochem Cytochem 49:369–378

Chapter 12

Peptide Labelling Strategies for Imaging Agents

Tuulia Huhtala, Janne Weisell, Jussi Rytkönen, and Ale Närvänen

Abstract

The main goal in modern biomedicine is to develop specific diagnostic and therapeutic agents for different diseases. Especially in cancer research tumor targeted molecules are the key factor in the development of new anti-tumor drugs. In addition, the early diagnosis of the disease is an important factor for a successful therapy. Synthetic peptides have been shown to be specific targeting agents for next generation diagnostic and therapeutic agents. Noninvasive in vivo imaging using targeting molecules provides modern method for the diagnosis of the pathological alterations like cancer. To evaluate the usefulness of a synthetic peptide for in vivo diagnostic purposes the preclinical biodistribution and targeting studies are essential. Today the widely used preclinical imaging modalities for the biodistribution and tissue alteration studies in experimental animals are single photon emission computed tomography (SPECT) and magnetic resonance imaging (MRI). Together with conventional histochemistry, the biodistribution and tissue/cell location can be determined. In this chapter we describe the conjugation and labelling methods of the peptides for histochemistry and for the molecular imaging with SPECT and MRI modalities.

Key words SPECT, Single photon emission computed tomography, MRI, Magnetic resonance imaging, Peptide, Labelling, Conjugation, Diagnostic, Noninvasive, Chelate

1 Introduction

Peptide mapping and phage display libraries produce new biologically active peptides with specific binding properties to target proteins such as receptors and proteases [1, 2]. Especially in vivo biopanning using phage libraries produce peptides with homing properties in living organisms [3]. Some of the identified receptors have been found from highly specialized tissues providing a possibility to use peptides for targeting [4]. Peptides with specific receptor binding domains have been used as carriers for different tracer and therapeutic molecules [5] as well as RNA/DNA molecules [6] or viruses [7]. These peptides serve as lead molecules for development of novel molecules for diagnostics and therapy [8].

Imaging technologies form an inseparable part of modern medicine and are becoming a major research focus globally. One of the basic challenges in future biotechnology and biomedicine is

Andrew E. Nixon (ed.), *Therapeutic Peptides: Methods and Protocols*, Methods in Molecular Biology, vol. 1088, DOI 10.1007/978-1-62703-673-3_12, © Springer Science+Business Media, LLC 2014

the localization of the pathogenic alterations by minimally invasive methods and specific therapy with targeting molecules. Especially successful cancer therapy often requires localization of the tumor and its possible metastases. Preclinical imaging in animal models is an important step towards clinical trials in humans. For preclinical molecular imaging and phenotyping purposes, several important in vivo modalities exist today. These are magnetic resonance imaging (MRI) and spectroscopy (MRS), single photon emission computed tomography (SPECT), positron emission tomography (PET), and computed tomography (CT), optical imaging and ultrasound imaging. Modern technology allows the co-registration of the images from different modalities, i.e., SPECT/CT or more sophisticated SPECT/CT/MRI images [9]. This provides accurate data on the dynamics and localization of the targeting molecules in vivo. Furthermore, the same peptides labelled with fluorescence tracers can be used in conventional and in vivo histochemistry. For in vivo histochemistry, labelled peptides are administered to living animal and are monitored postmortem in histological sections [10].

SPECT and MRI are widely used imaging modalities for the studies of new targeting molecules and tissue alterations in experimental animals. Our laboratory has used commercially available SPECT/CT and MRI equipments designed for small animal imaging and studied the biodistribution of USPIO labelled viruses [11] with MRI and glioblastoma homing peptide (Enbäck, unpublished data) with SPECT. The advantage of these modalities is the relative easy and cheap conjugation and labelling chemistry for the targeting molecules. For SPECT imaging there are two labelling strategies. When using radioactive ^{99m}Technetium or ^{111}Indium as tracers additional chelate is needed in the peptide. The most used chelates are diethylenetriaminepentaacetic acid (DTPA) and 1,4,7, 10-tetraazacyclododecane1,4,7,10-tetraacetic acid (DOTA) [12]. Iodine labelling requires paraaminobenzedene, and the peptide should contain additional tyrosine or its analogy. DTPA and DOTA chelates bind also Gadolinium, which is used as a contrast agent for MRI. The disadvantage of these contrast agents is relative low sensitivity. For MRI the most sensitive contrast agents are either small particles of iron oxide (SPIO) or ultra small particles of iron oxide (USPIO) which have diameters on average in the micrometer range and below 250 nm, respectively. The small size of USPIO makes them more suitable for in vivo imaging [13].

The main disadvantage in the use of peptides as diagnostic or therapeutic agents is their short life time due to the fast turn over via the kidneys. Conjugation of the peptides to polyethylene glycol polymer (PEG) has shown to prolong the circulation time leading to the better bioavailability [14, 15]. Peptides are synthesized mostly using solid phase peptide synthesis (SPPS) methodology. Using same chemistry conjugations and labelling could be done on resin.

In this chapter we introduce common instructions for the conjugation and the labelling methods for both on solid phase and for dissolved peptides.

2 Materials

2.1 Conjugation and Labelling of Peptides on Resin

1. Synthetic peptides may be synthetized in house or as custom service. The resin is stored at dry or for shorter periods in dimethyl formamide (DMF) at +4 °C. For the peptide design *see* **Notes 1** and **2**.
2. Organic solvents: Dichloromethane (DCM), DMF, and diethyl ether.
3. Coupling reagents: *O*-Benzotriazole-*N*,*N*,*N*′,*N*′-tetramethyluroniumhexafluorophosphate (HBTU) and *N*,*N*-*diisopropylethylamine* (DIPEA). For 1 ml of 0.3 M solutions of coupling reagents weight 0.115 g of HBTU and 0.039 g (55 μl) of DIPEA in DMF.
4. Cleavage solution: 95 % Trifluoroacetic acid (TFA), 3 % 1,2-ethanedithiol (EDT), 1 % triisopropylsilane, and 1 % water.
5. Deprotection solution : 20 % piperidine in DMF.
6. Kaiser test reagents: A: Phenol 4 g/1 ml abs. Ethanol. B: Kaliumcyanide 1 ml in 50 ml pyridine. C: Ninhydrine 1 g in 20 ml abs. Ethanol.
7. Chelates: Diethylenetriamine-*N*,*N*,*N*″,*N*″-tetra-*tert*-butyl acetate-*N*′-acetic acid DTPA(tBu4)). Tertbutyl protected DTPA chelate (B365, Macrocyclics). 1,4,7,10-Tetraazacyclododecane1,4,7-tris-*tert*-butyl acetate-10-acetic acid (B260, Macrocyclics).
8. Fluorescein isothiocyanate FITC, Biotin, Fmoc-PEG-OH (*see* http://www.irisbiotech.de/).
9. Empty PD10 columns, magnetic stirrer, dry path (100 °C) for class test tubes, centrifuge (4,000 ×*g*) and lyophilizer.

2.2 Radiochemistry

1. Radio isotopes: ^{99m}Technetium (^{99m}TcO4), half-life 6 h. Technetium should be delivered at the same day as it is used. ^{111}Indium ($^{111}InCl_3$), half-life 2.8 day. ^{123}I (Na^{123}I), half-life 13.3 h (*see* **Note 8**).
2. Labelling buffers: 200 mM sodium acetate buffer pH 5.0 (NaAc), 300 mM sodium citrate buffer pH 5.0 (NaCit), 10 mM sodium acetate buffer pH 5.
3. IODOGEN™ iodination tubes, Iodination buffer: 25 mM Tris buffer pH 7.0, 0.4 M NaCl.
4. Other reagents :Stannous (II) chloride, 10 mM HCl, Saline.
5. Needles, Nitrogen gas (>99.5 %), rubber sealed ampoule.

6. Instant Thin Layer Chromatography plates (ITLC, Silica Gel impregnated glass fiber sheets 20 cm × 20 cm).
7. Radioisotope calibrator (e.g., Cabintec CRC120).
8. HPLC unit *see* Subheading 2.4.

2.3 Ultra Small Particles of Iron Oxide (USPIO)

1. Primary amino derivatized USPIO particles, streptavidin-coated USPIO particles (*see* http://www.kiskerbiotech.com/).
2. Water soluble bifunctional reagent (maleinimido-N-hydroxysuccinimide, e.g., SulfoKMUS, *see* http://www.piercenet.com/).
3. High gradient magnet or Dialysis cassettes, MWCO 10,000 Da and Sepharose 6 (GE Healthcare, Sweden). Column length of 10.0 cm with diameter of 1.5 cm.
4. Multichannel spectrophotometer for microtitration plates with 405 nm filter.

2.4 HPLC

1. Normal gradient forming HPLC unit with C18 revere phase analytical (4 mm × 250 mm) and preparative (10 mm × 250 mm, e.g., Supelco C18, 5 μm) columns, UV and radiodetectors for HPLC, e.g., Bioscan BFC3100 for 125Iodine (<60 keV) and Bioscan BFC3200 for other radiotracers (>60 keV).
2. Running buffers. Analysis: A: 0.1 % TFA, 5 % acetonitrile (ACN) in water, B: 95 % ACN, 0.075 % TFA, 5 % water. Preparative: A: 0.1 % TFA in water, B: ACN. Iodine purification: 60 % methanol in water (isocratic run).

3 Methods

3.1 Conjugation of the Peptides with DTPA/DOTA Chelates, PEG and Biotin on Resin

For conjugations we are using DTPA(tBu)4, DOTA(tBu)3, Fmoc protected PEGacid (FmocPEGOH) and biotin (Fig. 1). All the molecules are conjugated via the free carboxylic acid forming amide bond between conjugated molecule and the free α-amino group of the peptide using HBTU as coupling reagent (*see* **Note 3**). The concentration of the peptide on the resin is approximated from the theoretical yield. The theoretical yield is calculated from the substitution of the resin (mmol of peptide/1 g of resin) and it is reported by the manufacturer. The following instructions are for 100–200 mg of resin.

3.1.1 On Resin Conjugation

1. Move the resin to the empty PD10 column and wash it with 5 ml of DMF. Plug the column and add small magnetic bar to the PD10 column. Hang the column over the magnetic stirrer.
2. Dissolve 10× molar excess of DTPA(tBu4), DOTA(tBu)3, FmocPEGOH or Biotin and HBTU and DIPEA. For example, using 100 mg of resin with substitution of 0.3 mmol/g

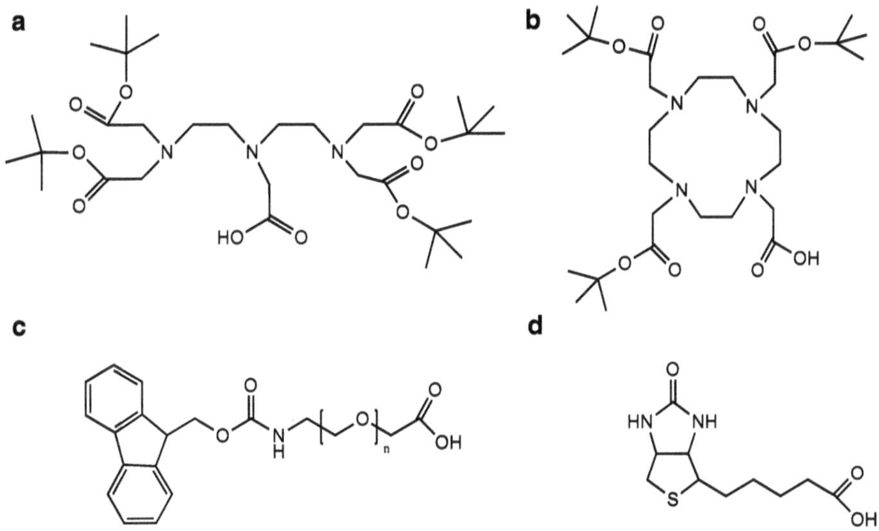

Fig. 1 Formulas of the tertButyl protected DTPA (**a**) and DOTA (**b**) and Fmoc protected PEG (**c**) for the on resin conjugation. Biotin (**d**) do not need any protection during the synthesis. DTPA and DOTA are conjugated via one of the carboxylic acid producing tetra- and tri-carboxylic acid forms, respectively

resin mix 1 ml of 0.3 M of chelate /PEG solution with 1 ml of 0.3 M HBTU and 0.3 M DIPEA solutions. For biotin use 0.03 M concentrations (*see* **Note 4**).

3. Incubate 10 min at room temperature and add the mixture on the resin.
4. Stir resin for 2 h at room temperature.
5. Unplug the column and wash the resin three times with 5 ml of DMF.
6. Take a small sample with the spatula and move it to the test tube. Add 50 μl of each Kaiser test solutions A, B, and C on the resin and incubate at 100 °C for 2 min. If the resin turns blue there are still free amino groups left and the coupling should be repeated. If the Kaiser test reagents stay yellow all the primary amines are coupled (*see* **Note 5**).
7. Wash the resin several times with DCM to remove DMF (*see* **Note 6**).
8. Plug the column carefully and add 2 ml of cleavage solution per 100 mg resin.
9. Cover the column with aluminum foil and stir 2 h at room temperature.
10. After stirring, unplug column carefully and let cleavage solution drop to 10 ml of ice-cold diethyl ether.
11. Wash the resin with 1 ml of 100 % TFA and let it drop to diethyl ether.

12. Incubate ether at 20 °C for 30 min, centrifuge the precipitate (4,000 × *g*) and decant ether.
13. Dry and reconstitute precipitate in water, freeze and lyophilize it.
14. Purify as described in Subheading 3.3.

3.1.2 Removal of Fmoc Protection Group

The primary amino group of PEG is protected with Fmoc during the conjugation. After PEG has been coupled to the peptide, it should be removed with base before the conjugation of tracer of chelate.

1. Add 3 ml of 20 % piperidine in DMF on resin and stir 5 min.
2. Wash resin with DMF and repeat the piperidine treatment.
3. Take a small sample with the spatula and move it to the test tube and make Kaiser test as described. If the color is yellow repeat piperidine treatment, if blue there are free primary amino groups for further couplings.

3.2 FITC Labelling

For the conjugation of fluorescein, we are using isothiocyanate derivate (FITC). It is coupled to the peptides and proteins via primary amines. In solid phase peptide synthesis the amino terminal amine of the peptide is not optimal target since in the most cases the cleavage from the resin with strong acid breaks the amide bond of the N-terminal amino acid in similar mechanism as in Edman chemistry. Instead of labelling amino terminal amine the additional lysine with special side chain protection group, methytrityl (Mtt), is synthesized to the peptide (*see* **Notes 1** and **2**). Mtt is removed with mild acid leaving all the other protection groups intact including Boc for lysine and arginine. Before the Mtt removal, the free amino terminus is blocked using acetylation.

3.2.1 Acetylation

1. Move the resin to the empty PD10 column and wash it with 5 ml of DMF. Plug the column and add small magnetic bar to the PD10 column. Hang the column over the magnetic stirrer.
2. To acetylate free amino terminus add 3 ml of 20 % acetic anhydrine in DMF on resin and stir for 1 min.
3. Remove anhydrine and add another 3 ml of 20 % anhydrine solution on resin and stir for 5 min.
4. Wash once with DMF.
5. Take a small sample with the spatula and perform the Kaiser test as described above. If the color is blue repeat the acetylation. If the color is yellow continue to Subheading 3.2.2.

3.2.2 Removal of Mtt Group

1. Wash the resin carefully with DCM to remove DMF.
2. Remove Mtt group from the additional lysine by adding 3 ml of 2 % TFA in DCM on resin and stir 2 min at room temperature. Repeat for four times.

3. Take a small sample with the spatula and perform the Kaiser test as described in Subheading 3.2.1. If the color is blue continue to Subheading 3.2.3.

3.2.3 FITC Conjugation

1. Wash resin with DMF.
2. Dissolve 10× molar excess of FITC to 3 ml of DMF and add it on the resin.
3. Stir resin 2 h at room temperature.
4. Unplug the column and wash the resin three times with 5 ml of DMF.
5. Take a small sample with the spatula and carry out Kaiser test as described above. If the color is blue repeat the conjugation.
6. If the Kaiser test reagents stay yellow all the primary amines are conjugated with FITC.
7. Wash the resin several times with DCM to remove DMF (*see* **Note 6**).
8. Plug the column carefully, cleave, precipitate and lyophilize as described in Subheading 3.1.1.
9. Purify the peptide as described in Subheading 3.3.

3.3 Purification and Analyses

After the lyophilization peptides are purified using conventional preparative reverse phase C18 column and 60 min gradient (0–90 % buffer B) 5 ml/min and monitored at 220 nm. After purification the pooled fractions are dried and dissolved in water to preferred concentration. Pooled fractions are analyzed with analytical HPLC using analytical reverse phase C18 column and 30 min gradient (0–95 % analytical buffer B) 1 ml/min monitored at 220 nm and verified with mass spectrometry (*see* **Note 7**).

3.4 Radiolabelling

For animal experiments variable radioactive doses are used for mouse and rat for imaging (Table 1.) The used peptide dose per animal for both species and for both radiolabels has varied 2–50 μg in our studies (*see* **Notes 8–10**).

Table 1
Typical radioactive doses (MBq) of different radiolabels for mouse and rat (*see* Note 10)

	^{111}Indium	^{99m}Technetium	^{123}Iodine
Mouse	3–5	20–50	10–40
Rat	5–10	40–150	30–80

3.4.1 Labelling DTPA Conjugate with ^{90m}Technetium

1. Pipet the peptide into glass ampoule with rubber sealing (*see* **Note 11**).
2. Mix the peptide with at least 50 μg of fresh stannous chloride solution dissolved in 10 mM HCl (maximum concentration 10 mg/ml) (*see* **Note 12**).
3. Seal the ampoule with rubber containing top and saturate the atmosphere in the ampoule with nitrogen for few minutes.
4. Inject ^{99m}technetium to syringe, and incubate for 10 min at room temperature.
5. Add saline to obtain the desired total volume and measure the radioactivity with the calibrator.
6. Measure the radiochemical purity by using ITLC on 13 cm × 1.5 cm strip and saline as a mobile phase.
   (a) Apply a drop of the labelled solution on the ITLC slide (~1 cm from the bottom of the slide) and eluate in a chamber as the mobile phase reaches approximately 1 cm from the top.
   (b) The labelled peptide stays at the injection place as the unlabeled technetium moves with the eluent.
   (c) Cut the slide in two equally length pieces between the applied drop and the eluate front.
   (d) Measure the activity of the upper (A) and lower (B) part separately, minus the background activity of both of the samples and calculate the radiochemical purity as follows:
   $B/(A+B) \times 100\ \%$.
   (e) The radiochemical purity should be at least 90 % for the in vivo imaging.
7. Pull the labelled peptide to syringe, measure the radioactivity, inject desired volume, and measure the activity of the syringe again for the calculation of injected dose.

3.4.2 Labelling DTPA Conjugate with Indium

1. Mix peptide and 200 mM NaAc buffer pH 5.0 to the desired concentration. Do not dilute acetate buffer more than 1:2.
2. Add radioactive indium chloride and incubate for 60 min at room temperature.
3. Add saline to obtain the desired total volume and measure the radioactivity with the calibrator.
4. Check the radiochemical purity as described in Subheading 3.4.1 (Fig. 2).

3.4.3 Labelling DOTA Conjugate with Indium

1. Mix peptide and 20 mM NaAc buffer pH 5.0 to the desired concentration. Do not dilute acetate buffer more than 1:2.
2. Add radioactive indiumchloride and incubate for 30 min at +96 °C in dry path.

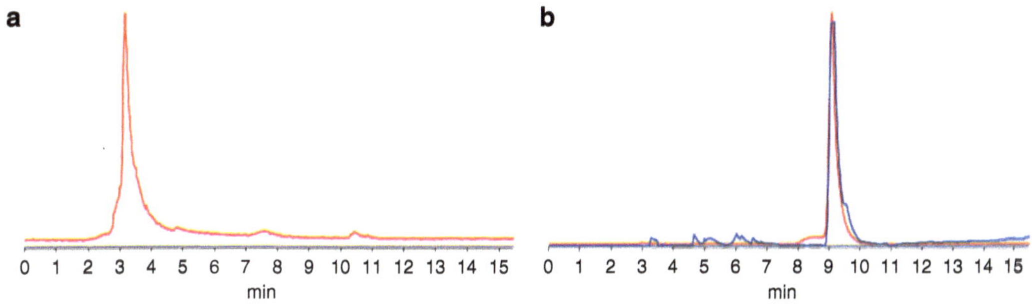

Fig. 2 HPLC chromatogram of free Indium (**a**) and radiolabelled DTPA-peptide conjugate (**b**). Peptide was labelled with 1.1 MBq of ^{111}Indium in 100 mM Acetate buffer for 30 min at RT. Free Indium and the peptide was analyzed with C18 reverse phase column using 15 min gradient (A: 5 % ACT, 0.1 % TFA in water, B 95 % ACN, 0.075 % TFA 5 % water) 1 ml/min and monitored at 220 nm and with radiodetector >60 keV. Radiochemical purity was >99 %

3. Remove the tube carefully from the hot plate and let to cool down at room temperature in a hood.
4. Add saline to obtain the desired total volume and measure the radioactivity with the calibrator.
5. Measure the radiochemical purity as described in Subheading 3.2.1.

3.4.4 Iodination of the Dissolved Peptide

Iodine labelling requires tyrosine or its analogue and in the peptide additional tyrosine is needed (*see* **Note 13**). Labelling with radioactive iodine is done in solution. Iodination is based on the oxidation of iodine using 1,3,4,6-tetrachloro-3α, 6α-diphenylglycouril. Since the oxidating reagent is insoluble in aqueous solutions, it remains adherent to the test tube allowing the oxidation without toxic reagents in solution.

1. Dissolve peptide to water of saline.
2. Wash IODOGEN tube with labelling buffer.
3. Add radioactive NaI solution to IODOGEN tube and incubate for 6 min. Swirl tube gently every 30 s.
4. Remove and add oxidated iodine to the peptide solution.
5. React for 10 min mixing gently every 30 s.
6. Purify labelled peptide with HPLC with preparative C18 reverse phase column using 60 % methanol in water 5 ml/min (*see* **Note 14**).
7. Collect and pool radioactive peptide fractions.
8. Evaporate methanol using nitrogen gas.
9. Measure the radiochemical purity as described in Subheading 3.2.1.

3.5 Conjugation of the Peptide on USPIO Particles

We have used two methods for conjugation of the peptides, via additional cysteine or using streptavidin–biotin method, where peptide is biotinylated via the amino terminus. The used chemistries are suitable for both SPIO and USPIO particles except that the size of the USPIO particles is typically 20–130 nm and requires special high gradient magnet for the separation. If the high gradient magnet is not available, dialysis or gel filtration can be used between the reaction steps. For the conjugation using additional cysteine water soluble bifunctional reagent (MBS) is used. It forms a bridge between the amino group of the particle and the side chain of cysteine. The amount of the free amino groups on the USPIO particles is reported by the manufacturer (*see* **Note 15**).

3.5.1 Streptavidin Derivatized USPIO Particles

1. Dilute USPIO beads to PBS.
2. Mix the biotinylated peptide with SAUSPIO.
3. Separate free peptide from the particles either with high field magnet or dialyzing
   (a) Magnetic separation: precipitate particles with the magnet and wash particles twice with PBS.
   (b) Dialysis: Put the solution into a 10,000 MWCO dialysis cassette or tube and dialyze against PBS over night. Change the buffer twice during the first 2 h.
4. Measure the concentration using turbidometric method (Subheading 3.5.3).

3.5.2 Conjugation via Additional Cysteine

1. Dilute USPIO beads to PBS.
2. Dissolve 10× molar excess of water soluble MBS in PBS and add it on USPIO particles Mix 30 min at room temperature.
3. Separate particles from free MBS either by the magnet or by gel filtration.
   (a) Magnetic separation: precipitate particles with the magnet and wash particles twice with PBS.
   (b) Gel filtration: Load column with Superose 6 in PBS. Run particles and collect fraction containing particles. If UV detector is available follow the fractions at 405 nm.
4. Mix 10× molar concentration of the peptide with particles and mix 60 min at room temperature.
5. Separate particles as described above.
6. Measure the concentration using turbidometric method (Subheading 3.5.3).

3.5.3 Turbidometric Measurement

1. Make a dilution series of the original stock of the USPIO particles in PBS and pipet 200 μl to the wells.
2. Measure the absorbance of the dilution series and the samples at 405 nm.

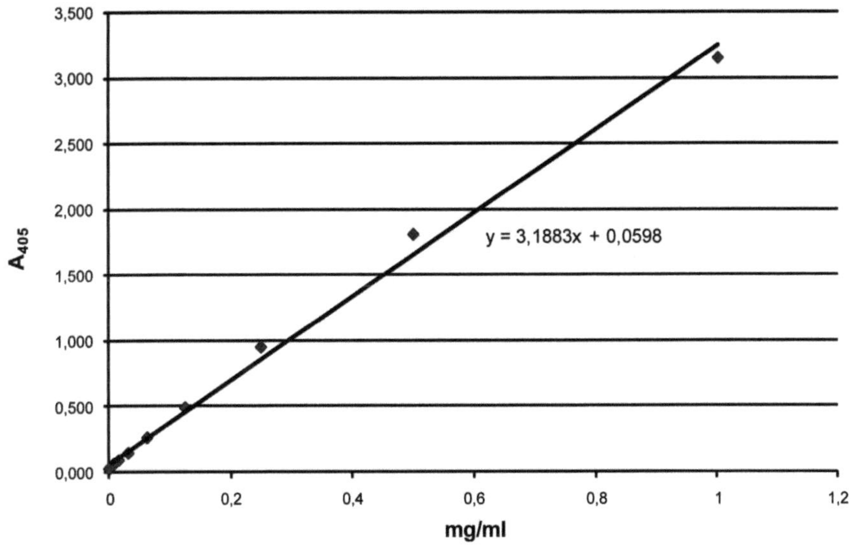

Fig. 3 Scatter blot of the USPIO standard curve. Particles were diluted 1:2 in PBS starting from 1 mg/ml. 200 μl of the samples were pipetted to the wells and measured at 405 nm

3. Calculate the standard curve and determine the concentrations of the samples using standard curve (Fig. 3).

4 Notes

1. The conjugation and labelling site may affect to the biological activity of the peptide. When chelates or additional amino acid are used the position in the sequence can be predetermined during the synthesis. Typically an additional amino acid is synthesized either at the amino or carboxyl terminal.
2. The conjugation or labelling site may be too near to the active site. Polyethylene glycol (PEG) is widely used as a spacer in biologically active molecules especially for in vivo use. Furthermore PEGylation may improve the bioavailability of the peptide (*see* **Note 10**). On the other hand if the length of the PEG is increased, the conjugation of chelates or biotin at the free amino group in PEG may be difficult.
3. Work all the time in the hood. All the reagents are toxic, strong acids or have strong odor.
4. Solubility of biotin is low in DMF (we use normally 10× lower concentrations than with amino acid, chelates, or PEG (0.03 M). For conjugation either volumes can be increased or repeated coupling steps are needed for complete biotinylation of the peptides.
5. If the amino terminal amino acid is proline the Kaiser test does not work and the free amino terminus can not be monitored. Perform an extra deprotection with 20 % piperidine in DMF.

6. Wash DMF carefully before TFA treatment, residual DMF reacts with TFA exothermally and may destroy the peptide.
7. Since DTPA and DOTA are negatively charged molecules, it may be difficult to analyze the conjugated peptide by mass spectroscopy in positive mode, if there is problems try negative mode.
8. Due to the low energy of ^{125}Iodine (39/60 keV) it is not suitable for in vivo imaging. Energies of the radiotracers used in imaging are 27/159/186 keV for ^{123}Iodine is 140.5 keV for ^{99m}Technetium and 171/245/416 keV for ^{111}Indium.
9. When using chelates we have adjusted the peptide dose and specific activity to avoid the purification of the peptide after the labelling. To optimize the labelling conditions *see* **Note 9**.
10. The bioavailability of the peptides varies due to the immunological response and turn over times. The in vivo peptide dose depends on the size of the experimental animal and the binding and stability properties of the peptide. Typically, due to the small size, most of the injected peptide is secreted via the kidneys during first few minutes. PEG has shielding properties against an immunological response and it increases the turn over and thus improves the bioavailability of the peptides. According to our experience, when too small of a dose of a peptide with relatively high specific activity is used the peptide will be metabolized too fast for imaging. However, too large of a dose of a peptide with relatively low specific activity may lead to the competition between the labelled and unlabelled peptide. Finally a high dose with a peptide with high specific activity may lead to a high background. Testing of the optimal ratio of peptide dose and specific activity for every investigated peptide is required.
11. Avoid phosphate containing solutions during technetium labelling since $SnCl_2$ precipitates with phosphates. Water or saline is recommended.
12. Stannous chloride solution should be used immediately after preparation due to fast oxidation of Sn^{II} to Sn^{IV}.
13. If the peptide sequence itself contains tyrosine, which is essential for the biological activity, the iodination approach cannot be used since the iodine atom will most probably destroy the activity of the tyrosine side chain.
14. In some cases the peptide is highly hydrophilic and may come out in the void with free iodine when using 60 % methanol. Before the labelling, the relative hydrophilicity can be tested using gradient of 0–60 % methanol.
15. For 130 nm USPIO particles, the substitution of free amino groups is about 50 mmol/g of particles (Kisker, PMN130) and concentration about 2.9×10^{12} particles/ml resulting 100,000 amino groups/particle.

Acknowledgment

Our original studies have been supported by National Technology Agency of Finland.

References

1. Enback J, Laakkonen P (2007) Tumor homing peptides: tools for targeting, imaging and destruction. Biochem Soc Trans 35:780783
2. Koistinen H, Närvänen A, Pakkala M, Hekim C, Mattsson JM, Zhu L, Laakkonen P, Stenman UH (2008) Development of peptides specifically modulating the activity of KLK2 and KLK3. Biol Chem 389:633642
3. Koivunen E, Arap W, Rajotte D, Lahdenranta J, Pasqualini R (1999) Identification of receptor ligands with phage display peptide libraries. J Nucl Med 40:883888
4. Shadidi M, Sioud M (2003) Selective targeting of cancer cells using synthetic peptides. Drug Resist Updat 6:363371
5. Huang PS, Oliff A (2001) Drug targeting strategies in cancer therapy. Curr Opin Genet Dev 11:104110
6. Antopolsky M, Azhayev A (2001) Synthesis of peptideoligonucleotide phosphorothioate conjugates by convergent or stepwise solid-phase strategies. Nucleosides Nucleotides Nucleic Acids 20:539550
7. Turunen MP, Puhakka HL, Koponen JK, Hiltunen MO, Rutanen J, Leppanen O, Turunen AM, Närvänen A, Newby AC, Baker AH, YläHerttuala S (2002) Peptideretargeted adenovirus encoding a tissue inhibitor of metalloproteinase1 decreases restenosis after intravascular gene transfer. Mol Ther 6: 306312
8. Langer M, BeckSickinger AG (2001) Peptides as carrier for tumor diagnosis and treatment. Curr Me Chem Anticancer Agents 1:7193
9. Lee JD, Huang CH, Weng YH, Lin KJ, Chen CT (2007) Improved accuracy of brain MRI/SPECT registration using a two cluster SPECT normalization algorithm and a combinative similarity measure: application to the evaluation of Parkinson's disease. Ann Nucl Med 21:197207
10. Laakkonen P, Porkka K, Hoffman JA, Ruoslahti E (2002) A tumor homing peptide with a targeting specificity related to lymphatic vessels. Nat Med 8:751755
11. Räty JK, Liimatainen T, Wirth T, Airenne KJ, Ihalainen TO, Huhtala T, Hamerlynck E, VihinenRanta M, Närvänen A, YläHerttuala S, Hakumaki JM (2006) Magnetic resonance imaging of viral particle biodistribution in vivo. Gene Ther 13:14401446
12. Kwekkeboom D, Krenning EP, de Jong M (2000) Peptide receptor imaging and therapy. J Nuc Med 41:17041713
13. McCarthy JR, Kelly KA, Sun EY, Weissleder R (2007) Targeted delivery of multifunctional magnetic nanoparticles. Nanomed 2:153167
14. Lee H, Jang IH, Ryu SH, Park TG (2003) N-terminal site specific monoPEGylation of epidermal growth factor. Pharm Res 20:818825
15. Roberts MJ, Bentley MD, Harris JM (2002) Chemistry for peptide and protein PEGylation. Adv Drug Deliv Rev 54:459476

Chapter 13

Peptide Optimization and Conjugation Strategies in the Development of Molecularly Targeted Magnetic Resonance Imaging Contrast Agents

Andrew F. Kolodziej, Zhaoda Zhang, Kirsten Overoye-Chan, Vincent Jacques, and Peter Caravan

Abstract

Peptides are highly selective, high-affinity ligands for a diverse array of disease targets, but suitably derivatizing them for application as diagnostic or therapeutic agents often presents a significant challenge. Covalent modification with metal chelates frequently results in decreased binding affinity, so a variety of strategies must be explored to find suitable locations for modification and facile peptide conjugation chemistries that maintain or enhance binding affinity. In this chapter, we present a paradigm for systematically optimizing peptide binding and determining the favorable sites and methods for peptide conjugation. This strategy is illustrated by two case studies of peptide-based targeted gadolinium contrast agents: EP-2104R for diagnosis of thrombosis and EP-3533 for diagnosis of cardiac perfusion and fibrosis. Two different architectures for the peptide–metal complex conjugation were designed: EP-2104R contains a total of four gadolinium (Gd) chelates linked at the N- and C-termini, whereas EP-3533 is derivatized with three Gd chelates, two on the N-terminus and one on a lysine side chain. Detailed protocols are provided for two Gd chelate conjugation methods.

Key words Magnetic resonance imaging, Targeted contrast agent, Peptide conjugation, Peptide optimization

1 Introduction

Molecularly targeted contrast agents are emerging as powerful tools in diagnostic medicine. These bifunctional reagents are composed of a targeting moiety coupled to an image-enhancing moiety. The image-enhancing moiety increases the contrast of the image in order to provide specific diagnostic information and may be detectable by any of several diagnostic imaging techniques such as magnetic resonance imaging (MRI) [1], gamma scintigraphy [2–4] (e.g., PET = positron emission tomography or SPECT = single-photon emission computed tomography), near-infrared light imaging [5, 6], and ultrasound [7, 8]. In binding to a protein up-regulated

Andrew E. Nixon (ed.), *Therapeutic Peptides: Methods and Protocols*, Methods in Molecular Biology, vol. 1088,
DOI 10.1007/978-1-62703-673-3_13, © Springer Science+Business Media, LLC 2014

or associated with a cell type or a disease state, the targeting group confers selective binding or uptake of the agent in a tissue or at a site of disease, thereby providing diagnostic specificity that is in addition to any anatomical and functional information provided by the imaging modality. Peptide-based contrast agents are an important class of targeted agents owing in part to the availability of peptides specific to clinically relevant targets either from natural sources or derived from phage display or other peptide-based library screening [9–12]. Determining how and where to attach the image-enhancing moiety to the peptide while maintaining affinity to the target, however, remains a key technical hurdle. This challenge is compounded in the design of MR contrast agents as the weaker signal sensitivity of MRI often necessitates conjugation of multiple metal complexes to sufficiently amplify the signal and enable robust detection [1].

We used peptides in the development of two molecularly targeted gadolinium-based MR contrast agents. EP-2104R, a bright spot imaging agent for diagnosis of vascular thrombosis, comprises a fibrin-specific peptide for binding to blood clots linked to four Gd chelates for image contrast [13]. EP-3533 is composed of a collagen-specific peptide conjugated to three Gd chelates for imaging fibrosis, specifically resulting from myocardial infarction, where collagen levels are elevated [14]. We describe in this chapter a systematic approach to choosing optimal sites for chelate conjugation and peptide optimization which resulted in two unique molecular architectures of the final clinical candidates.

1.1 Peptide Optimization and Conjugation Strategies

Phage display libraries of cyclic, disulfide-cross-linked peptides were used to derive peptide leads specific to fibrin and collagen. In these libraries, the cysteine loop was varied in length from six amino acids ($X_3CX_4CX_3$) to ten amino acids ($X_3CX_8CX_3$) and was flanked by three amino acids on the N- and C-termini [15]. All amino acids except cysteine were allowed at each position. Cyclic libraries were chosen because the conformational restraints imposed by the disulfide generally result in higher affinity peptides, presumably due to reduced entropic costs of binding to the protein target [16, 17]. Nevertheless, the derived peptides bound their respective protein targets with modest affinity (low μM) and required further optimization to increase binding affinity to a level suitable for in vivo efficacy (*see* **Note 1**). Sites amenable to conjugation with a metal chelate also needed to be identified, and the loss of binding affinity frequently accompanying conjugation further increased the affinity requirements for the parent peptide. The strategy we followed for optimizing an initial peptide lead and for determining the most favorable sites for chelate conjugation is outlined in Fig. 1. The case studies below illustrate how these strategies were followed in the optimization of peptides specific to fibrin and collagen.

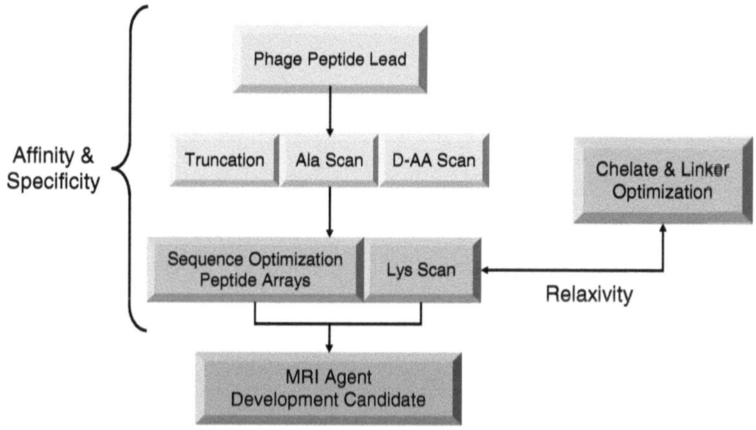

Fig. 1 MRI contrast agent optimization process. Lead characterization and initial SAR are obtained from alanine and D-amino acid scans. Further sequence optimization is conducted by testing peptide arrays and favorable sites for conjugation are identified via a lysine scan. Chelate and linker design and optimization should be examined early in lead optimization to determine the effect of conjugation on peptide affinity and to assess the correlation between peptide and conjugate affinity

N- and C-terminal truncations of the peptide were first tested to determine the minimum sequence required to retain binding. In parallel, amino acid scans were performed to define the structural and functional requirements at each position. An alanine scan wherein each amino acid is individually changed to alanine bluntly probes whether a particular side chain contributes to interaction with the target [18]. Sites relatively insensitive to change are generally more amenable to conjugation or to further optimization; similarly, required amino acid side chains can be further optimized, but with more conservative substitutions such as closely related analogs. The alanine scan also yields a gross assessment of peptide specificity. Peptides with a high proportion of positions that tolerate alanine, particularly in the more structurally constrained portion within the cysteine loop, may bind via nonspecific interactions which are not readily optimized (*see* **Note 2**).

Complementary information is obtained from a D-amino acid scan. By changing the side chain orientation, this scan also identifies side chains important for binding (*see* **Note 3**). In some instances, as in the development of EP-2104R, a gain in affinity may be realized. The change in stereochemistry also results in perturbations to the peptide backbone conformation, and in the context of cyclic peptides, D-amino acid scans can provide information regarding turn conformation. Finally, D-amino acids are more resistant to hydrolysis by peptidases and their inclusion can increase in vivo stability. Sites identified in the alanine and D-amino acid scans (*see* **Note 4**) are then subjected to further optimization

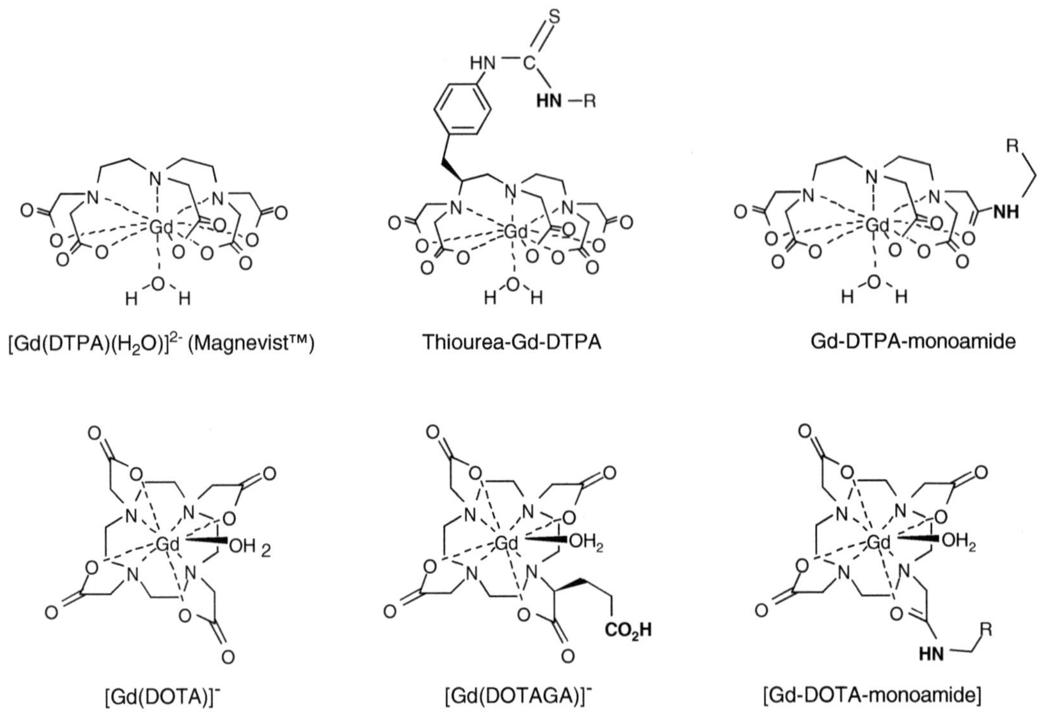

Fig. 2 Gd-DTPA and Gd-DOTA chelates and derivatives

through peptide array strategies to systematically optimize each position alone and in combination with other substitutions. The translation of these improvements to the final metal chelate-conjugated product must be verified (*see* **Note 5**).

The alanine scan and peptide optimization process also identifies sites that are potentially amenable to conjugation. This may be accomplished more directly with a lysine scan where the peptide products are conjugated to Gd-DTPA at each lysine and (optionally) at the N- or C-terminus. This strategy was used to quickly evaluate sites in the design of EP-3533 [19]. In addition to side chain modification, N- and C-terminal modifications were incorporated in the design of EP-2104R due to their advantages for maintaining optimal binding affinity and for increasing in vivo stability. In both cases, terminal lysine or 2,4-diaminobutyric acid (dab) moieties served as diamine linkers for adding two Gd chelates.

Selection of an appropriate Gd chelate offers a range of choices for the ligand and coupling chemistries. Gd-DTPA and Gd-DOTA are approved MRI agents with high kinetic and thermodynamic stability [13, 20–22], and were used in the program described here (Fig. 2). Coupling of the chelate to the peptide via an amide bond is the most common conjugation strategy given the ease of synthesis and commercial availability of suitably functionalized chelates. Chelates may be linked to the peptide by converting a ligand acetate to an amide, although this is less preferred. Chelates

Table 1
Structure–activity data for fibrin-binding peptides

		Amino acid											
Peptide	**K_i (μM)**	**1**	**2**	**3**	**4**	**5**	**6**	**7**	**8**	**9**	**10**	**11**	**12**
Fbn-01	2.2	Q	W	E	C	P	Y	G	L	C	W	I	Q
Fbn-02	2.9	Q	F	H	C	P	Y	D	L	C	H	I	L
Fbn-03	0.9		W	E	C	P	Y	G	L	C	W	I	Q
Fbn-04	1.5			E	C	P	Y	G	L	C	W	I	Q
Fbn-05	3.1				C	P	Y	G	L	C	W	I	Q
Fbn-06	2.8		W	E	C	P	Y	G	L	C	W	I	
Fbn-07	1.5		F	H	C	P	Y	D	L	C	H	I	L
Fbn-08	0.75		F	H	C	Hyp	Y	D	L	C	H	I	L
Fbn-09	0.37		F	H	C	P	Y(3-Cl)	D	L	C	H	I	L
Fbn-10	0.26		F	H	C	Hyp	Y(3-Cl)	D	L	C	H	I	L
Fbn-11	0.15		W	E	C	Hyp	Y(3-Cl)	G	L	C	W	I	Q
Fbn-12	0.16		Y	E	C	Hyp	Y(3-Cl)	G	L	C	Y	I	Q
Fbn-13	0.12		Y	e	C	Hyp	Y(3-Cl)	G	L	C	Y	I	Q

K_i values were obtained in a fluorescence polarization assay measuring displacement of fluorescein-labeled Fbn-01 from DD(E), a soluble fibrin fragment produced by limited plasminolysis of the fibrin polymer [23]. Amino acid abbreviations: *e* D-glutamate, *Y(3-Cl)* 3-chlorotyrosine, *Hyp* 4-hydroxyproline

derivatized via conversion of a single acetate (Gd-DTPA-monoamide and Gd-DOTA-monoamide) generally result in a less stable complex with a slower water exchange rate and consequently lower relaxivity. These problems are circumvented by attaching a pendant reactive moiety to the chelate backbone. Gd-ITC-DTPA, used in the collagen peptide study, reacts with an amine to form a thiourea-Gd-DTPA adduct with improved stability. Similarly, the chelate Gd-DOTAGA contains a pendant carboxylic acid that may be coupled to an amine without derivatizing an acid group involved in Gd chelation.

1.2 Case Study 1: EP-2104R, a Fibrin-Targeted MR Contrast Agent

Cyclic peptide libraries were screened for binding sequences against fibrin [23]. Two sequences, Fbn-01 and Fbn-02, were obtained from a $X_3CX_4CX_3$ library (where X = all 19 amino acids, excluding cysteine), as shown in Table 1. Their sequence similarity strongly suggested that the residues within the cysteine loop would be required for activity with some flexibility in position 7, that aromatic residues were preferred in positions 2 and 10, and that a hydrophobic side chain was preferred in position 11.

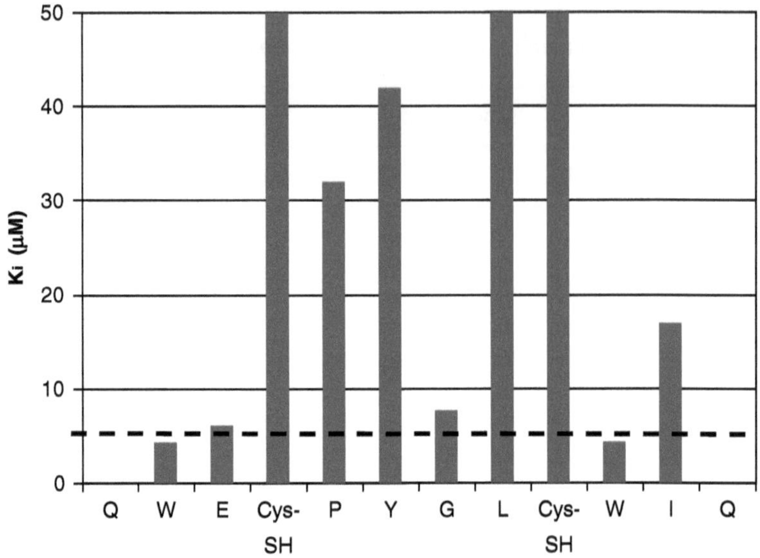

Fig. 3 Peptide affinity when an amino acid in peptide QWECPYDLCWIQ (Fbn-01) is replaced by L-ala. Cys was not replaced with L-ala; values for Cys-SH indicate the affinity of the peptide after disulfide reduction with TCEP; N- and C-terminal glutamine residues were not probed. K_i values were obtained in a fluorescence polarization assay measuring displacement of fluorescein-labeled Fbn-01 from DD(E), a soluble fibrin fragment produced by limited plasminolysis of the fibrin polymer [23]. Full-scale values indicate $K_i > 50$ μM

Truncation studies with Fbn-01 (Table 1) indicated that the N-terminal amino acid was not required, but that removal of the C-terminal glutamine reduced affinity twofold. Fortuitously, removal of the N-terminal glutamine improved activity twofold, and affinity was similarly increased by removal of the N-terminal tryptophan from Fbn-02 (Fbn-07). Further truncation of the peptide from the N- (Fbn-04, Fbn-05) or C-terminus (Fbn-06) resulted in progressive loss of affinity, and the peptide Fbn-03 was identified as the minimum required peptide sequence. The modest changes in affinity for N- and C-terminal truncations indicated that these areas were not providing important interactions with fibrin and might be good sites for chelate conjugation.

The alanine scan data for Fbn-01 were consistent with the sequence convergence observed in the phage display isolates (Fig. 3). Residues within the cysteine loop, Pro5, Tyr6, and Leu8, were all required. Substitution of Gly7 was tolerated, although a similar substitution in Fbn-02 reduced affinity tenfold (data not shown). In addition, alanine replacement of Ile11 resulted in tenfold loss of affinity. Changes to the flanking residues resulted in small losses in affinity, in keeping with the results from the truncation studies indicating their peripheral involvement in target interactions.

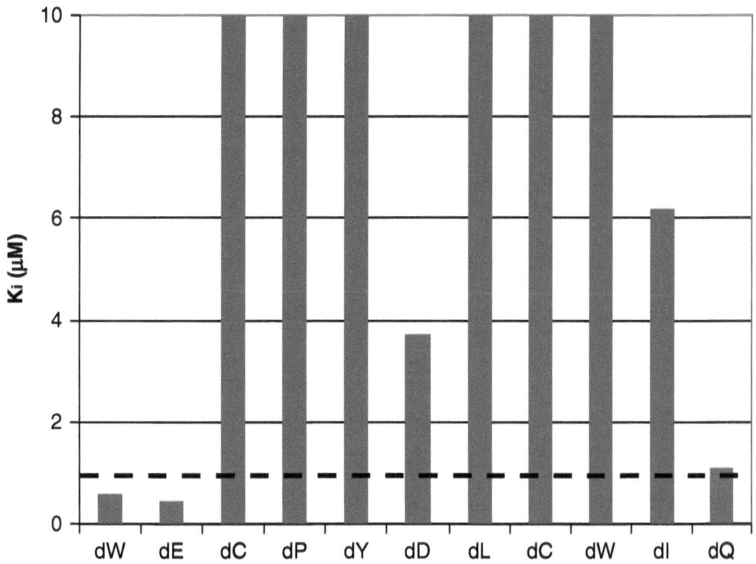

Fig. 4 Peptide affinity when an amino acid in peptide WECPYDLCWIQ (Fbn-03) is replaced with the corresponding D-amino acid. K_i values were obtained as described in Fig. 3. The dashed line indicates the K_i for Fbn-03. Full-scale values indicate K_i > 10 μM

The requirement for the intraloop residues eliminated them from consideration for chelate conjugation, and this conclusion was further supported by the high sensitivity of binding activity to even subtle modification at these positions. The cysteine disulfide was also required for binding. While alanine was not introduced at the Cys positions, reduction of the disulfide with tris(2-carboxyethyl) phosphine (TCEP) destroyed binding.

The results of a D-amino acid scan supported the findings from the alanine scan, and further highlighted the intraloop amino acid requirements (Fig. 4). Positions proximal to the cysteine loop were also sensitive to D-amino acid changes, but in divergent ways. The conformational constraints imposed by the disulfide hold the residues in orientation for binding, and D-amino acids introduced proximal to the cysteines perturb the backbone conformation. The introduction of D-Trp at position 10, a site tolerant to alanine substitution and therefore not a critical point of interaction, substantially reduced affinity. The bulky side chain presented in an opposite orientation may have caused a conformational change to the peptide backbone that abrogated peptide binding. On the other hand, a D-Glu at the N-terminal Glu3 flanking the cysteine loop increased binding affinity two-fold, and this change was incorporated into the final agent design. A small improvement in binding affinity was also observed with D-Trp at position 1. This modification was not combined with the higher binding D-Glu substitution in the final peptide as this resulted in a slight loss of affinity. Finally, the more

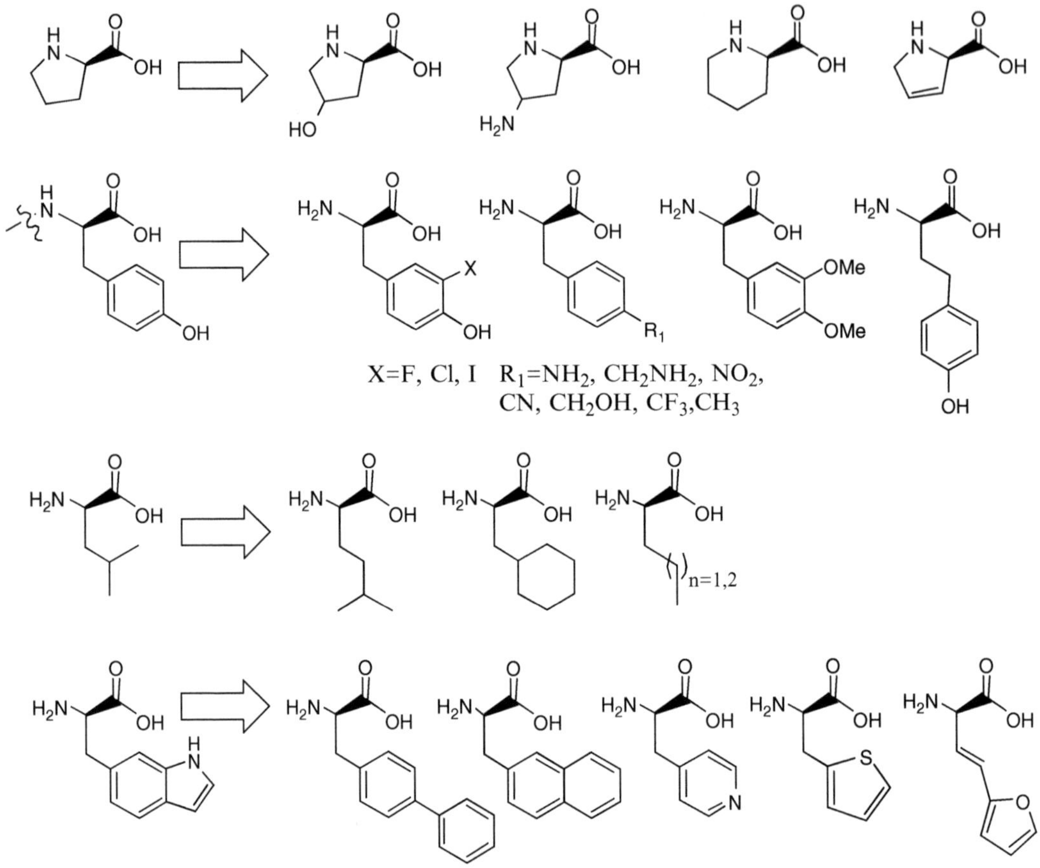

Fig. 5 Examples of unnatural and natural amino acid analogs of proline, tyrosine, leucine, and tryptophan used in peptide array optimization

substantial modifications of the cysteine loop conformation introduced with the mixed D-Cys-L-Cys disulfide loops resulted in pronounced decreases in affinity.

The Ala and D-amino acid scans identified residues important for binding and this information served to guide further peptide optimization. Residues important for binding, Pro5, Tyr6, and Leu8, were further probed with conservative replacements using proline, tyrosine/phenylalanine, and hydrophobic amino acid analogs (examples of introduced natural and unnatural amino acids are shown in Fig. 5). Peptide arrays with single or double substitutions were synthesized and tested using Fbn-07 as the parent sequence (*see* **Note 6**). Replacement of Pro5 with 4-hydroxyproline (Fbn-08) and Tyr6 with 3-chlorotyrosine (Fbn-09) resulted in two- and four-fold affinity improvements, respectively, and when combined improved affinity nearly 5.7-fold (Fbn-10). These substitutions translated to the Fbn-03 sequence and together improved binding affinity of this peptide sixfold (Fbn-11). Hydrophobic substitutions to Ile11, including valine, resulted in loss of binding affinity.

This conserved residue was therefore retained. Arrays of natural (e.g., His, Tyr, Phe) and unnatural amino acid substitutions at the aromatic residues Trp2 and Trp10 were also investigated. These changes did not result in a marked increase in peptide affinity; however the peptide with a disubstitution of Tyr for Trp2 and Trp10 (Fbn-12) bound with equal affinity to fibrin and offered synthetic advantages. In total, five substitutions were incorporated into the EP-2104R peptide (Fbn-13) which was improved 18-fold in binding affinity compared to the original phage-derived sequence. To synthesize EP-2104R, the peptide was conjugated to four Gd-DOTA-like moieties. EP-2104R bound to human fibrin at two equivalent sites with $K_d = 1.7 \pm 0.2$ μM. Given the 14-fold loss in affinity, peptide optimization was clearly essential to successfully obtain a clinical imaging agent with sufficient affinity to enable clot detection. Since the protein target fibrin is present in high concentration in blood clots (~10–100 μM), this affinity was adequate to obtain high fibrin and thrombus binding in vivo [25–27]. The compound also maintained high selectivity for fibrin over fibrinogen (over 100-fold) and serum albumin (over 1,000-fold), two potentially interfering off-target proteins present in plasma at high concentrations. This selectivity was obtained by removing phage from the library during panning with affinity to these proteins present in plasma at high concentration, and was maintained throughout the lead optimization process.

1.3 EP-2104R Conjugation Strategy

EP-2104R was designed to balance the need for a large Gd payload with affinity and in vivo stability requirements. A minimum of four Gd chelates per peptide was required to achieve adequate signal contrast, necessitating either multiple sites of attachment on the peptide, conjugation to branched linkers carrying multiple Gd complexes, or a combination of these strategies. While there were no attractive conjugation sites within the peptide, the N- and C-termini were tolerant to truncation and substitution. Three architectures were contemplated for linking four Gd complexes to the peptide: N-terminal derivatization with a polyamine scaffold bearing four Gd complexes; a similar C-terminal conjugate; or N- and C-terminal derivatization with diamine linkers bearing two Gd complexes. The N- or C-terminal single-site conjugates were not pursued because these compounds were susceptible to rapid C- or N-terminal degradation to inactive fragments by plasma and microsomal proteases, and in vivo stability was inadequate for systemic delivery of the agent. A combination of N- and C-terminal derivatizations, however, protected both of these protease-sensitive sites.

In order to introduce a multiplicity of chelates, the EP-2104R peptide was conjugated on both the N- and C-termini of the peptide to two Gd-DOTAGA(R)-2-(4,7,10-tris(carboxymethyl)-1,4,7,10-tetraazacyclododecan-1-yl)pentanedioic acid (Fig. 6; EP-2104-08 is an activated ester of DOTAGA) chelate moieties joined by a bis

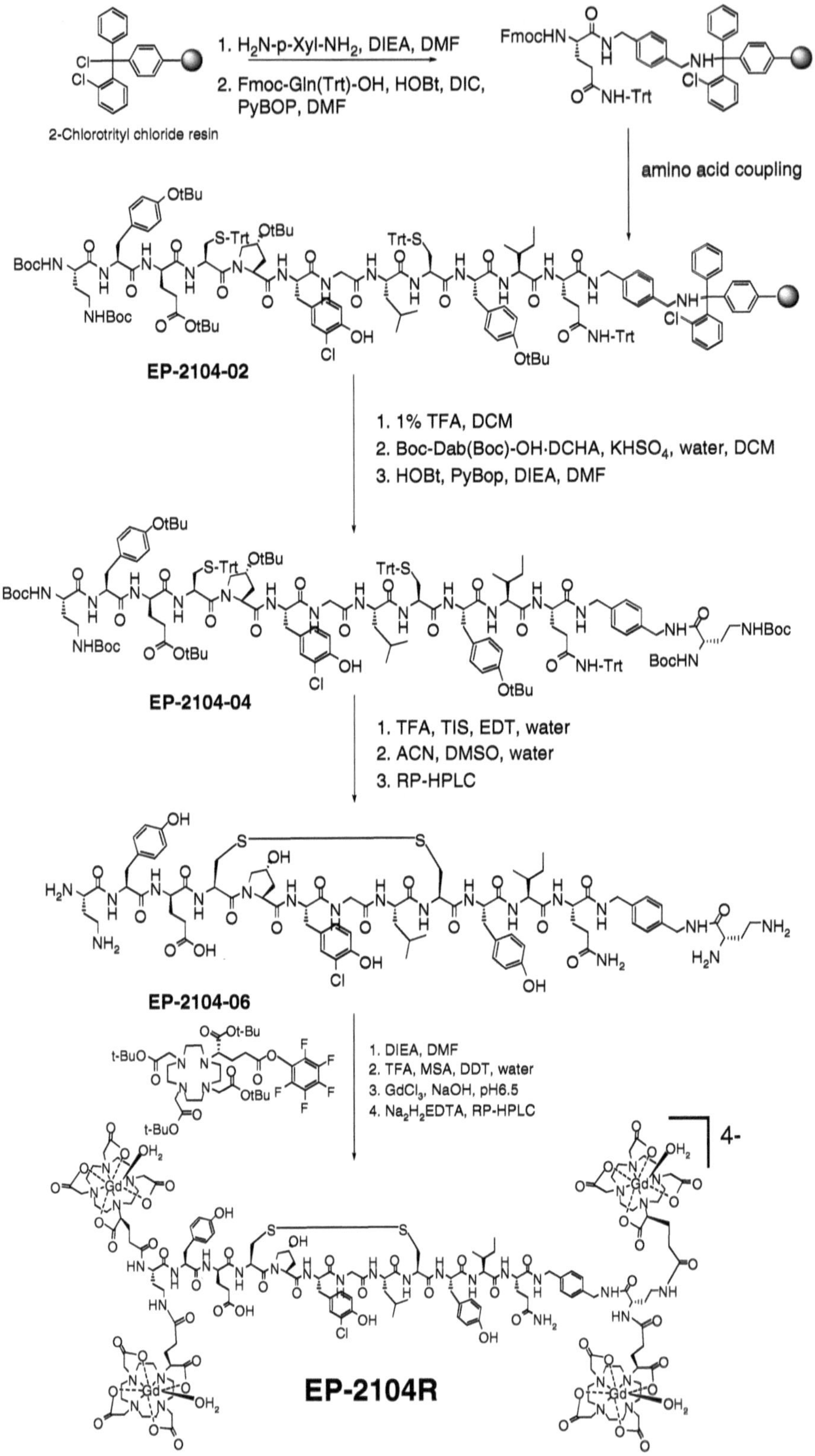

Fig. 6 Synthesis of EP-2104R. The peptide was synthesized on a 2-chlorotrityl chloride resin via a *p-xylylenediamine* linker, then converted to a peptide tetraamine by coupling of Boc-Dab(Boc)-OH. Activated DOTAGA-OPfp was coupled to the deprotected peptide tetraamine to yield EP-2104R

amine linker, dab. The DOTAGA ligand features a pendant acid moiety for amide bond coupling to the peptide. This avoids coupling to an acid group involved in Gd^{3+} chelation, and results in a more stable ligand–metal complex with superior relaxivity properties [13, 28]. Synthesis of the final product was accomplished by first converting the C-terminus to an amino terminus using a *p*-xylylenediamine linker (*see* **Note 7**). In the method illustrated here (Fig. 6), 2-chlorotrityl chloride resin was derivatized with *p*-xylylenediamine. The peptide was extended from this linker using standard fluorenylmethyloxycarbonyl (Fmoc) coupling conditions and was capped at the N-terminus with the *N-tert-butyloxycarbonyl* (Boc)-protected dab moiety Boc-Dab(Boc)-OH. The protected peptide was cleaved from the resin using 1 % trifluoroacetic acid (TFA) in dichloromethane (DCM), and the C-terminal Boc-Dab(Boc)-OH was then added to the peptide in solution. The protecting groups were then removed using a stronger acidic cocktail (TFA, triisopropylsilane, ethylenedithiol, water), and the peptide was cyclized in a dimethylsulfoxide (DMSO) solution. The activated pentafluorophenyl ester of the chelating ligand was synthesized and subsequently coupled to the four primary amine groups in solution. The *t*-butyl protecting groups on the ligands were hydrolyzed in acid solution, and the fully deprotected intermediate was chelated with Gd. The desired product was purified from the crude reaction mixture by reverse-phase chromatography (*see* **Note 8**).

1.4 Case Study 2: EP-3533, a Collagen-Targeted MR Contrast Agent

The disulfide-linked 16 amino acid cyclic peptide $GQ^{1}W^{2}H^{3}C^{4}T^{5}T^{6}R^{7}F^{8}P^{9}H^{10}H^{11}Y^{12}C^{13}L^{14}Y^{15}G^{16}$ was identified by phage display against human type I collagen [14]. A glycine was introduced at the N-terminus as the original phage-derived peptide had an N-terminal glutamine that underwent cyclization to pyroglutamate. A plate-based binding assay was used to measure collagen binding as a percent bound (under conditions of 5 μM collagen and peptide). An alanine scan of the peptide (Fig. 7) identified several residues critical for binding, including Trp2, Phe8, Pro9, and Tyr15, and several residues where a broad diversity of substitutions could be explored, including Gln1, His3, Thr5, Arg7, His10, Leu14, and Gly16. Reduction of the cysteine disulfide also markedly reduced collagen binding (<15 % bound), indicating that the cyclic structure was required to maintain an active conformation. D-Amino acid replacements systematically resulted in dramatic loss of activity in all positions except for Gln1, Thr6, and Tyr12 (Fig. 8). Additional D-amino acids were tested at the Thr6 and Tyr12 positions to explore the effect of different functionalities (Arg, Gln, Ser, Thr, and Tyr), but these did not yield significant advantage.

Development of a collagen-targeted MRI contrast agent from this peptide posed several challenges for obtaining a functional conjugate with adequate gadolinium payload. First, it would be necessary to identify sites for chelate attachment within the peptide

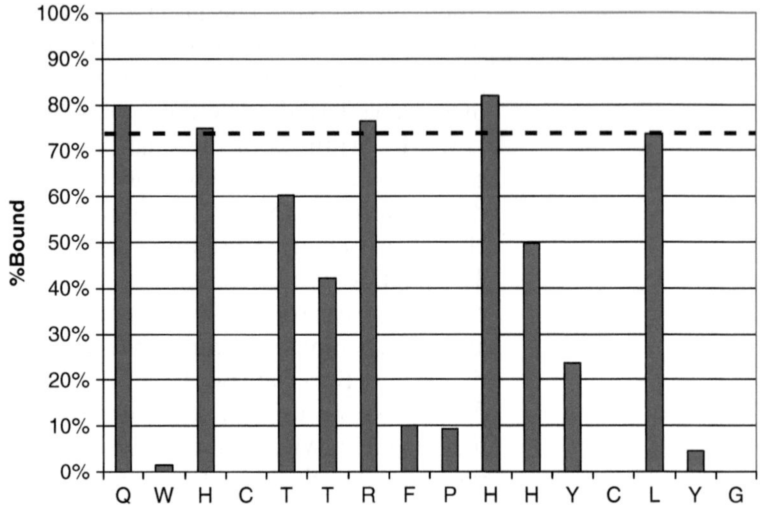

Fig. 7 The fraction of peptide GQWHCTTRFPHHYCLYG bound to collagen when an amino acid is substituted by L-Ala. The *dashed line* shows binding of the parent peptide. N- and C-terminal glycine residues were not probed

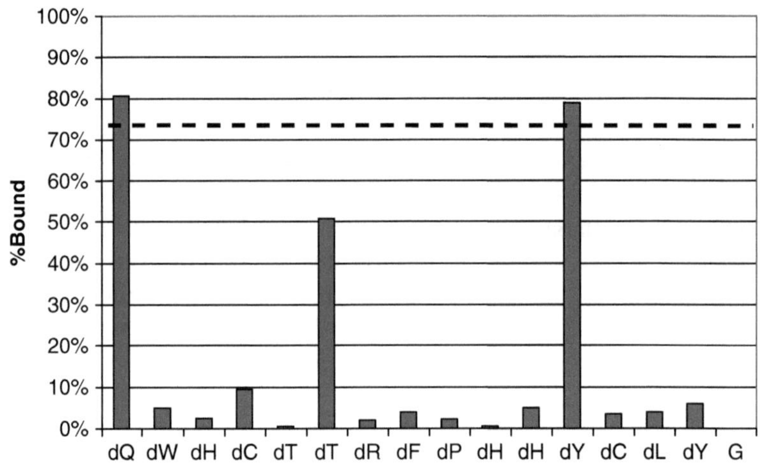

Fig. 8 The fraction of peptide bound to collagen when an amino acid in peptide GQWHCTTRFPHHYCLYG is replaced with the corresponding D-amino acid. *Dashed line* indicates the fraction bound of the parent peptide

or to introduce a scaffold carrying multiple Gd moieties. Early studies using a chelate conjugation strategy similar to that used for EP-2104R resulted in conjugates with reduced activity. A bis Gd-DTPA conjugate of the peptide GQWHCTTYFPHHYCLYG derivatized at the N-terminus and at the C-terminus (C-terminus was derivatized via a *p*-xylylenediamine linker) bound modestly to collagen (48 %). Additional experiments indicated that C-terminal modification was responsible for loss of binding activity, while the

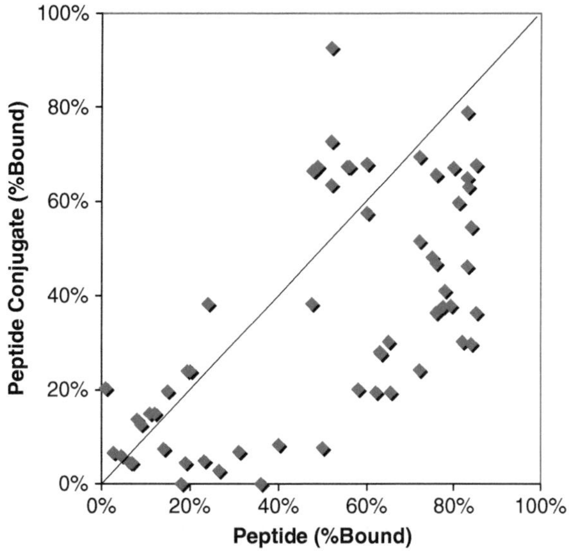

Fig. 9 Fraction bound of N-terminal Gd-ITC-DTPA-conjugated peptides compared to their corresponding parent sequence. Peptides contain single or multiple amino acid replacements of the sequence GQWHCTTRFPHHYCLYG

N-terminus was more tolerant to modification. It also became apparent that at least three Gd chelates would be required to provide sufficient contrast for imaging fibrosis or cardiac perfusion. The problem of Gd chelate conjugation, however, could not be addressed independently from the peptide optimization process. In contrast to the fibrin peptide conjugates, binding affinity of collagen-targeted peptide chelate conjugates did not correlate well with that of the parent peptides, as indicated by a comparison of the binding data for 56 peptides and their N-terminal Gd-DTPA derivatives (Fig. 9). While conjugation did not affect affinity for some peptides, a pronounced loss of binding was observed for a substantial proportion of the peptides. This behavior was not predictable, i.e., was largely without any relation to the position or the nature of introduced amino acid substitutions.

The poor translation of affinity from peptide to conjugate necessitated direct testing of the N-terminal Gd-DTPA conjugates in order to more productively guide lead optimization. To facilitate screening, we developed a method to quickly generate and test these derivatives on a small scale [19]. Purified peptide was reacted with two equivalents per peptide amine of Gd-isothiocyanatobenzyl DTPA (Gd-ITC-DTPA). 2-(4-Isothiocyanatobenzyl)-diethylenetriaminepentaacetic acid (ITC-DTPA) is a commercially available DTPA derivative which reacts with primary amines to yield a thiourea-linked benzyl DTPA conjugate. Coupling DTPA to the peptide requires a rather difficult subsequent purification of the peptide ligand adduct prior to chelation with Gd. Therefore, we first formed the Gd chelate of ITC-DTPA, keeping the isothiocyanate moiety intact, and then

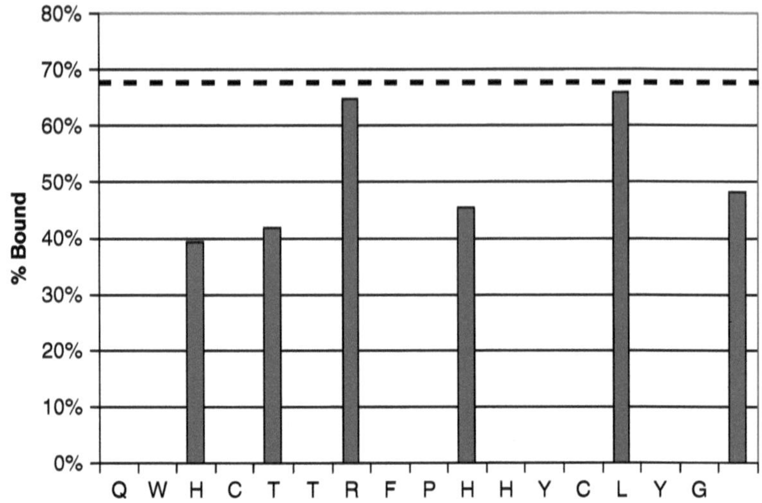

Fig. 10 Fraction bound of N-terminal GdDTPA-GQWHCTTRFPHHYCLYG substituted at select positions by GdDTPA-lysine. *Dashed line* is the binding of the conjugate without GdDTPA-lysine

reacting it the peptide. After overnight reaction, unreacted Gd-ITC-DTPA was quenched with excess ethylenediamine. Rather than purify and isolate each peptide, the reaction mixture was assayed directly for binding. The assay method conveniently involved incubation of the peptide mixture with gelled collagen immobilized (by drying) on a microtiter plate. The bound concentration was obtained as the fractional difference in peptide concentration before and after incubation with collagen. Since the peptide was quantified using HPLC, the peptide conjugate concentration was readily determined even in the presence of undesired side products or small amounts of unreacted peptide.

Based on the alanine scan results, a lysine scan was performed whereby two Gd-DTPA chelates were introduced using the Gd-ITC-DTPA chemistry, one at the N-terminus, and a second on the ε-NH_2 of lysine [19]. Lysine was introduced at each position where alanine substitution was tolerated (Fig. 10). While the chelate was tolerated at all positions tested, including the C-terminus, peptides substituted at Arg7 and Leu14 retained the highest level of binding.

Sites identified as tolerant to alanine substitution were also targeted for systematic introduction of natural and unnatural amino acids. A series of conjugates with 2 Gd-DTPA chelates at the N-terminus was prepared: Gd-DTPA-G-K(G-Gd-DTPA)-W-H-C-X^5-T-X^7-F-P-H-H-Y-C-X^{14}-Y-X^{16} (Gly was added to both α- and ε-NH_2 groups of the N-terminal lysine). Binding data for these peptides are shown in Table 2. The Arg7 → Tyr, Leu14 → Val, and C-terminal Gly16 → Aromatic substitutions significantly improved binding affinity of the bis DTPA peptide conjugates.

Table 2
Structure–activity data for collagen-binding bis(Gd-DTPA)–peptide conjugates

X^5	X^7	X^{14}	X^{16}	% Bound
T	L	L	G	31
T	Y	V	G	78
Y	Y	L	G	55
Y	Y	V	G	91
Y	Y	V	Y	83
T	Y	L	Y	72
T	Y	L	Bip	78

Table 3
Structure–activity data for collagen-binding tris(Gd-DTPA)–peptide conjugates

Compound	% Bound	Peptide sequence
EP-3533	85	GK(G)WHCTTKFPHHYCLYBip
EP-3539	81	KKWHCYTYFPHHYCVYG
EP-3566	76	GK(G)WHCYTYFPHHYCLYGK

Sites of conjugation are underlined

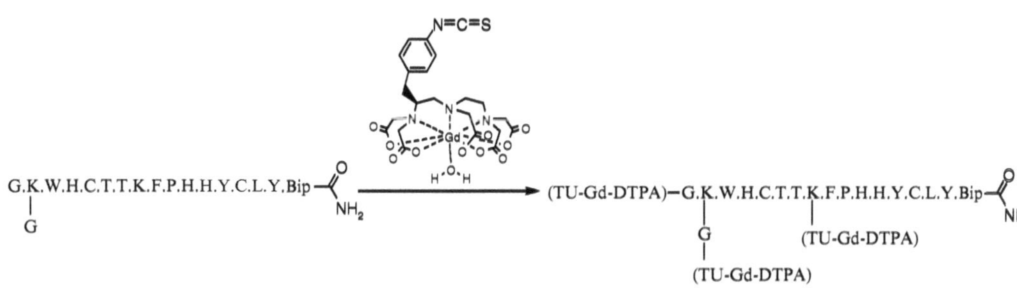

Fig. 11 Synthesis of the tris GdDTPA conjugate EP-3533 by reaction of peptide with Gd-ITC-DTPA to yield a thiourea-linked benzyl Gd-DTPA (TU-Gd-DTPA; *see* Fig. 2 for structure)

The results of the lysine scan were then combined with the structure–activity data to yield a series of peptides incorporating three Gd-DTPA moieties while retaining excellent collagen binding affinity (Table 3). EP-3533 (Fig. 11) with 4-biphenylalanine (Bip) at X^{16}, Gd-DTPA-substituted lysine at X^7, and two Gd-DTPA moieties at the N-terminus was 85 % bound to collagen under the assay conditions. Based on the highest binding peptide, chelates were added to the two Lys ε-NH_2 and the N-terminal NH_2 groups of the peptide KKWHCYTYFPHHYCVYG to form the tris Gd-DTPA conjugate EP-3539 which was 81 % bound to collagen. C-terminal lysine derivatization generally reduced binding of the peptide compared to alternative sites, although the best such peptide, EP-3566, a tris Gd-DTPA conjugate, bound appreciably to collagen (76 %). Considerable flexibility in peptide conjugation strategies was realized from these studies, but the interdependent effects on binding accompanying changes at even distal sites within the peptide, whether involving chelate conjugation or amino acid substitutions, highlight the need to proceed cautiously in agent optimization. In some cases, combinations of favorable substitutions reduced binding, or alternatively certain substitutions conferred improvements in affinity only in combination with other specific changes.

1.5 Summary

Peptide-based MRI contrast agent design requires optimization of peptide affinity by strategies that also enable conjugation of a chelate to the peptide. We recommend a stepwise process (Fig. 1) to identify sites amenable to optimization or to conjugation. Conjugation will likely impact affinity and DMPK properties, and for MR-based probes, this challenge may be compounded by the need to introduce multiple Gd chelates. This chapter gave examples of modifying the peptide, the conjugation position, and the linker to improve affinity, metabolic stability, and Gd payload. Similar approaches could be (and have been by others) taken to improve other DMPK properties [e.g., minimize plasma protein binding, reduce renal retention, alter elimination pathway (hepatic vs. renal)], affinity, or agent contrast-enhancing properties (relaxivity) [29–31].

2 Materials

The ligand ITC-DTPA was purchased from Macrocyclics (Dallas, TX) as the tris–hydrochloride salt.

2-Chlorotritylchloride resin (substitution 0.98 mmol/g) was purchased from Novabiochem (San Diego, CA). Protected amino acids which include Fmoc-Ala-OH, Fmoc-Asp(OtBu)-OH, Fmoc-Cys(Trt)-OH, Fmoc-D-Glu(OtBu)-OH, Fmoc-Gln(tBu)-OH, Fmoc-Gly-OH, Fmoc-His(Trt)-OH, Fmoc-Ile-OH, Fmoc-Leu-OH, Fmoc-Lys(Fmoc)-OH, Fmoc-Lys(Boc)-OH, Fmoc-4-hydroxyproline(tBu)-OH, Fmoc-Pro-OH, Fmoc-Thr(tBu)-OH, Fmoc-Tyr(tBu)-OH, and Fmoc-Trp(Boc)-OH were purchased from Novabiochem (San Diego, CA). Fmoc-3-chlorotyrosine-OH and Fmoc-Bip-OH (4-biphenylalanine) were purchased from RSP amino Acids (Shirley, MA). Boc-Dab(Boc)-OH·dicyclohexylamine (DCHA) salt was purchased from Bachem (Torrance, CA). *p-Xylylenediamine*, diisopropylethylamine (DIEA), diisopropylcarbodiimide (DIC), benzotriazol-1-yl-oxytripyrrolidinophosphonium hexafluorophosphate (PyBOP), polystyrene-carbodiimide (PS-DCC), 1-hydroxybenzotriazole (HOBt), and all other reagents were purchased from Sigma-Aldrich. All solvents were purchased from J.T. Baker. A solid-phase peptide synthesis vessel having a medium-porosity fritted glass resin support with side port and lower Teflon stopcocks for filtration under positive pressure (CG-1861) was purchased from Chemglass (Vineland, NJ).

3 Methods

3.1 Synthesis of EP-2104R

All reactions and procedures were carried out at room temperature, unless otherwise specified.

3.1.1 Step 1: Linker loading and First Amino acid Coupling

1. Swell 2-chlorotritylchloride resin (2.0 g; substitution 0.98 mmol/g) in DCM (10 mL/g). Wash the resin with dimethylformamide (DMF) three times (3 × 20 mL), and then suspend in 20 mL DMF.
2. Add 3.5 eq of *p*-xylylenediamine to the suspension. Adjust the pH of the mixture to 7 with DIEA, and agitate the reaction for 2 h.
3. Add 20 mL of a solution of DMF/methanol/DIEA (17:2:1), and agitate the reaction for 30 min. Wash the resin with DMF (3 × 20 mL).
4. Suspend the 1,4-bisaminomethylbenzyl-2-chlorotritylchloride resin in 20 mL DMF. Add Fmoc-Gln(Trt)-OH (3.5 eq), DIC (5.0 eq), and 1-hydroxybenzotriazole (HOBt) (2.5 eq) to the reaction vessel, and agitate for 2 h. Monitor the reaction by Kaiser [32] and chloranil tests [33] (*see* **Notes 9** and **10**).
5. Wash the resin with DMF (3 × 20 mL) and then DCM (3 × 20 mL). Dry the resin under reduced pressure to give EP-2104-01 [Fmoc-Gln(Trt)-1,4-bisaminomethylbenzyl-2-chlorotritylchloride resin]. Determine the substitution level of the resin by UV spectrometry (*see* **Note 11**).

3.1.2 Step 2: Peptide Extension

The following general procedure is used to attach each subsequent amino acid.

1. Wash the resin with DMF (3 × 20 mL), and then add 20 % piperidine in DMF (20 mL) to remove the N-terminal Fmoc protecting group while agitating the reaction for 30 min.
2. Wash the resin with DMF (3 × 20 mL) and then 5 % HOBt (w/v) in DMF (1 × 20 mL). Suspend the resin in DMF (20 mL) and add the Fmoc-amino acid (1.5 eq), DIC (3.0 eq), and

HOBt (1.5 eq). Monitor the reaction by Kaiser and chloranil tests, and wash the resin with DMF (3×20 mL) and DCM (3×20 mL) upon reaction completion.

3. Repeat this procedure for each amino acid in the following order: Fmoc-Ile-OH, Fmoc-Tyr(tBu)-OH, Fmoc-Cys(Trt)-OH, Fmoc-Leu-OH, Fmoc-Gly-OH, Fmoc-Tyr(3-Cl)-OH, Fmoc-Hyp(tBu)-OH,Fmoc-Cys(Trt)-OH,Fmoc-D-Glu(OtBu)-OH, Fmoc-Tyr(tBu)-OH, and Boc-Dab(boc)-OH (*see* **Note 12**).
4. Dry the elongated peptide on the resin under reduced pressure to give EP-2104-02.

3.1.3 Step 3: Peptide Cleavage from Resin

1. Suspend EP-2104-02 in a solution of 1 % trifluoroacetic acid (TFA) in DCM (20 mL). Filter the resin and wash with DCM (2×20 mL). Adjust the pH of the filtrate to 4–5 with DIEA (note: it is important to neutralize the TFA to avoid concentration of TFA and lowering of the pH which will remove the protecting groups on the peptide), and evaporate the filtrate solution under reduced pressure.
2. Add 0.5 % potassium hydrogen sulfate solution ($KHSO_4$) (50 mL) to precipitate the peptide. Filter the solid, wash with water (2×20 mL), and isopropyl ether (2×20 mL), and then dry under reduced pressure to give EP-2104-03.

*3.1.4 Step 4: dab Addition to C-Terminus (See **Note 13**)*

1. Dissolve Boc-dab(Boc)-OH·DCHA (1.2 g; 2.4 mmol; 1.2 eq) in DCM (100 mL) and wash with 0.5 % $KHSO_4$ (100 mL) to remove DCHA. Evaporate the organic layer under reduced pressure to give the potassium salt as a solid.
2. Dissolve EP-2104-03 and the solid (1.2 eq) in DMF (50 mL) and cool the solution to 0 °C. Add PyBOP (1.2 eq) and HOBt (1.2 eq). Adjust the pH to 6.5–7 with DIEA. Stir the reaction for 2 h, and monitor progress by HPLC (*see* **Note 14**).
3. Pour the reaction mixture into a 0.5 % $KHSO_4$ solution (150 mL) to precipitate the peptide. Filter the solid, wash with water (2×100 mL), and dry under reduced pressure.
4. Dissolve the dried solid in methanol (50 mL) and add sodium carbonate (1 eq). Stir the reaction for 1 h, and monitor by HPLC. Pour the reaction into a 2 % $KHSO_4$ solution (150 mL) to precipitate the peptide. Filter the solid, wash with water (3×50 mL) and isopropyl ether (3×50 mL), and dry under reduced pressure to give EP-2104-04.

3.1.5 Step 5: Peptide Deprotection

1. Dissolve EP-2104-04 in a solution of TFA, triisopropylsilane, ethanedithiol, and water (92.5:2.5:2.5:2.5, ~15 mL/g EP-2104-04). Stir the reaction for 1 h, and monitor progress by HPLC.

2. Evaporate the reaction solution to ~10 mL under reduced pressure, and add isopropyl ether (100 mL) to precipitate the deprotected peptide.
3. Filter the solid, wash with isopropyl ether (2 × 50 mL), and dry under reduced pressure to give EP-2104-05.

3.1.6 Step 6: Peptide Disulfide Cyclization

1. Dissolve EP-2104-05 in a solution of DMSO:acetonitrile:water (15:10:75) (1 g/100 mL). Stir the reaction for 24 h, and monitor cyclization by HPLC.
2. Dilute the reaction mixture with an equal volume of water. Pump the solution onto a reverse-phase preparative HPLC column (Kromasil C18, 100 Å, 10 μm, 5 cm × 25 cm) equilibrated with 95:5 A:B (consisting of A: 0.1 % TFA in water; B: 0.1 % TFA in acetonitrile). Hold starting conditions for 5 min to remove DMSO. Elute the peptide from the column with the following steps: 0.1 % TFA-buffered acetonitrile/water gradient (5–20 % B over 5 min; 20–35 % B over 25 min; 35–95 % B over 3 min; wash for 10 min at 95 % B, column re-equilibration at 5 % B for 10 min). Collect fractions and analyze for purity by analytical HPLC. Combine fractions with purity greater than 98 % and lyophilize to give EP-2104-06.

3.1.7 Step 7: Chelate Activation for Coupling to Peptide

The following describes the synthesis of the pentafluorophenyl ester of the chelating ligand that is subsequently reacted with peptide EP-2104-06. The synthesis of the chelating ligand [2-(*R*)-2-(4,7,10-tris-(*t*-butylcarboxymethyl)-1,4,7,10-tetraazacyclododec-1-yl)-pentanedioic acid 1-*t*-butyl ester] (DOTAGA(OtBu)$_4$-OH) has already been described. Other chelates can alternatively be used (*see* **Note 15** for commercially available analogs).

1. Dissolve DOTAGA(OtBu)4-OH (6 eq, 4.3 g) and pentafluorophenol (1.2 eq) in DCM (100 mL). Add polystyrene-carbodiimide (PS-DCC; 1.2 eq; 1.23 mmol/g loading capacity) and shake the reaction on an orbital shaker for 2 h. Monitor the reaction by HPLC (*see* **Note 16**).
2. Filter and rinse the resin with DCM (2 × 50 mL). Evaporate the filtrate containing the product [DOTAGA(OtBu)$_4$-OPfp] (EP-2104-08) under reduced pressure. Dry the glassy solid under reduced pressure.

3.1.8 Step 8: Conjugation of Peptide EP-2104-06 to [(DOTAGA(OtBu)$_4$-OPfp)]

1. Dissolve 3.4 g peptide EP-2104-06 in 100 mL DMF, and add the activated ligand EP-2104-08 (4 eq) as a solid to the reaction. Adjust the pH to 6.5 with DIEA. Stir the reaction for 1 h.
2. Add additional DOTAGA(OtBu)$_4$-OPfp (0.5 eq) to the reaction mixture. Adjust the pH to 6.5 with DIEA and stir the reaction for 0.5 h.

3. Again, add to the reaction additional DOTAGA(OtBu)$_4$-OPfp (0.5 eq), adjust the pH to 6.5 with DIEA, and stir the reaction for 0.5 h. Finally, add 0.3 eq DOTAGA(OtBu)$_4$-OPfp to the reaction mixture, adjust the pH to 6.5 with DIEA, and stir for another 1 h. Monitor reaction progress by HPLC.
4. Add 200 mL saturated sodium chloride to the reaction to precipitate the conjugated peptide. Stir the suspension for 30 min to complete precipitation, and filter the solid. Wash the solid with water (150 mL) and dry under reduced pressure to give EP-2104-09 as an off-white solid.

3.1.9 Step 9: DOTAGA(OtBu)$_4$ Deprotection

1. Dissolve EP-2104-09 in a cocktail of TFA, dodecanethiol, methanesulfonic acid, and water (92.5:2.5:2.5:2.5; ~15 mL/g EP-2104-09). Stir the reaction for 1 h, and monitor reaction progress by HPLC.
2. Add cold ethyl acetate to precipitate the deprotected product. Filter the solid, wash with isopropyl ether, and dry under reduced pressure to give EP-2104-10 as an off-white solid.

3.1.10 Step 10: Gd Chelation

1. Add EP-2104-10 (7.1 g; 2 mmol) in portions to a solution of water (150 mL) while maintaining a pH of 6.5 using 1 N sodium hydroxide (NaOH) (80 mL). Add gadolinium (III) chloride hexahydrate ($GdCl_3 \cdot 6H_2O$, 4 eq) to the solution and adjust the pH to 6.5 with 1 N NaOH (20 mL). Stir the reaction for 1 h at room temperature, adjust the pH to 6.5 as necessary, and monitor reaction progress by HPLC.
2. Neutralize residual gadolinium (III) chloride with ethylenediaminetetraacetic acid (EDTA, 0.25 eq). Stir the reaction for 15 min while adjusting the pH to 6.5 with 1 N NaOH. Monitor the neutralization reaction to minimize excess EDTA by testing with xylenol orange (*see* **Note 17**).
3. Isolate and purify the product from solution by reverse-phase chromatography (~3.5 g product/run). Divide the chelated peptide solution into two equal portions with water (50 mL). Pump each portion onto a preparative HPLC column (C4 Kromasil, 100 Å, 10 μm, 5 cm × 25 cm) equilibrated with acetonitrile:water (2:98). Hold starting conditions for 5 min. Initiate elution using the following gradient routine at a flow rate of 100 mL/min: 2 to 18 % ACN over 25 min; 18–50 % ACN over 1 min; and 50–80 % ACN over 7 min. Detect peaks at 220 nm. Analyze purified fractions by analytical HPLC. Combine high-purity fractions (>98 %) and lyophilize to give the EP-2104R as a white powder.
4. Analysis of EP-2104R yielded the following data: electrospray ionization-time-of-flight mass spectrometry: m/z expected $[C_{154}H_{212}ClGd_4Na_5N_{34}O_{56}S_2 + H + Na]^{2+}$ 2,151.5173, found

2,151.5958, expected $[C_{154}H_{212}ClGd_4Na_5N_{34}O_{56}S_2+2Na]^{2+}$ 2,162.5082, found 2,162.5650. UV: λ (ε) 192 nm (327,000 cm^{-1} M^{-1}), 219 nm (60,600 cm^{-1} M^{-1}), 277 nm (5,700 cm^{-1} M^{-1}). Chiral amino acid analysis showed that all amino acids were >99.5 % in the L form except Glu which was analyzed as 49.7 % D-Glu. This is expected from the D-Glu in the molecule and the L-Gln which is hydrolyzed to Glu in the analytical method.

3.2 Peptide Labeling with Gd-DTPA-ITC

All reactions and procedures were carried out at room temperature, unless otherwise specified.

3.2.1 Step 1: Preparation of Gd-DTPA-ITC Solution

1. Dissolve DTPA-ITC 1.72 g (2.65 mmol) in 10 mL of distilled deionized water (ddH_2O) in a 100 mL beaker with stirring. Insert a pH electrode to monitor the reaction (*see* **Note 18**).
2. Adjust pH to 6 by addition of 1 M NaOH (~13 mL).
3. Add $GdCl_3{\cdot}6H_2O$ (781 mg, 2.1 mmol) with stirring and readjust pH to 6 with 1 M NaOH (~6 mL). Add an additional 186 mg (0.55 mmol) of $GdCl_3{\cdot}6H_2O$ and readjust the pH to 6 with 1 M NaOH (~2 mL). This resulted in one typical preparation in a final volume of 43.6 mL, and a concentration of 59.6 mM Gd-DTPA-ITC (*see* **Note 19**).
4. Check reaction completion by analytical HPLC on a Kromasil C18 column, 50×4.6 mm, 3.5 μm, developed with a binary solvent system. Mobile phase A is 50 mM ammonium formate; mobile phase B is a solution of acetonitrile and 50 mM ammonium formate (9:1, v/v). The gradient is initiated at 5 % B for 2 min, and then 5–40 % B over 4 min followed by a 3-min ramp to 95 % B and re-equilibration at 5 % B for 3 min. Flow rate is 0.8 mL/min (*see* **Note 20**).

3.2.2 Step 2: Preparation of Peptide–Gd-DTPA Conjugate (See **Note 21**)

1. Suspend purified cyclic peptide (0.05 mmol) containing *N* primary amines in 10 mL pH 9 borate buffer (100 mM). Acetonitrile may be used to dissolve the peptide if aqueous solubility is poor; the reaction has been run successfully in up to 40 % acetonitrile.
2. Add Gd-ITC-DTPA solution (59.6 mM; 3.35 mL for two amines) in twofold excess (stoichiometry is calculated as 2×N amines×mmol peptide) and stir reaction overnight (*see* **Note 22**).
3. Purify by analytical or preparative HPLC. *Preparative method*: Kromasil C18 column, 250×20 mm. Mobile phase A is 50 mM aqueous ammonium acetate; mobile phase B is acetonitrile. The gradient is initiated at 2 % B for 5 min, and then 2–20 % B in 5 min and 20–45 % B over 15 min, followed by a 4-min ramp to 95 % B over 4 min, a ramp to 2 % B in 4 min, and re-equilibration at 2 % B over 6 min. Flow rate is 20 mL/min.

4 Notes

1. Phage display against fibrin was conducted for four rounds, a high degree of sequence selection. Analysis of phage isolates at earlier rounds of panning may have yielded greater sequence diversity, although a greater proportion of lower affinity peptides would in principle be present. In the collagen screen, panning was conducted for two rounds and yielded 140 diverse sequences, but only 3 of these yielded high-affinity peptides.
2. A sequence that was abandoned in the collagen project tolerated alanine at six of eight positions in the cysteine loop. Binding activity of peptides in this series increased with the degree of peptide aggregation rather than through modulation of specific interactions with the protein target.
3. Fully D-amino acid-substituted peptides are frequently excellent negative controls for specificity studies, and have advantages over "scrambled" peptide controls where the sequence of the parent peptide is haphazardly rearranged. If binding is specific to the protein target, the change in chirality should abolish all binding. The use of the same amino acid sequence conserves other properties of the peptide such as solubility, nonspecific binding, aggregation state, charge alignment, and the hydrophobic profile that are altered when scrambled peptides are used.
4. Achiral glycine is frequently tolerant to D-amino acid substitution, and an array of D-side chain functionalities can be tested as a strategy for improving affinity.
5. Several other additional scans can be considered, including *N*-methyl scans (introduction of a hydrophobic group in orientation opposite to the side chain; perturbation of the backbone conformation), α-amino isobutyric acid (modulation of helical conformation) [34], β-amino acid (modulation of cysteine loop size and peptide loop conformation), aspartic acid (negatively charged amino acid scan), leucine (hydrophobic scan), or tyrosine (aromatic). These scans can be considered depending on the flexibility of the sequence to substitution. In cases where considerable diversity in the sequence is tolerated or where functional roles of particular residues are poorly defined, one or more of these scans can quickly establish structural and functional preferences for optimal binding.
6. The vast majority of substitutions in both the fibrin and collagen projects resulted in affinity loss. It is recommended to first perform single substitutions and then test double substitutions with tolerated changes to determine synergies.
7. Resins are commercially available with a choice of preloaded diamine linkers from which the peptide may be extended.

Additional linkers that may be considered include *m-xylylenediamine*, 1,2-ethylenediamine, and 2,2′-oxydiethanamine. We selected the *p*-xylylenediamine linker as the affinity of the conjugates tended to be approximately fivefold higher than with these other alternatives.

8. The method described here is a more general method, but other synthetic approaches, specific to this sequence, are feasible, as previously published [13].
9. The protocol for the Kaiser test to determine qualitatively the presence of a primary amine is as follows:
   (a) Prepare a solution of ninhydrin in ethanol (5 g/100 mL).
   (b) Dissolve 80 g of liquefied phenol in 20 mL ethanol.
   (c) Prepare 20 mL of 0.001 M potassium cyanide (WARNING: very toxic; take extreme caution, use protective clothing, and prepare in hood) in water.
   (d) Add 2 mL 0.001 M potassium cyanide to 98 mL pyridine (WARNING: toxic and unpleasant odor; harmful if inhaled or absorbed through skin; use protective clothing and handle in hood).
   (e) Wash a few resin beads three times with ethanol (several mL).
   (f) Transfer beads to a glass test tube and add two drops of each of the solutions above in order.
   (g) A positive Kaiser test is blue.
10. The chloranil test is used for detecting secondary amines associated with proline:
   (a) Prepare test solutions: 2 % acetaldehyde in DMF and 2 % *p*-chloranil in DMF.
   (b) Add a drop of each solution to 1–5 mg of resin, and leave at room temperature for 5 min.
   (c) A positive chloranil test is blue.
11. To determine the substitution level of the first amino acid added to the resin:
   (a) Weigh ~5 mg washed, dry resin into a 10 mL flask.
   (b) Add 2 mL 2 % diazabicyclo[5.4.0]undec-7-ene (DBU) in DMF and agitate gently for 30 min to cleave the peptide from the resin.
   (c) Dilute solution to 10 mL with acetonitrile.
   (d) Remove 2 mL of solution and dilute to 25 mL with acetonitrile.
   (e) Prepare a reference by following the above steps without resin.

(f) Place diluted solution in cuvette (allow any suspended resin to settle) and record A304, the absorbance at 304 nm (use acetonitrile as blank). The substitution level, calculated in mmol/g is
$[(A304_{sample} - A304_{reference}) \times 16.4]/\text{mg resin}$

12. The absence of lysine side chains simplified the protection scheme required for synthesis of EP-2104R. For peptides requiring a lysine ε-amine for activity, it is recommended that the lysine-containing portion be synthesized as a fragment and then coupled to the remaining portion and/or that carbobenzyloxy (Cbz) groups be used to protect the lysine side chain.
13. The dab linkers could be added to both the N- and C-termini in a single step, although this was found to result in 10–15 % racemization of the linker.
14. Excess Boc-Dab(Boc)-OH reacted with 3-chlorotyrosine which was coupled without side chain protection to the peptide, as side chain-protected 3-chlorotyrosine is not commercially available. Any ester that is formed, however, is hydrolyzed during the work-up in **step 4** of the Boc-Dab(Boc)-OH addition.
15. Other compounds could also be used to couple to the peptide at this step. For example, acid side chain-protected or unprotected DOTA-NHS-ester (DOTA N-hydroxysuccinimide ester) [35]; DTPA tetra (*t*-butyl ester); DOTA tris (*t*-butyl ester); and DOTA-TFP-ester (tetrafluorophenol ester) are all available from Macrocyclics Inc. (www.macrocyclics.com). Additional solid-phase synthetic strategies can be found in references 30 and 31.
16. DCC immobilized on solid support (PS-DCC) is useful for avoiding carryover of contaminating DCC in the subsequent coupling step while providing a rapid and convenient filtration work-up. Unwanted reaction of excess pentafluorophenol with side chain acid groups on the peptide is also thereby minimized.
17. To perform a photometric titration of chelatable equivalents using the xylenol orange test:
    (a) Prepare 0.02 M xylenol orange solution in 50 mM sodium acetate buffer, pH 4.85.
    (b) Prepare solutions of 5.0 and 0.5 mM $Gd(NO_3)_3$ in water.
    (c) Dissolve the ligand in water to a concentration greater than 1 mM.
    (d) Adjust the pH to ~6 using 1 M NaOH (if the ligand was isolated under acid conditions) and record the volume of solution.
    (e) Pipette 10 μL of ligand solution and 1.2 mL of xylenol orange solution into a 1.5 mL quartz cuvette and mix.

(f) Place cuvette into a UV-Vis spectrophotometer set at 572 nm and zero the reading.

(g) Add 10 μL aliquots of 5.0 mM and then 0.5 mM $Gd(NO_3)_2$, depending on the expected ligand solution concentration, until a positive absorbance is observed. Record the absorbance value (Abs-I).

(h) A positive absorbance represents the end of the titration. To estimate how much excess $Gd(NO_3)_2$ is present, add 1.2 mL 0.02 M xylenol orange to a 1.5 mL quartz cuvette. Zero the absorbance at 572 nm. Add 10 μL 5 mM $Gd(NO_3)_2$ solution and record the absorbance (Abs-II).

(i) The concentration of chelatable equivalents is given by the equation

$$\text{Ligand Conc} = \frac{5\times\left(\text{Vol 5mmGd}\left(NO_3\right)_3\right) + 0.5\times\left(\text{Vol 0.5mm Gd}\left(\text{NO}_3\right)_3\right)}{10\mu\text{L}} - \left[5\times\frac{\text{Ab-I}}{\text{Ab-II}}\right]$$

18. To avoid decomposition of ITC-DTPA to a benzylamine, and its coupling to excess ITC-DTPA, NaOH addition should be performed with care so as not to exceed pH 6.
19. A considerable volume of 1 M NaOH is required to deprotonate the acid form of the ligand (2–2.5 eq), ligand bound HCl (3 eq), and to neutralize HCl released from $GdCl_3$ (3 eq). Rinsing of vial and pH electrode with ddH_2O between additions of NaOH or $GdCl_3{\cdot}6H_2O$ results in additional volume. Weigh reaction vessel before and after to calculate the volume of liquid (subtract weight of solid reagents).
20. The Gd-ITC-DTPA solution can be stored frozen and is stable for at least one month. Small-volume aliquots are recommended to avoid freeze/thaw cycles that reduce activity.
21. The reaction proceeds more efficiently in a microwave synthesizer. Peptide (0.05 mmol) containing N primary amines is suspended in 10 mL of pH 7.5 phosphate buffer (200 mM Pi). Gd-ITC-DTPA solution (59.6 mM) is added in excess (2 × N amines × 0.05 mmol peptide), typically 1–5 mL of solution. The mixture is heated to 80 °C for 20 min using an Emrys Optimizer microwave synthesizer, and cooled to room temperature before purification.
22. The indicated (two) equivalents of Gd-ITC-DTPA used in the reaction assume that the peptide potency is 100 %. In practice, the potency is typically 50 % due to salt and water incorporated in the peptide during work-up, and therefore the number of equivalents added is probably closer to 4. No attempt was made to quantify peptide potency for each reaction.

Acknowledgements

We would like to thank Thomas McMurry and Phil Graham for their leadership and scientific contributions to the development of EP-2104R and EP-3533. We also acknowledge the many contributions of colleagues at EPIX Pharmaceuticals to the development of these strategies and protocols, including John Amedio, Jaclyn Chasse, Biplab Das, Qing Deng, Stephane Dumas, Matthew Greenfield, Steffi Koerner, Richard Looby, Shrikumar Nair, Luhua Shen, Wei-Chuan Sun, and Stephan Zech. Discovery of the peptide leads was enabled by collaborations with David Buckler, Bob Ladner, Dan Sexton, and Charles Wescott at DYAX Corporation.

References

1. Caravan P (2006) Strategies for increasing the sensitivity of gadolinium based MRI contrast agents. Chem Soc Rev 35:512–523
2. Ladner RC (1999) Polypeptides from phage display. A superior source of in vivo imaging agents. Q J Nucl Med 43:119–124
3. Lever JR (2007) PET and SPECT imaging of the opioid system: receptors, radioligands and avenues for drug discovery and development. Curr Pharm Des 13:33–49
4. Reubi JC, Maecke HR (2008) Peptide-based probes for cancer imaging. J Nucl Med 49:1735–1738
5. Nahrendorf M, Sosnovik DE, Weissleder R (2008) MR-optical imaging of cardiovascular molecular targets. Basic Res Cardiol 103: 87–94
6. Pham W, Choi Y, Weissleder R, Tung CH (2004) Developing a peptide-based near-infrared molecular probe for protease sensing. Bioconjug Chem 15:1403–1407
7. Villanueva FS (2008) Molecular imaging of cardiovascular disease using ultrasound. J Nucl Cardiol 15:576–586
8. Klibanov AL (2006) Microbubble contrast agents: targeted ultrasound imaging and ultrasound-assisted drug-delivery applications. Invest Radiol 41:354–362
9. Newton J, Deutscher SL (2008) Phage peptide display. Handb Exp Pharmacol 185(pt 2): 145–63
10. Aina OH, Liu R, Sutcliffe JL, Marik J, Pan CX, Lam KS (2007) From combinatorial chemistry to cancer-targeting peptides. Mol Pharm 4:631–651
11. Lam KS, Kiu R, Miyamoto S, Lehman AL, Tuscano JM (2003) Applications of one-bead-one-compound combinatorial libraries and chemical microarrays in signal transduction research. Acc Chem Res 36:370–377
12. Uchiyama F, Tanaka Y, Minari Y, Tokui N (2005) Designing scaffolds of peptides for phage display libraries. J Biosci Bioeng 99:448–456
13. Overoye-Chan K, Koerner S, Looby RJ, Kolodziej AF, Zech SG, Deng Q, Chasse JM, McMurry TJ, Caravan P (2008) EP-2104R: a fibrin-specific gadolinium-based MRI contrast agent for detection of thrombus. J Am Chem Soc 130:6025–6039
14. Caravan P, Das B, Dumas S, Epstein FH, Helm PA, Jacques V, Koerner S, Kolodziej A, Shen L, Sun WC, Zhang Z (2007) Collagen-targeted MRI contrast agent for molecular imaging of fibrosis. Angew Chem Int Ed Eng 46:8171–8173
15. Sato AK, Sexton DJ, Morganelli LA, Cohen EH, Wu QL, Conley GP, Streltsova Z, Lee SW, Devlin M, DeOliveira DB, Enright J, Kent RB, Wescott CR, Ransohoff TC, Ley AC, Ladner RC (2002) Development of mammalian serum albumin affinity purification media by peptide phage display. Biotechnol Prog 18:182–192
16. Ladner RC (1995) Constrained peptides as binding entities. Trends Biotechnol 13: 426–430
17. Lowman HB, Chen YM, Skelton NJ, Mortensen DL, Tomlinson EE, Sadick MD, Robinson ICAF, Clark RG (1998) Molecular mimics of insulin like growth factors 1 (IGF-1) for inhibiting IGF-1:IGF-1 binding protein interactions. Biochemistry 37:8870–8878
18. Cunningham BC, Wells JA (1989) High-resolution epitope mapping of hGH-receptor interactions by alanine scanning mutagenesis. Science 44:1081–1085

19. Caravan P, Das B, Deng Q, Dumas S, Jacques V, Koerner SK, Kolodziej A, Looby RJ, Sun WC, Zhang Z (2009) A lysine walk to high relaxivity collagen-targeted MRI contrast agents. Chem Commun (Camb) 4:430–432
20. Eisenwiener KP, Powell P, Macke HR (2000) A convenient synthesis of novel bifunctional prochelators for coupling to bioactive peptides for radiometal labelling. Bioorg Med Chem Lett 10:2133–2135
21. Kumar K, Chang CA, Francesconi LC, Dischino DD, Malley MF, Gougoutas JZ, Tweedle M (1994) Synthesis, stability, and structure of gadolinium(III) and yttrium(III) macrocyclic poly(amino carboxylates). Inorg Chem 33:3567–3575
22. Pulukkody KP, Normann TJ, Parker D, Royle L, Brouan CJ (1993) Synthesis of charged and uncharged complexes of gadolinium and yttrium with cyclic polyazaphosphinic acid ligands for in vivo applications. J Chem Soc Perkins Trans 2:605–620
23. Kolodziej AF, Nair SA, Graham P, McMurry TJ, Wescott W, Sexton DJ, Ladner RC (2012) Fibrin specific peptides derived by phage display: characterization of peptides and conjugates for imaging. Bioconjug Chem 23:548–556
24. Moskowitz KA, Budzynski AZ (1994) The (DD)E complex is maintained by a composite fibrin polymerization site. Biochemistry 33:12937–12944
25. Spuentrup E, Buecker A, Katoh M, Wiethoff AJ, Parsons EC Jr, Botnar RM, Weisskoff RM, Graham PB, Manning WJ, Günther R (2005) Molecular magnetic resonance imaging of coronary thrombosis and pulmonary emboli with a novel fibrin-targeted contrast agent. Circulation 111:1377–1382
26. Spuentrup E, Botnar RM, Wiethoff AJ, Ibrahim T, Kelle S, Katoh M, Ozgun M, Nagel E, Vymazal J, Graham PB, Günther RW, Maintz D (2008) MR imaging of thrombi using EP-2104R, a fibrin-specific contrast agent: initial results in patients. Eur Radiol 18:1995–2005
27. Sirol M, Aguinaldo JG, Graham PB, Weisskoff R, Lauffer R, Mizsei G, Chereshnev I, Fallon JT, Reis E, Fuster V, Toussaint JF, Fayad ZA (2005) Fibrin-targeted contrast agent for improvement of in vivo acute thrombus detection with magnetic resonance imaging. Atherosclerosis 182:79–85
28. Levy SG, Jacques V, Zhou KL, Kalogeropoulos S, Schumacher K, Amedio JC, Scherer JE, Witowski SR, Lombardy R, Koppetsch K (2009) Development of a multigram asymmetric synthsis of 2-(R)-2-(4,7,10-Tris tert-butylcarboxymethyl-1,4,7,10-tetraazacyclododec-1-yl)-pentanedioic acid, 1-tert-butyl ester, (R)-tert-Bu_4-DOTAGA. Org Process Res Dev 13:535–542
29. Liu S (2008) Bifunctional coupling agents for radiolabeling of biomolecules and target-specific delivery of metallic radionuclides. Adv Drug Deliv Rev 60:1347–1370
30. De Leon-Rodriguez LM, Kavacs Z (2008) The synthesis and chelation chemistry of DOTA-peptide conjugates. Bioconjug Chem 19: 391–402
31. Yoo B, Page MD (2007) Peptidyl molecular imaging contrast agents using a new solid phase peptide synthesis approach. Bioconjug Chem 18:903–911
32. Kaiser E, Colescott RL, Bossinger CD, Cook PI (1970) Color test for detection of free terminal amino groups in the solid-phase synthesis of peptides. Anal Biochem 34: 595–598
33. Vojkovsky T (1995) Detection of secondary amines on solid support. Pept Res 8:236–7
34. Karle IL, Balaram P (1990) Structural characteristics of alpha-helical peptide molecules containing Aib residues. Biochemistry 29: 6747–56
35. Lewis MR, Kao JY, Anderson AL, Shively JE, Raubitschek A (2001) An improved method for conjugating monoclonal antibodies with N-hydroxysulfosuccinimidyl DOTA. Bioconjug Chem 12:320–324

Chapter 14

Evaluation of Prenylated Peptides for Use in Cellular Imaging and Biochemical Analysis

Joshua D. Ochocki, Urule Igbavboa, W. Gibson Wood, Edgar A. Arriaga, Elizabeth V. Wattenberg, and Mark D. Distefano

Abstract

Protein prenylation involves the addition of a farnesyl (C15) or geranylgeranyl (C20) isoprenoid moiety onto the C-terminus of approximately 2 % of all mammalian proteins. This hydrophobic modification serves to direct membrane association of the protein. Due to the finding that the oncogenic protein Ras is naturally prenylated, several researchers have developed inhibitors of the prenyltransferase enzymes as cancer therapeutics. Despite numerous studies on the enzymology of prenylation in vitro, many questions remain about the process of prenylation in living cells. Using a combination of flow cytometry and confocal microscopy, we have shown that synthetic fluorescently labeled prenylated peptides enter a variety of different cell types. Additionally, using capillary electrophoresis we have shown that these peptides can be detected in minute quantities from lysates of cells treated with these peptides. This method will allow for further study of the enzymology of protein prenylation in living cells.

Key words Peptide, Lipid modification, Posttranslational modification, Prenylation, Farnesyl, Cell-penetrating peptide

1 Introduction

An important posttranslational modification that occurs on approximately 2 % of all mammalian proteins [1] is called protein prenylation. Protein prenylation involves the addition of a hydrophobic isoprenoid moiety (15 carbon farnesyl or 20 carbon geranylgeranyl) to the C-terminal cysteine of proteins that contain a "CAAX" box motif [2–4]. In this case, "A" represents an aliphatic amino acid and "X" represents a directing residue that controls whether a protein is farnesylated or geranylgeranylated [5]. There is considerable interest in this field due to the finding that several oncogenic proteins, including Ras, are naturally prenylated. Many researchers have developed potent inhibitors of the prenyltransferase enzymes

Andrew E. Nixon (ed.), *Therapeutic Peptides: Methods and Protocols*, Methods in Molecular Biology, vol. 1088,
DOI 10.1007/978-1-62703-673-3_14, © Springer Science+Business Media, LLC 2014

Farnesylated CAAX peptides:

1a: (VM)
A_1=V, A_2=I, X=M

1b: (VS)
A_1=V, A_2=L, X=S

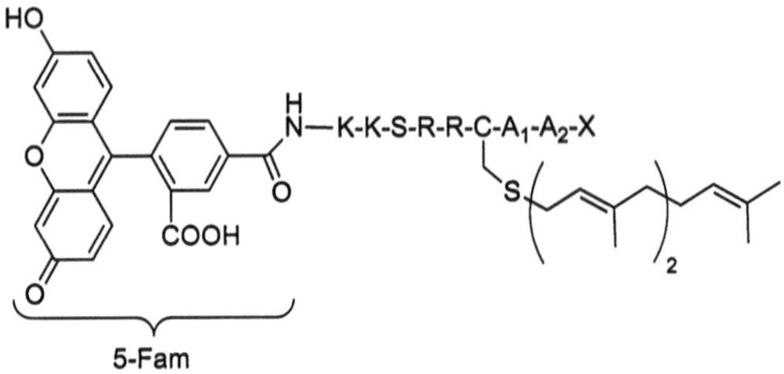

Fig. 1 Synthetic, farnesylated, and fluorescently labeled cell-penetrating peptides. For detailed synthesis of these peptides, *see* ref. 16

as anticancer therapeutics [6, 7], which have progressed to various stages of clinical trials [8–10]. In addition to studying the prenyltransferase enzymes for anticancer therapies, new light has recently been shed on the involvement of protein prenylation in several other important diseases such as Alzheimer's and Parkinson's disease [11–13]. Eckert and co-workers have shown that the levels of farnesyl and geranylgeranyl diphosphate (the isoprenoid substrates for the protein prenyltransferases) are elevated in the brains of Alzheimer's patients [14]. Additionally, Liu et al. have shown that inhibition of the farnesylated protein UCH-L1 may be a therapeutic strategy for slowing the progression of Parkinson's disease [15].

While in vitro research has contributed significantly to the understanding of protein prenylation, it is clear that an enhanced understanding of the enzymology of prenylation in vivo is required to develop enhanced therapeutics. We have synthesized farnesylated peptides that are N-terminally labeled with a fluorescein derivative (5-carboxyfluorescein, 5-Fam), based on the C-terminal sequence of the naturally prenylated protein CDC42 [16, 17]. A variety of cell types have been shown to take up these synthetic peptides, and thus allow studies in living cells to be performed. In accordance with this finding, we have recently described the synthesis of several peptides, as well as their applications in cellular studies using flow cytometry and confocal microscopy. This chapter describes the methods involved in the use of these farnesylated peptides (Fig. 1) to quantify cellular uptake and study localization patterns in living cellular systems, from both immortalized cell lines and primary neuron cells. Additionally, micellar electrokinetic capillary chromatography (MEKC) is used to detect subnanomolar levels of the peptides in lysates of cells treated with them.

2 Materials

2.1 Cell Culture

1. Dulbecco's Modified Eagle Medium (DMEM) (Invitrogen, Carlsbad, CA) supplemented with 10 % fetal bovine serum (FBS) (Intergen, Purchase, NY).
2. F12K medium (ATCC, Manassas, VA) supplemented with 10 % horse serum (HS) (Fischer Scientific, Hampton, NH) and 5 % FBS.
3. Solutions of trypsin (10×, 2.5 %) and versene (Invitrogen, Carlsbad, CA) (*see* **Notes 1** and **2**).
4. Mouse β-nerve growth factor (β-NGF) (Raybiotech, Norcross, GA) was dissolved at 100 μg/mL in phosphate-buffered saline (PBS, *see* **Note 3**) and stored in single-use aliquots at −80 °C. Working solutions are prepared by dilution to 100 ng/mL in F12K medium supplemented with 1 % HS.
5. Poly-D-lysine hydrobromide (mol. wt. > 300,000, Sigma Aldrich, St. Louis, MO) was dissolved at 0.1 mg/mL in PBS and stored in aliquots at −20 °C.

2.2 Primary Cell Isolation and Culture

1. Incubation medium: Neuronal basal medium (Gibco/BRL, Bethesda, MD) supplemented with 10 % FBS (HyClone, Ogden, UT), 1 % penicillin/streptomycin (100 U/mL), 2 mM GlutaMax (Gibco), and 10 % B-27 (Gibco) or 10 % Gem 21 (Gemini Bio Prod).
2. Dissecting medium: Same as **item 1**, but supplemented with 50 μg/mL Fungin (purchased as a 10 mg/mL solution, InvivoGen, San Diego, CA).
3. 1× TrypLE Express (Invitrogen, Carlsbad, CA).
4. 5-Fluoro-2-deoxyuridine and uridine (Sigma Aldrich, St. Louis, MO).
5. 70 μm cell strainer (BD Biosciences, Franklin Lakes, NJ).
6. 15 mL polypropylene centrifuge tubes with a conical bottom, sterilized (herein referred to as "15 mL centrifuge tube," BD Biosciences, Franklin Lakes, NJ).
7. Dissecting scissors and forceps.
8. C57BL/6J mice (The Jackson Laboratory, Bar Harbor, Maine).

2.3 Flow Cytometry

1. 12×75 mm polypropylene round bottom test tube (BD Biosciences, Franklin Lakes, NJ).
2. 15 mL centrifuge tubes.

2.4 Confocal Microscopy

1. Culture dishes (35 mm) fitted with a glass bottom (14 mm diameter, No. 1.5 coverslip) (MatTek Corporation, Ashland, MA).

2. Wheat germ agglutinin Alexa Fluor 594 conjugate plasma membrane stain (Invitrogen, Carlsbad, CA) was dissolved at 5 mg/mL in PBS and stored in aliquots at −20 °C.
3. Hoechst 34850 nuclear stain (Invitrogen, Carlsbad, CA) was dissolved at 5 mg/mL in water and stored in aliquots at −20 °C.

2.5 Capillary Electrophoresis

1. Fused silica capillary tubing (50 μm ID, 363 μm OD, 20 μm CT) (Polymicro Technologies, Phoenix, AZ).
2. Run buffer composed of 100 mM sodium dodecyl sulfate (SDS, MP Biomedicals, Solon, OH), 25 mM sodium tetraborate (Fisher Chemical, Fair Lawn, NJ), and 3 M urea (Sigma Aldrich, St. Louis, MO) (*see* **Note 4**).

3 Methods

The peptides used in this study (Fig. 1) were previously synthesized as described in ref. 16. These peptides were first evaluated for their ability to be taken up by immortalized HeLa cells using flow cytometry, which was followed by visualizing their localization inside cells using confocal microscopy. Several cell types were tested using these methods to establish the versatility of peptide uptake. Having established that these peptides were naturally cell penetrating, we sought to explore their utility in primary cultured cells using mouse neurons. Neurons were chosen due to the potential relevance of protein prenylation in Alzheimer's disease (*see* Subheading 1). Because these peptides were demonstrated to exhibit the uptake in both immortalized and primary cells, we next sought to analyze the peptides from cell lysates using a modification of capillary electrophoresis. MEKC coupled with laser-induced fluorescence (LIF) detection is used to separate components in the cell lysate and detect any fluorescent peptide present.

3.1 Preparation of Immortalized Cells

1. HeLa, NIH/3T3, D1TNC1 astrocyte, and PC-12 cells are passaged when nearing complete confluence with trypsin/versene to continue the growth of maintenance cultures. All cell lines are maintained in DMEM supplemented with 10 % FBS except PC-12, which are maintained in F12K medium supplemented with 10 % HS and 5 % FBS (*see* **Note 5**).
2. HeLa, NIH/3 T3, and D1TNC1 astrocyte cells are seeded in 35 mm glass-bottomed culture dishes (*see* **Note 6**) at 7.8×10^3 cells/cm^2 and grown to approximately 50 % confluence (typically 24 h). PC-12 cells are seeded at 3.9×10^3 cells/cm2 and grown to approximately 20 % confluence (typically 24 h). Prior to peptide incubation, PC-12 cells are differentiated into neurons (*see* **Note 7**).

3.2 Isolation and Maintenance of Primary Cells

1. Coat 35 mm glass bottom culture dishes with polylysine (*see* **Note 5**).
2. Sterilize dissecting scissors, forceps, and all metallic equipment with 95 % ethanol.
3. 17- to 18-day gestated C57BL/6J mice were asphyxiated with CO_2, cleaned with 95 % ethanol, and the stomach cut open with scissors. The uterus was removed and the embryos were transferred into a 10 cm^2 dish on ice. The head was snipped just below the ear and the opening at the bottom of the head was widened by cutting on both the left and right side with a small scissors. Carefully apply force with forceps on the frontal portion of the head to remove the brain. The brains were collected in a 15 mL centrifuge tube containing 10 mL dissecting medium on ice.
4. Under the microscope, the cerebellum and brain stem were removed while the cerebral cortices were collected for primary neuronal preparation.
5. Mince the cortices using a small scissor and suspend in TrypLE Express in a 15 mL centrifuge tube. The suspension was defragmented by pipetting up and down (two to three times) using a 1 mL pipetman pressed against the bottom of the tube. The cell suspension was placed in a 10 cm^2 dish, covered, and placed in an incubator at 37 °C with 5 % CO_2 for 7 min.
6. Neutralize the suspension with 2 mL FBS and 3 mL neuronal basal medium containing Fungin (50 μg/mL). Pipet the solution repeatedly (5–10 times) to dislodge clumps, filter through a 70 μm cell strainer, and centrifuge in a 15 mL centrifuge tube at $130 \times g$ for 5 min.
7. Aspirate the medium from the pellet and re-suspend in 10 mL neuronal basal medium containing Fungin (50 μg/mL). Seed the cells at 0.5×10^6 cells/cm^2 and place in an incubator at 37 °C with 5 % CO_2 and 80 % humidity.
8. After 1 h, remove the medium, replace it with freshly warmed incubation medium, and place back in the incubator.
9. 48 h later, replace the medium with fresh incubation medium containing 5-fluoro-2-deoxyuridine (0.75 mg/mL) and uridine (1.75 mg/mL) to inhibit any mitotic processes of astrocytes or glia. Incubate with this medium for an additional 48 h.
10. Replace the medium with fresh incubation medium and replace every 48 h until the primary cells are ready to use.

3.3 Analyzing Cells Using Flow Cytometry

1. After reaching appropriate confluency, the cells are rinsed with PBS (2×, 1 mL each) and serum-free DMEM is added (or F12K with 1 % HS in the case of PC-12).

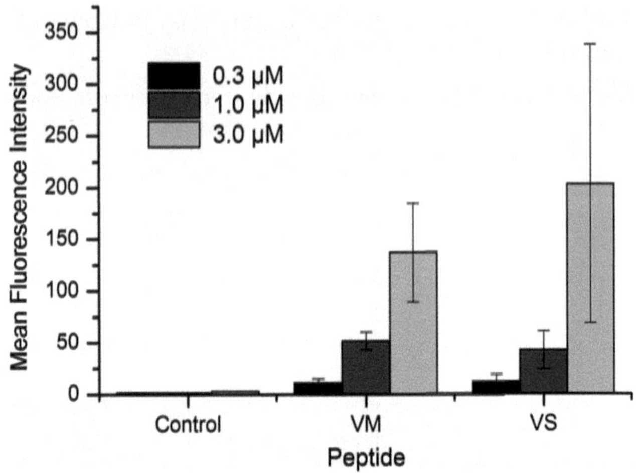

Fig. 2 Flow cytometry data for two farnesylated peptides incubated with HeLa cells at 37 °C for 1 h at varying concentrations. The "control" in this case consists of HeLa cells that were not treated with peptide. After peptide incubation, cells were trypsinized for 15 min to remove membrane-bound peptide. Each bar represents the geometric mean fluorescence intensity of 10,000 cells. Experiments performed in triplicate with the results expressed as the mean fluorescence intensity ± standard deviation

2. The farnesylated peptide is added to the culture medium to reach the desired concentration, generally between 0.3 and 3 μM.
3. Incubate the cells with the peptide for 1 h at 37 °C and 5 % CO_2, then rinse the cells twice with PBS, and trypsinize for 15 min (0.2 mL of trypsin/versene solution, *see* **Note 2**). Trypsinization normally proceeds for 5 min; however, a 15-min time is used in this case to remove any peptide bound to the outer cell membrane and thus allow measurement of only internalized peptide.
4. Add 1.8 mL complete media to the cells and transfer to a 15 mL centrifuge tube. Centrifuge cells at 100 × *g* for 5 min at room temperature.
5. Aspirate medium from cell pellet and re-suspend cells in 2 mL PBS. Transfer cells to a 12 × 75 mm round-bottom test tube for flow cytometry analysis.
6. The mean fluorescence intensity of 10,000 cells is counted using a BD FACS Calibur flow cytometer. Typical results of this analysis are shown in Fig. 2.

3.4 Localization Studies with Confocal Microscopy

1. Once the cells have reached confluence in glass-bottom dishes, rinse the cells with PBS (2×, 1 mL each) and add serum-free DMEM (or F12K with 1 % HS in the case of PC-12).

2. Add the desired peptide to the culture medium to reach the desired concentration, generally either 1 or 3 μM, and incubate at 37 °C and 5 % CO_2 for 2 h (*see* **Note 8**).
3. During the final 20 min of incubation, add Hoechst 34580 nuclear dye to a final concentration of 1 μg/mL. Additionally, for the final 10 min of incubation, add wheat germ agglutinin Alexa Fluor 594 plasma membrane stain to a final concentration of 5 μg/mL.
4. Rinse the cells twice with PBS (1 mL each) and place cells back in serum-free DMEM (or F12K with 1 % HS in the case of PC-12).
5. The cells are imaged live using an Olympus FluoView 1000 inverted confocal microscope with a 60× objective. Excitation at 405 nm induces fluorescence of Hoechst nuclear dye (emission 461 nm), excitation at 488 nm induces peptide fluorescence (from 5-Fam, emission 519 nm), and excitation at 543 nm induces wheat germ-Alexa Fluor 594 plasma membrane fluorescence (emission 618 nm). Collection of all fluorescence channels occurs simultaneously using the confocal laser scanning microscope (*see* Figs. 3 and 4 for sample pictures).

3.5 Micellar Electrokinetic Capillary Electrophoresis

1. Before performing MEKC, the sample must first be prepared. This sample consists of HeLa cells that have been treated with the *VS* peptide and lysed.
2. Grow and treat HeLa cells with peptide following the steps in Subheading 3.3, **1–6**.
3. Aspirate the medium from the pellet, re-suspend the cell pellet in 2 mL PBS, and centrifuge at 100 × *g* for 5 min again (to rinse the pellet and remove all medium).
4. Add 150 μL of MEKC run buffer (100 mM SDS, 25 mM borate, 3 M urea, pH 9.3) to the cell pellet to immediately lyse the cells. The sample will become thick and viscous. This sample is frozen at −80 °C until analysis with MEKC.
5. Cut a 35 cm length of fused silica capillary and burn a detection window approximately 12 cm from one end of the capillary.
6. Feed capillary through the detector end of the capillary cartridge (Beckman) and position the detector window in the opening for the detector; replace white plastic clips with red seals for coolant flow.
7. Using the metal template (provided with instrument), cut the capillary to the correct length and attach the detector pieces for the LIF detector.
8. Insert capillary and cartridge into the instrument and condition the capillary (*see* **Note 9**).

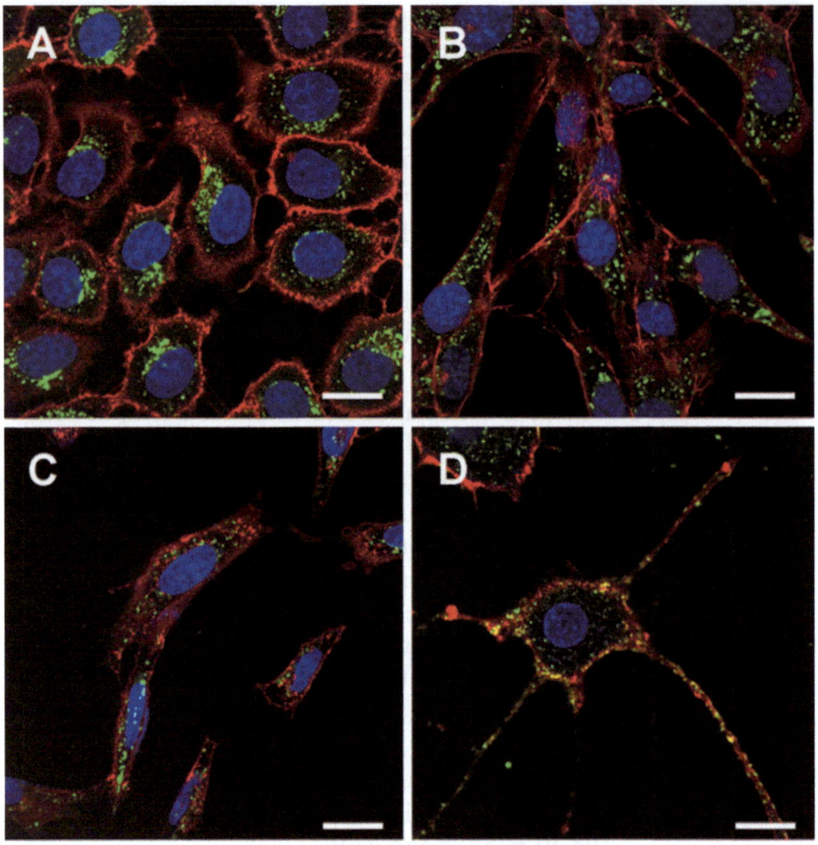

Fig. 3 Confocal microscopy images of different live immortalized cells after incubation with peptide VS at either 1 or 3 μM for 2 h. (**a**) HeLa cells treated at 3 μM. (**b**) NIH/3T3 fibroblasts treated at 1 μM. (**c**) D1TNC1 astrocytes treated at 1 μM. (**d**) PC-12 differentiated neurons treated at 1 μM. Hoechst 34850 was used to stain the nucleus *blue* and wheat germ agglutinin Alexa Fluor 594 conjugate was used to stain the plasma membrane *red*. The peptide is visualized as *green* fluorescence. Size bar represents 20 μm

9. After capillary conditioning, it is ready for use. Set up the sample(s) in the sample tray and run. For each sample, injection occurs at a pressure of 0.3 psi for 5 s and separation occurs for 12 min at 12.0 kV.
10. The data is exported as an ascii file and graphed using Origin graphing software (*see* Fig. 5 for typical data that is obtained).

4 Notes

1. Unless stated otherwise, all solutions should be prepared in water that has a resistivity of at least 15 MΩ cm. This standard is referred to as "water" in this text.
2. The trypsin (10×) and versene were purchased separately from Invitrogen and stored at −20 and 4 °C, respectively. Prior to

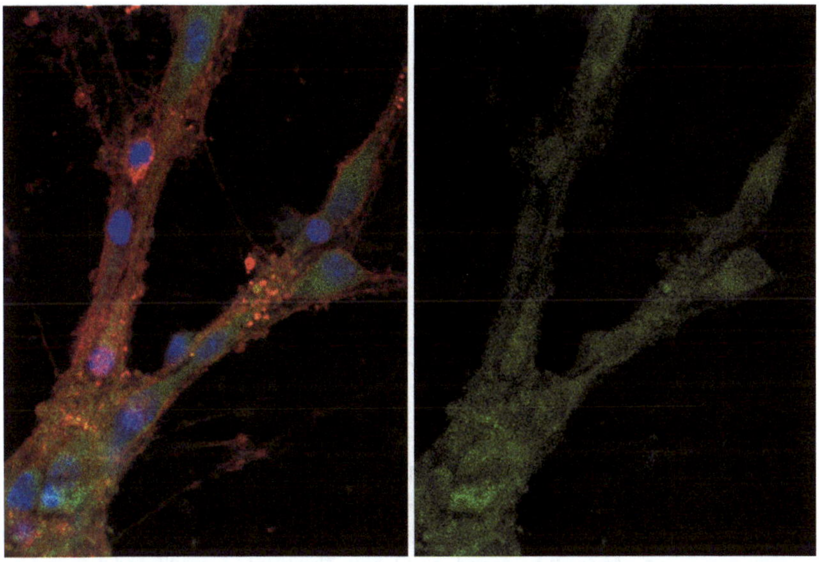

Fig. 4 Confocal microscopy images of primary mouse neurons after incubation with peptide VS at 1 μM for 2 h. The *left panel* is an overlay of all fluorescence channels, while the *right panel* represents only the *green* channel for clarity. Hoechst 34850 was used to stain the nucleus *blue* and wheat germ agglutinin Alexa Fluor 594 conjugate was used to stain the plasma membrane *red*. The peptide is visualized as *green* fluorescence. Size bar represents 20 μm

use, trypsin was added to versene to reach a final concentration of 0.063 % trypsin.

3. PBS is composed of 137 mM NaCl, 2.7 mM KCl, 4.3 mM Na_2HPO_4, and 1.47 mM of KH_2PO_4. The solution is pH adjusted to 7.4, autoclaved, and stored at 4 °C for future use.
4. After preparing the capillary electrophoresis run buffer at the specified concentration, the pH is adjusted to 9.3 and the buffer is filtered through a 0.2 μm PTFE acrodisc syringe filter (Sigma Aldrich, St. Louis, MO).
5. PC-12 cells are maintained and seeded on polylysine-coated culture dishes. To coat the dishes, a thin layer (enough solution to completely coat the bottom of the plate) of a 0.1 mg/mL poly-D-lysine hydrobromide solution is added to the plates and allowed to sit at RT overnight. The following day, the plates are rinsed twice with PBS and are ready for use.
6. Glass bottom culture dishes (MatTek) are used if confocal microscopy will be performed. If flow cytometry is desired, regular plastic 35 mm culture dishes are used.
7. After seeding the PC-12 cells to 20 % confluence in 35 mm glass-bottom dishes, the cells are serum starved for 24 h with F12K+1 % HS. Addition of F12K medium containing 1 % HS and 100 ng/mL of β-NGF initiated neuron differentiation.

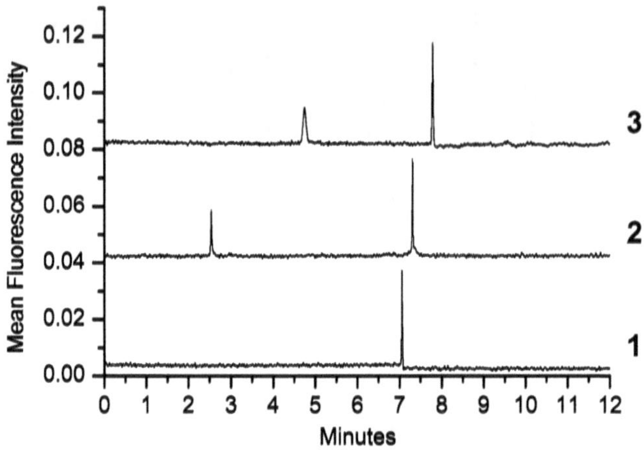

Fig. 5 Micellar electrokinetic capillary chromatography (MEKC) of HeLa lysate samples. *Electropherogram 1* represents peptide VS injected alone at 25 nM as a standard. In *electropherogram 2*, HeLa cells were treated with peptide VS at 1 μM for 2 h, rinsed, and lysed. The lysate was diluted 1:2 and analyzed with MEKC. *Electropherogram 3* represents the same lysate (also diluted 1:2) that has been spiked with the peptide standard (VS) at a concentration of 25 nM. The co-migration of the large peak at approximately 7.5 min verifies the presence of the VS peptide in the cell lysate. The peak from the lysate at 2.5 min in *electropherogram 2* and 4.6 min in *electropherogram 3* represents a possible proteolytic fragment of peptide VS (this peak is not present in a lysate sample not treated with peptide (data not shown)). The average tR_M of the peptide peak is as follows: *Electropherogram 1*: 7.46 ± 0.29 min. *Electropherogram 2*: 7.22 ± 0.45 min. *Electropherogram 3*: 7.48 ± 0.46 min (standard deviation represents at least three trials)

Incubation with β-NGF proceeded for 6–8 days until maximum neurite growth was observed; the medium was replaced every 2 days. Once the cells begin differentiating, they are quite fragile and easily disturbed. When giving cells fresh medium, leave a small amount of the previous medium behind so that the cells do not become exposed to the air as they will rapidly dry out.

8. We have discovered that longer peptide incubation times lead to increased cellular uptake. In order to study the peptide localization with microscopy, 2-h peptide incubation is used to increase the uptake.
9. To condition the capillary, rinse at 20 psi for 8 min with the following solutions, in order: 0.1 M HCl, water, 0.1 M NaOH, and run buffer.

References

1. Epstein WW, Lever D, Leining LM, Bruenger E, Rilling HC (1991) Quantitation of prenylcysteines by a selective cleavage reaction. Proc Natl Acad Sci U S A 88:9668–9670
2. Benetka W, Koranda M, Eisenhaber F (2006) Protein prenylation: an (almost) comprehensive overview on discovery history, enzymology, and significance in physiology and disease. Monatsh Chem 137:1241–1281
3. Casey PJ (1992) Biochemistry of protein prenylation. J Lipid Res 33:1731–1740
4. Clarke S (1992) Protein isoprenylation and methylation at carboxyl-terminal cysteine residues. Annu Rev Biochem 61:355–386
5. Zhang FL, Casey PJ (1996) Protein prenylation: molecular mechanisms and functional consequences. Annu Rev Biochem 65:241–269
6. Agrawal AG, Somani RR (2009) Farnesyltransferase inhibitor as anticancer agent. Mini-Rev Med Chem 9:638–652
7. Sousa SF, Fernandes PA, Ramos MJ (2008) Farnesyltransferase inhibitors: a detailed chemical view on an elusive biological problem. Curr Med Chem 15:1478–1492
8. Blum R, Kloog Y (2005) Tailoring Ras-pathway-inhibitor combinations for cancer therapy. Drug Resist Updates 8:369–380
9. Karp JE, Lancet JE (2007) Development of farnesyltransferase inhibitors for clinical cancer therapy: focus on hematologic malignancies. Cancer Invest 25:484–494
10. Zhu K, Hamilton AD, Sebti SM (2003) Farnesyltransferase inhibitors as anticancer agents: current status. Curr Opin Investig Drugs 4:1428–1435
11. Cole SL, Vassar R (2006) Isoprenoids and Alzheimer's disease: a complex relationship. Neurobiol Dis 22:209–222
12. Florent-Bechard S, Desbene C, Garcia P, Allouche A, Youssef I, Escanye M-C et al (2009) The essential role of lipids in Alzheimer's disease. Biochimie 91:804–809
13. Ostrowski SM, Wilkinson BL, Golde TE, Landreth G (2007) Statins reduce amyloid-beta production through inhibition of protein isoprenylation. J Biol Chem 282: 26832–26844
14. Eckert GP, Hooff GP, Strandjord DM, Igbavboa U, Volmer DA, Muller WE et al (2009) Regulation of the brain isoprenoids farnesyl- and geranylgeranylpyrophosphate is altered in male Alzheimer patients. Neurobiol Dis 35:251–257
15. Liu Z, Meray RK, Grammatopoulos TN, Fredenburg RA, Cookson MR, Liu Y et al (2009) Membrane-associated farnesylated UCH-L1 promotes alpha-synuclein neurotoxicity and is a therapeutic target for Parkinson's disease. Proc Natl Acad Sci U S A 106: 4635–4640
16. Ochocki JD, Igbavboa U, Wood WG, Wattenberg EV, Distefano MD (2010) Enlarging the scope of cell-penetrating prenylated peptides to include farnesylated CAAX box sequences and diverse cell types. Chem Biol Drug Des 76:107–115
17. Wollack JW, Zeliadt NA, Mullen DG, Amundson G, Geier S, Falkum S et al (2009) Multifunctional prenylated peptides for live cell analysis. J Am Chem Soc 131:7293–7303

References

Chapter 15

Ultraviolet Absorption Spectroscopy of Peptides

Mangala R. Liyanage, Kunal Bakshi, David B. Volkin, and C. Russell Middaugh

Abstract

UV absorption spectroscopy is commonly used with peptides for determining concentration and enzyme activity, but high-resolution UV spectra can also provide information on peptide secondary and tertiary structure and association behavior. New developments using temperature- and cation-dependent high-resolution second derivative absorption methods can also provide information concerning peptide dynamics. Data from several low-resolution spectroscopic techniques, including UV absorption, can be combined to generate an overall picture of peptide structure as a function of environmental conditions.

Key words Ultraviolet, UV absorption, Near UV, Far UV, Peptide, Conformational integrity, Stability

1 Introduction

1.1 Spectroscopic Characterization of the Conformational Integrity and Stability of Peptides

Although peptides are commonly characterized analytically by their primary sequence, conformational aspects of their solution behavior (e.g., a peptide's secondary and tertiary structures) often govern many of their functional and stability properties. Thus, the experimental determination of various aspects of higher order structural information is vital to the elucidation of structure–activity relationships of peptides. High-resolution structure determination techniques such as NMR, X-ray crystallography, and molecular dynamic (MD) simulations have been widely used for this purpose. Certain practical limitations, however, such as the need for large sample quantities, irrelevant analysis conditions, and complex data analysis can limit the general utility of these high-resolution methods. To circumvent these problems, a number of alternative, lower resolution spectroscopic techniques are available to provide complementary analyses. Advantages of these lower resolution techniques include their ready availability and their direct applications

Mangala R. Liyanage and Kunal Bakshi have contributed equally to this manuscript.

Andrew E. Nixon (ed.), *Therapeutic Peptides: Methods and Protocols*, Methods in Molecular Biology, vol. 1088, DOI 10.1007/978-1-62703-673-3_15, © Springer Science+Business Media, LLC 2014

to a wide variety of environmental perturbations such as temperature, pH, and ionic strength. These useful spectroscopic methods include UV-visible absorbance and fluorescence spectroscopy, far- and near-UV circular dichroism (CD), and infrared methods such as Fourier transform infrared (FTIR) spectroscopy. This chapter is the first in a series of four chapters that describe each of these techniques as applied to the characterization of peptides.

Secondary structural analysis of peptides (and other biomolecules such as larger proteins) has commonly relied on far-UV CD and FTIR techniques while information pertaining to tertiary structure and more subtle structural alterations has often been obtained using fluorescence, UV absorbance, and near-UV CD spectroscopies. In addition to secondary structure and more subtle aspects of conformational behavior, peptide association (aggregation) and dissociation can also be monitored using complementary data available during data acquisition by these techniques. For example, fluorescence (light scattering), UV absorbance (optical density measurements in non-absorbing spectral regions), and FTIR (intermolecular β-structure formation) can all be used for this purpose.

Despite the lower resolution of these methods, they have emerged as important sources of higher order structural information about peptides, especially when they are combined into self-consistent pictures of structure as a function of environmental conditions (e.g., temperature and solution pH). For example, thermal and pH perturbation studies performed using these techniques have been combined to create multidimensional vector representations (empirical phase diagrams, EPDs) of the physical states of peptides. As discussed in more detail in Chapter 18 in this series, these EPDs can provide a comprehensive, yet practical and useful, picture of peptide structure and behavior in solution with a relatively small investment of time and money.

1.2 Ultraviolet Absorption Spectroscopy

Ultraviolet absorption spectrophotometers are one of the most versatile and ubiquitous analytical instruments in the modern protein and peptide chemistry laboratories. Despite the widespread use of UV absorption spectroscopy (UV-AS) for simple quantitative applications such as concentration determination and enzyme assays, the full potential of this analytical technique is often not exploited. Measurement of a high-resolution protein/peptide UV absorption/optical density spectrum (190–400 nm) can be acquired in a very short period of time (seconds to minutes) and is capable of providing simultaneous information on peptide secondary structure (180–220 nm), tertiary structure (240–300 nm), and association (aggregation)/dissociation behavior (light scattering above 300 nm). Minimal or no sample preparation, rapid spectral acquisition, low cost, and robust and readily available instrumentation all make UV-AS an ideal technique for studying various aspects of peptide structure and behavior.

2 Materials

1. A research-grade UV-visible spectrophotometer. Instruments can be obtained from suppliers such as Agilent (Santa Clara, CA); Jasco Inc. (Easton, MD); Varian Inc. (Palo Alto, CA); and Perkin Elmer (Waltham, MA), and are available in a variety of configurations for different wavelength ranges and sensitivity. For applications with peptide samples, UV spectrophotometers are typically equipped with both (a) a combination of deuterium and tungsten lamps (wavelength range 190–800 nm) as a light source, and (b) a photomultiplier tube (PMT) or diode array-based detector. Other choices include single/dual-beam designs and scanning, high-throughput microplate reader spectrophotometers. For high-resolution work, a diode array instrument is often preferred as discussed below.
2. A sample container for measurement (a "cuvette") is chosen based on sample volume, path length, and concentration. Quartz or quarasil cuvettes are used in the UV region for transparency and are available in 10–0.01 mm path lengths with corresponding volumes. Black-walled, matched, variable path length, and disposable cuvettes are also available. For high-throughput microplate spectrometers, 6- to 1,024-well microplates and strip wells are available in both UV-opaque (glass or plastic) or UV-transparent (e.g., quartz) forms. The path length in a microplate is usually determined by sample height and may change between samples. Most modern microplate spectrometers, however, provide a software feature to correct the effective path length to 1 cm.

3 Methods

3.1 Preparation of Instrument and Sample Cuvettes

1. Turn on the spectrophotometer and light sources. Light sources should be allowed to warm up and stabilize for at least 30 min prior to measurements.
2. Turn on the temperature controller and set the desired temperature.
3. The cuvettes must be thoroughly cleaned prior to use by soaking in a strong acid and/or base or a detergent solution and thoroughly rinsed with copious amounts of distilled water. Cuvettes should then be allowed to dry in an oven or air-dried prior to use. This cuvette cleaning step is especially important for peptide samples due to the possibility of nonspecific adsorption of peptides to the inner surface of the cuvette.
4. Always carefully wipe off the outside of cuvette windows using a non-scratching material to remove any residual liquid and

dust to clear the optical faces of any fingerprint smudges or stains prior to inserting the cuvette into the sample holder.

5. It is a good practice to run a wavelength scan without anything in the sample chamber, to test for any instrumentation errors. The spectrophotometer should then be zeroed with the sample buffer in the cuvette under identical conditions to those of the sample to be examined.

3.2 Sample Preparation

1. For experimental purposes, it is preferred to have a sample concentration such that the absorbance is in the range of 0.05–1.0. Although modern instruments sometimes permit absorbance as high as 3–4 to the quantitatively examined, this possibility should be carefully tested. A concentration-dependent study using serial dilutions of the peptide sample is initially recommended to ensure that the absorbance falls within the linear range of the instrument under the specific conditions employed (*see* **Notes 1** and **2**).
2. Aqueous peptide samples can be prepared by their dissolution or dilution in water or buffer containing aqueous solution. Water is a good solvent for UV spectroscopy because it has no significant absorbance far into the UV region. Peptide samples can also be prepared in a wide variety of buffers to better control the solution pH. A preferred buffer should not itself absorb in the UV wavelength range of interest. Buffers that are well suited for this purpose include tris, citrate, imidazole, and phosphate.
3. The optical density and absorbance of the peptide samples should ideally be minimal above 320 nm. An optical density significantly above zero in this region is usually indicative of light scattering effects due to the presence of higher order species (e.g., aggregates, multimers). Light scattering effects can be minimized by centrifugation or filtration of the sample beforehand to remove impurities like precipitates. Alternatively, by extrapolating log/log plots of optical density (O.D.) vs. wavelength from non-absorbing regions into the near-UV region and subsequent subtraction, the effects of light scattering can also be corrected for mathematically. Most modern spectrophotometers contain built-in algorithms for this purpose. Light scattering effects can also be minimized by the use of derivative analysis (*see* **Note 3**).

3.3 Data Acquisition

1. The sample should be loaded into a clean and dry cuvette, and filled with a volume sufficient to cover the height of the aperture through which the light beam passes. Care should be taken to make sure that no air bubbles or meniscus are present in the light path. The cuvette should be capped or sealed especially during thermal experiments to minimize sample evaporation.

Microtiter plates can be covered with a variety of filmlike materials for the same purpose.

2. Modern UV spectrophotometers permit acquisition to be performed in a variety of different modes such as spectral acquisition (wavelength scanning), quantitation (concentration dependence), kinetic (time mode) analysis, or thermal perturbation. These considerations vary sufficiently from one instrument to another that the reader is referred to the individual operating manuals.
3. Depending on the information sought, one can choose the experiment to be performed. For example, (a) at fixed wavelength for concentration determination, (b) high-resolution UV spectral measurements incorporating derivative analysis to provide information on major or subtle changes in peptide conformation and/or dynamics, or (c) turbidity measurements in a kinetic (time based) mode to study peptide aggregation behavior. Typical parameters to be adjusted for a scanning spectrometer include wavelength, slit widths, and temperature. For peptides, a wavelength range of 190–400 nm (or higher) is convenient at 0.1–0.5 nm increments and 1 second or more integration times. Longer integration times (e.g., 30 s) may be used if the signal-to-noise ratio is poor. Note that longer exposure times to UV light may lead to photodegradation in some samples. In the case of diode array instruments, the entire spectrum is obtained simultaneously, so the major variable is the total time of acquisition. Acquisition at fixed wavelengths in dispersive instruments can also be chosen for applications like concentration determination, enzyme activity assays, and/or time-dependent aggregation studies. For thermal perturbation studies in which the temperature is increased in a stepwise manner, it is a good idea to let the sample equilibrate at each temperature increment before collecting spectra to ensure that equilibrium has been reached. This temperature equilibration can be checked for by the constancy of the absorbance value with time. We find that increasing the temperature by 2–3 °C intervals is usually satisfactory for these types of measurements.
4. A diode array instrument is particularly useful for peptide structural studies due to its high wavelength reproducibility and precision (since no moving parts are involved), rapid acquisition times (generally less than 1 s), and high precision of the absorbance value at the individual detector diodes (typically with 0.5–1.0 nm spacing). The high precision and reproducibility of diode array instruments permit even minor shifts in peak positions or peak intensities to be investigated with confidence. In addition, fast acquisition rates also facilitate kinetic studies. Because the entire spectrum is obtained during a diode array experiment, the wavelength range (or fixed wavelength) to be

studied is selected after data acquisition. The integration time generally used ranges from 0.5 to 20 s, and can be adjusted to increase or decrease the signal-to-noise ratio.

3.4 Data Analysis and Interpretation

3.4.1 Zero-Order Data

The raw data collected from a typical UV spectrometer is a plot of degree of absorption (AU) or optical density (OD) on the *Y*-axis versus wavelength range on the *X*-axis. This is also referred to as a zero-order spectrum. The OD is the sum of the absorption and other extinction processes such as light scattering, and OD values should be so indicated when light scattering effects are present. The extinction coefficient is sometimes used in place of absorbance or optical density.

Peptide absorbance in UV spectra originates largely from three sources: the peptide bond, aromatic amino acid side chains, and prosthetic groups and metals. The strongest of the three sources is usually the backbone peptide bond, which absorbs in the far-UV range (180–220 nm) with a maximum near 190 nm ($\pi \rightarrow \pi^*$) and a weaker shoulder at 220 nm ($n \rightarrow \pi^*$). The side chains of Asp, Glu, Asn, Gln, Arg, and His also absorb weakly in the far-UV region, but most peptide spectra are dominated by the more intense peptide bond absorbance (*see* **Note 4**). Interpretation of data in the far-UV region, however, can be complicated because of possible interference from commonly used buffer components, because of the presence of dissolved oxygen below 200 nm, and by different amino acid side chains all absorbing in the same wavelength range. Purging the spectrophotometer with nitrogen can alleviate the oxygen interference problem and careful choice of other solutes can minimize their contribution, but the use of circular dichroism to examine the same electronic transitions with better resolution has essentially relegated far-UV absorption measurements for secondary structure analysis to primarily historical interest.

The absorption bands in the near-UV region (250–300 nm) originating from the aromatic side chains of tryptophan, tryosine, and phenylalanine are the most commonly used signals in peptide studies. The most intense absorption band in the near-UV region is from the indole group of tryptophan and is centered near 280 nm. A comparatively weaker absorbance at 276 nm is observed due to the phenol side chain of tyrosine with shoulders at about 267 and 280 nm. The weakest absorbance results from the benzene side chain of phenylalanine in the 250–270 nm region and manifests significant vibrational structure. Some representative examples of zero-order near-UV adsorption spectra from three model peptides are shown in Fig. 1. In this example, three different peptides that differ in their content of aromatic amino acid residues display different zero-order near-UV adsorption spectra (angiotensin I contains one Tyr and Phe, α-melanocyte-stimulating hormone peptide contains one Trp, Tyr, and Phe, while substance P contains two Phe residues; *see* Fig. 1a, c, e, respectively). These

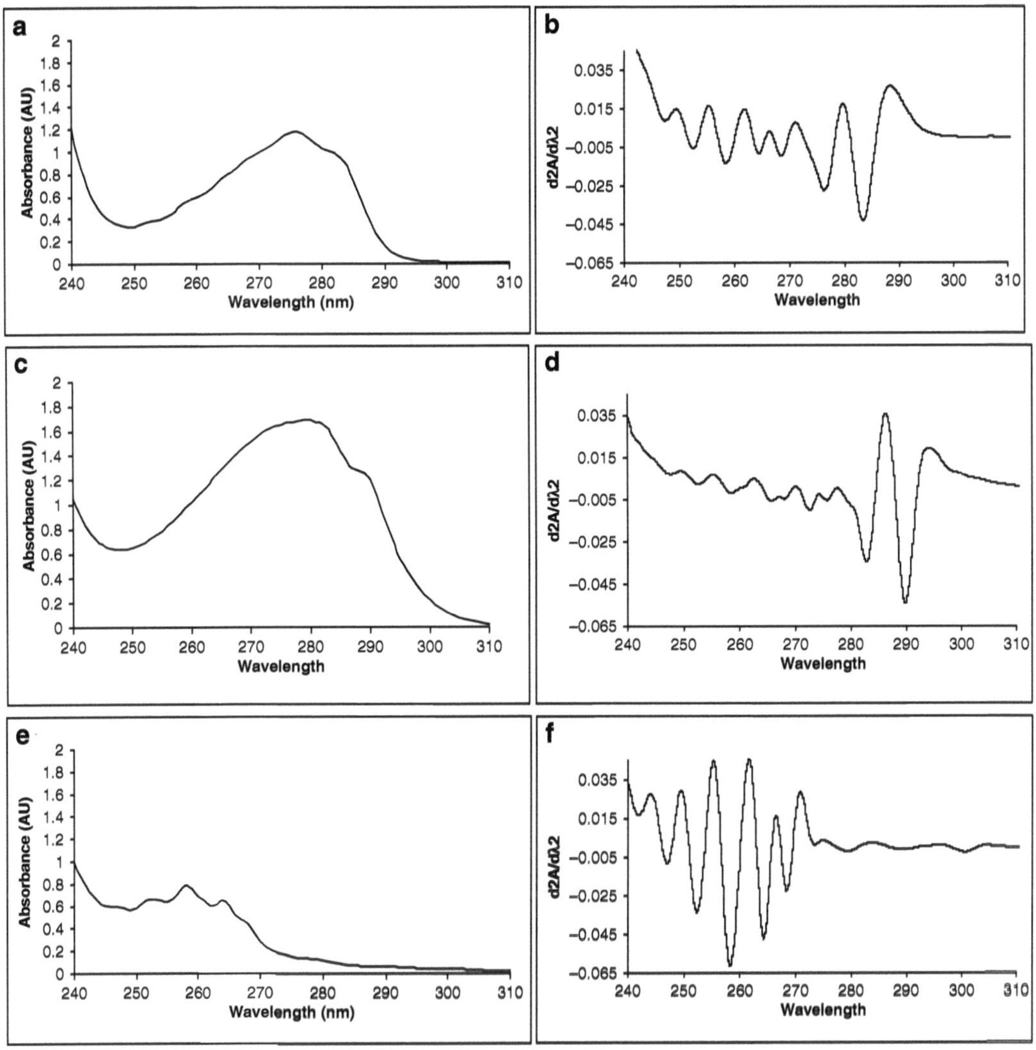

Fig. 1 Zero-order and second-derivative near-UV absorption spectra of three model peptides. (**a**, **b**) Angiotensin I contains single Tyr and Phe residues and displays a zero-order (**a**) and second-derivative (**b**) spectrum showing both the Phe peaks and two additional peaks from Tyr ca. 276.6 and 283.7 nm. (**c**, **d**) α-Melanocyte-stimulating hormone (α-MSH) contains one each of all three aromatic residues and displays a zero-order (**c**) and second-derivative (**d**) spectrum showing both the Phe and Tyr peaks as well as strong absorbance peaks from Trp at 280 and 290 nm. Note that peaks from Trp and Tyr overlap ca. 284 nm, and therefore, this is a combination peak to which both amino acid residues contribute. Similarly, some overlap of Phe and Tyr peaks is seen in the 260 nm region. (**e**, **f**) Substance P contains two Phe residues and displays a distinctive zero-order (**e**) and second-derivative (**f**) spectrum which includes absorption between 245 and 270 nm and significant vibrational fine structure, with peaks at 247.3, 256.6, 264.4, and 268.5 nm

aromatic side chain absorbance bands are very sensitive to their surrounding microenvironment and thus are potentially capable of reporting even subtle structural perturbations of changes in their environment. The exception will be peptides in which these side chains are completely exposed to solvent and lack any significant intramolecular contacts. In experimentally acquired UV spectra (zero-order spectra) of peptides, however, these bands are often present as a single poorly resolved, broad-absorption peak in the 250–300 nm range and thus are commonly used for relatively simple applications like concentration determination, enzymatic assays, or detection of major structural changes (which are less likely to be observed in small peptides).

3.4.2 Concentration Determination

For peptide concentration determinations, absorbance at or near 280 nm is used. According to the Beer–Lambert law (Eq. 1), absorbance is directly proportional to path length (l), concentration (c), and extinction coefficient (ε), permitting any one parameter to be determined with knowledge of other two:

$$A = \varepsilon c\, l \tag{1}$$

The peptide extinction coefficient can be experimentally determined using a variety of methods such as dry weight or amino acid analysis. It can also be theoretically calculated with a high degree of accuracy based on the known extinction coefficients of the absorbing residues present [1–6]. If higher order structure is present in a peptide, an average extinction coefficient determined from an analysis of proteins of known extinction coefficient can be used (*see* **Note 5**). Corrections may be required for light scattering effects if present (*see* **Note 6**).

3.4.3 Derivatization

In the case of smaller peptides, where often only one or two such aromatic side chains are present, the near-UV absorbance bands may be well resolved even in zero-order spectra. In such cases, raw data plots can be used to study peptide structure perturbations. In peptides that contain multiple aromatic residues, these overlapping contributions can be at least partially resolved by calculating higher order (first through fourth) derivatives of the zero-order UV absorbance spectrum. Derivatization is the most common data processing procedure used to analyze raw UV spectra. Taking the derivative of a UV absorption spectrum is the simple process of calculating the slope or the gradient of an absorption band. This gradient ($dA/d\lambda$) is plotted against wavelength (λ) to produce a first-order derivative plot. An iteration of this process leads to higher order derivatives ($d^nA/d\lambda^n$). A popular method to calculate derivatives uses the Savitzky–Golay method, in which a window of data filter points is chosen and a polynomial is fitted using a least squares method. This procedure can be used to smooth the data and thus partially compensate for the loss of signal-to-noise ratio during the derivatization process. The number of data points and the order of polynomial can be varied to control the degree of smoothing.

3.4.4 Second Derivatives

We generally find that the second derivative is the most useful spectral form since peaks continue to be peaks although they are now present in a negative direction. Typically, for second derivative calculations, a 9-point data filter and a fifth-order Savitzky–Golay polynomial work well. To further enhance resolution, splining of the raw derivative spectra can be performed in which *n* points are interpolated between two consecutive data point for further smoothing. This is especially useful for diode array-derived spectra in which the diode array itself sets the spacing of the raw data points. This process can result in resolution as high as 0.01 nm under optimal conditions. Some representative examples of second-derivative near-UV adsorption spectra from three model peptides are shown in Fig. 1. In this example, three different peptides that differ in their content of aromatic amino acid residues display different second-derivative near-UV adsorption spectra (angiotensin I contains one Tyr and Phe, α-melanocyte-stimulating hormone contains one Trp, Tyr, and Phe, while substance P contains two Phe residues; *see* Fig. 1b, d, f, respectively).

3.4.5 Structural Changes

The resultant second (or other)-derivative spectra can be used to sensitively monitor changes in peak position and peak intensity ratios. Changes in these values as a function of various external variables (e.g., temperature, pH, ionic strength, peptide, and solute concentration) can be used to monitor peptide structural alterations. In general, shifts to lower wavelengths (blue shifts) correspond to an increase in polarity of aromatic side chain microenvironments while shifts to higher wavelengths suggest that such side chains are moving to less polar regions (note that this effect is in the opposite direction observed for intrinsic fluorescence spectra for the same amino acid side chain residues; see below). These trends can potentially be used to monitor subtle to large peptide conformational changes. Derivative peak intensities can also be measured and can be directly correlated to the number of phenylalanine, tyrosine, and tryptophan residues in the peptides. Derivative peak positions are relatively insensitive to broad underlying spectral components and thus can at least partially obviate the need for minor light scattering corrections.

3.4.6 Cation–Pi Probing of Peptide Dynamics

Another application of UV absorption spectroscopy applied to the study of peptides involves a methodology to examine protein structure and dynamics using second-derivative peak shifts of aromatic residues upon interaction with cations of varying sizes (e.g., Li^+, Na^+, Cs^+). This method is based on the observation that cation–pi interactions between aromatic side chains and certain metals can produce changes in the magnitude and direction of peak positions in aromatics which are dependent on protein charge and the solvent accessibility of the aromatic groups. In general, smaller cations ($Li^+ < Na^+ < Cs^+$ based on their unhydrated radii and $Cs^+ < Na^+ < Li^+$ based on their hydrated size) can more easily diffuse into occluded sites, producing larger shifts. Thus, this method has the potential to probe dynamic aspects of peptide structure although simple interpretation of such data is complicated by the possibility of specific interaction between cations and other sites on the peptides [7].

3.4.7 Temperature-Induced Peak Shifts as a Probe of Protein Dynamics

Temperature-dependent shifts in the position of second-derivative peaks of aromatic residues can also be used to probe peptide dynamics. This approach is based on the sensitivity of aromatic peaks to the dielectric properties of the surrounding microenvironment and solvent penetration effects due to protein dynamic motions. Experimental evidence suggests a direct correlation between the hydration of buried aromatic residues and the slope of a plot of peak shifts versus temperature [8]. Multiple signals are available for each of the aromatic side chains, making this potentially a very powerful technique.

3.4.8 Peptide Aggregation

Besides providing secondary and tertiary structure information concerning peptides, an optical density spectrum is also capable of providing significant information on protein association/dissociation phenomena. This application of UV spectroscopy is especially useful because peptide aggregates may scatter significant light that can be independently detected above 320 nm, a region where the intrinsic chromophores of proteins and peptides do not show any absorbance (unless they possess non-peptide chromophores). The scattering of light of wavelength less than 1/20 of the scattered light by peptide aggregate species can lead to an increase in optical density that is inversely proportional to approximately the fourth power of the incident wavelength light. Larger particles display a more complex dependence of scattering on their size. Increasing optical density in the 320–500 nm region is often used as an indication of peptide aggregation.

3.4.9 Kinetic-Based Studies

Studies of peptide aggregation using kinetic-based turbidity measurements are both useful and informative as well as adaptable to high-throughput methodologies. The shape of kinetic-based turbidity curves can provide insight into the underlying mechanism of aggregation. Lag times, rate of aggregation, as well as the extent of the aggregation can all be used in this regard [9–13] (*see* **Note 7**). Turbidity studies using absorbance/optical density-based microplate readers are commonly used to identify stabilizing excipients (e.g., inhibitors of aggregation) during preformulation studies of peptide-based pharmaceuticals.

4 Notes

1. Calibration of UV-visible spectrophotometers should be performed regularly.
2. For low-concentration samples, peptides can be lost from solution by adsorption to cuvette walls. This loss can sometimes be minimized by using high-ionic-strength solutions, or the addition of small quantities (~0.01 %) of either a nonionic detergent like Brij 35 or an inert polymer, which competes for surface adsorption sites.
3. As the size of particles in a sample becomes larger, light scattering corrections may not be sufficient. Absorption flattening is a phenomenon in which high local concentrations of peptides (due to their aggregation) create an inhomogeneous distribution of sample chromophores in the light path. Absorption flattening typically leads to reduced signal intensities and red shifts in absorption bands and may be erroneously interpreted as peptide structural changes [13]. The use of short-path-length cuvettes or disruption of aggregates (reduction in size) by a technique such as sonication can often be used to reduce or eliminate this artifact.
4. Cysteine, methionine, and histidine residues also absorb weakly in the near-UV region, but their spectral contribution is often obscured by the much stronger absorbance of the aromatic residues. For peptides in which aromatic amino acid residues are absent or present in very low amounts, as is often the case in smaller peptides, the more weakly absorbing groups may be detectable.
5. Extinction coefficient (ε) at 280 nm of peptides with higher order structures can be accurately calculated (to ±2 % in most cases) using established Eq. 2 or 3 that allow determination of peptide extinction coefficient (ε) when they exist in some type of folded state:

$$\varepsilon(280\,nm) = 5{,}500\left(n_{Trp}\right) + 1{,}490\left(n_{Tyr}\right) + 125\left(n_{cystine}\right) \quad (2)$$

$$\varepsilon(280\,nm) = 5{,}540\left(n_{Trp}\right) + 1{,}480\left(n_{Tyr}\right) + 134\left(n_{cystine}\right) \quad (3)$$

where n is the number of indicated residues in the peptide.

For small peptides with little or no structure, the extinction coefficient can be calculated using Eq. 4 established for unfolded proteins [4]:

$$\varepsilon(280\,nm) = 5{,}690 n_{Trp} + 1{,}280\left(n_{Tyr}\right) + 120\left(n_{cystine}\right) \quad (4)$$

where n is the number of indicated residues in the peptide.

6. A light scattering correction at 280 nm for concentration determination can be achieved by using Eq. 5:

$$OD_{280} = 10^{(2.5\log OD_{320}} - 1.5\log OD_{350}) \quad (5)$$

where OD_{280} is the optical density at 280 nm and OD_{320} and OD_{350} are OD values at 320 and 350 nm, respectively. Another approach in which second (or higher order)-derivative spectra of aromatic residues that are largely insensitive to the underlying bands and obey Beer–Lambert law may also be used [5, 6].

7. Many peptides display a characteristic lag time prior to the onset of aggregation during turbidity studies. The presence of a lag phase is usually considered to be suggestive of a critical nucleation event. Nucleation-based aggregation is highly dependent upon sample concentration, with an increase in sample concentration often resulting in shortening of the lag phase. An empirical relationship between protein concentration (C) and nucleation lag phase (t_d) was demonstrated by Hofrichter et al. [10]. According to these studies, the length of lag phase (t_d) is proportional to (C^{-N}), where N is the number of monomers in the critical nucleus. A number of studies [11, 12], however, have also indicated that aggregates too small to be detectable by turbidity measurements may be present early in the aggregation process, which significantly complicates such analyses. Thus, optical density-based turbidity measurements should be complemented by other methods to address this issue.

References

1. Pace CN, Vajdos F, Fee L, Grimsley G, Gray T (1995) How to measure and predict the molar absorption coefficient of a protein. Protein Sci 4:2411–2423
2. Mach H, Middaugh CR, Lewis RV (1992) Statistical determination of the average values of the extinction coefficients of tryptophan and tyrosine in native proteins. Anal Biochem 200(1):74–80
3. Edelhoch H (1967) Spectroscopic determination of tryptophan and tyrosine in proteins. Biochemistry 6:1948–1954
4. Gill SC, von Hippel PH (1989) Calculation of protein extinction coefficients from amino acid sequence data. Anal Biochem 182:319–326
5. Weiser WE, Pardue HL (1983) Evaluation of multi-wavelength derivative spectra for quantitative applications in clinical chemistry. Clin Chem 29(9):1673–1677
6. Mach H, Middaugh CR (1993) Measuring protein spectra in the presence of light scattering. Biotechniques 15(2):240–242
7. Lucas LH, Ersoy BA, Kueltzo LA, Joshi SB, Brandau DT, Thyagarajapuram N, Peek LJ, Middaugh CR (2006) Probing protein structure and dynamics by second-derivative ultraviolet absorption analysis of cation-π interactions. Protein Sci 15(10):2228–2243
8. Esfandiary R, Hunjan JS, Lushington GH, Joshi SB, Middaugh CR (2009) Temperature dependent 2nd derivative absorbance spectroscopy of aromatic amino acids as a probe of protein dynamics. Protein Sci 18(12):2603–2614
9. Wegner A, Savko P (1982) Fragmentation of actin filaments. Biochemistry 21(8):1909–1913
10. Hofrichter J, Ross PD, Eaton WA (1976) Supersaturation in sickle cell hemoglobin solutions. Proc Natl Acad Sci U S A 73(9): 3035–3039
11. Tomski SJ, Murphy RM (1992) Kinetics of aggregation of synthetic beta-amyloid peptide. Arch Biochem Biophys 294:630–638
12. Snyder SW, Ladror US, Wade WS, Wang GT, Barrett LW (1994) Amyloid-beta aggregation: selective inhibition of aggregation in mixtures of amyloid with different chain lengths. Biophys J 67:1216–1228
13. Schneider AS (1973) Analysis of optical activity spectra in turbid biological suspensions. Methods Enzymol 27:751–767

Chapter 16

Fluorescence Spectroscopy of Peptides

Mangala R. Liyanage, Kunal Bakshi, David B. Volkin, and C. Russell Middaugh

Abstract

In peptides, steady-state fluorescence can be used to measure the intrinsic fluorescence of Trp, Tyr, and Phe residues, as well as associated dyes, which can change upon exposure to different environmental conditions. This technique can be thus used to detect changes in the conformational states of peptides. Time-resolved fluorescence can also be used to study fast motions of peptides and their interactions with fluorescence dyes. Data from several low-resolution spectroscopic techniques, including fluorescence, can be combined to generate an overall picture of peptide structure as a function of environmental conditions.

Key words Fluorescence, Steady state, Time resolved, Peptide, Conformational integrity, Stability

1 Steady-State Fluorescence

1.1 Introduction

Due to the sensitivity of the indole side chain of tryptophan (Trp) to its local environment, including properties such as solvent polarity, viscosity, and pH, the intrinsic fluorescence intensity and peak positions of Trp residues in peptides can change upon exposure to different environmental conditions. This polarity dependence of Trp can also be used to study conformational changes of Trp-containing peptides. Although tyrosine and phenylalanine residues are also intrinsic fluorophores, their relative insensitivity towards local environmental changes makes them less useful in such studies. The exception is peptides that lack Trp residues and contain only Phe or Tyr residues. In such cases, their fluorescence is not obscured by the dominant Trp emission and their reduced environmental sensitivity may still be of some use. As peptides change their structure, those containing aromatic residues may see shifts in their fluorescence wavelength maxima as well as changes in intensity and fluorescence lifetime. In general, shifts to the blue (lower

Mangala R. Liyanage and Kunal Bakshi have contributed equally to this chapter.

Andrew E. Nixon (ed.), *Therapeutic Peptides: Methods and Protocols*, Methods in Molecular Biology, vol. 1088, DOI 10.1007/978-1-62703-673-3_16, © Springer Science+Business Media, LLC 2014

wavelength) indicate the movement of aromatic side chains to a less polar (more hydrophobic) environment, while shifts to the red (to higher wavelength) indicate movement to more polar surroundings. When completely exposed to an aqueous environment, Trp residues typically fluoresce in the 348–355 nm range. Specific interactions with other side chains as well as solutes can further modify the position of this emission maximum (*see* **Note 1**).

1.2 Materials

1. A spectrofluorometer equipped with a turreted single/multi-position or queued temperature-controlled cell holder and a xenon lamp is preferred. (Major instrument vendors include Photon Technology International, Varian, OLIS, Horiba, Jasco, etc.)
2. A set of 10 mm × 10 mm, 1 mm × 10 mm, and front-face (triangular) geometry quartz cuvettes (Starna, Helma, etc.). (*See* **Note 2**).
3. Software dedicated to fluorescence data acquisition and analysis. For example, Felix software for a PTI fluorometer. In addition, software packages such as Excel, GRAMS, and Origin are typically used for data processing.

1.3 Methods

1.3.1 Instrument Preparation

1. The light source (e.g., xenon lamp) is first turned on (prior to turning on the rest of the components of the fluorometer). Subsequently, the power to the monochromator assembly, PMT module, control electronics, and temperature control are turned on. Water circulation and/or Peltier control should be started prior to initiating the experiment.
2. Excitation and emission slit widths are set to optimum bandwidths, which are governed by the absorbance in the near-UV excitation region. Detector saturation can be avoided by starting with narrow slit settings and gradually increasing them until the emission intensity reaches the desired level. As with most spectroscopy experiments, the sensitivity is increased as the slit is opened but resolution is correspondingly decreased. Because of the wide width of the emission peak of the aromatic side chains, preference will often be to optimize instrument settings in terms of sensitivity (*see* **Note 3**).
3. For the exclusive excitation of Trp, the excitation wavelength is set at about 295 nm (>95 % Trp emission) or above, with emission scans typically monitored from 300 to 400 nm. The scattering peak can also be captured by widening this range to 290–390 nm. For Tyr-containing peptides, the optimal excitation is achieved at 280 nm with emission analyzed from 290 to 390 nm. For Phe-containing peptides, settings are typically excitation at 250–260 nm with emission monitored from 265 to 365 nm. A spectral resolution of 1 nm and 1-s integration time are often initially chosen, but can be further optimized as demanded

by a particular system during data acquisition (*see* **Note 3**). For example, analysis of Phe fluorescence may require much wider slit width settings due to its weak emission. An example of typical fluorescence spectra from three model peptides is shown in Fig. 1. In these examples, three different peptides that vary in their content of aromatic amino acid residues display quite different fluorescence emission spectra (angiotensin I contains one Tyr and Phe residue, α-melanocyte-stimulating hormone contains one Trp, Tyr, and Phe residue, while substance P contains two Phe residues; *see* Fig. 1a–c, respectively). As expected, the fluorescence spectra of angiotensin I is dominated by Tyr emission (ca. 303 nm), α-melanocyte-stimulating hormone by Trp emission (ca. 350 nm), and substance P by Phe emission (ca. 300–320 nm).

4. Fluorescence scans can be collected as a function of varying stress conditions, including temperature, pH, peptide concentration, and ionic strength. For example, thermal perturbation of structure may be followed by collecting spectra over a wide temperature range (e.g., 10–90 °C) at 2–3 °C increments (*see* **Note 4**). A water bath-cooled Peltier system is ideal for this purpose. Samples should be allowed to equilibrate at each temperature for a given period of time (typically 2–5 min) to permit temperature equilibrium to be reached. This equilibration can be confirmed by the constancy of the fluorescence emission signal. Note that fluorescence emission decreases intrinsically in a curvilinear fashion with temperature due to the temperature dependence of non-radiative processes.

1.3.2 Sample Preparation

1. Quartz cuvettes with the desired path lengths (typically 10 mm) are filled with an adequate volume of peptide yielding an optical density below 0.1 (preferably below 0.05) to avoid inner filter (i.e., self-absorption) effects. The cuvette is placed in the sample cell holder in such a way that fluorescence can be observed at right angles from the exciting light in a centrally illuminated cuvette. When dealing with samples of higher optical densities (i.e., high-concentration peptides), shorter path length cuvettes or other geometric positioning of the cuvettes (or different cuvette geometries) can be employed (*see* **Note 5**).
2. Cuvettes need to be vigorously cleaned after each experiment especially after thermal studies, which may leave peptide deposited on the inner surface of the cuvette. As described in Chapter 15 for UV absorption measurements, sample cells should be rinsed with water and then soaked in a strong acid solution for 1 h. Other extreme conditions (base, detergents, organic solvents, peptidases, etc.) may also be necessary. Cuvettes should finally be rinsed with nano pure water and stored in a safe, dust-free location to maintain cleanliness.

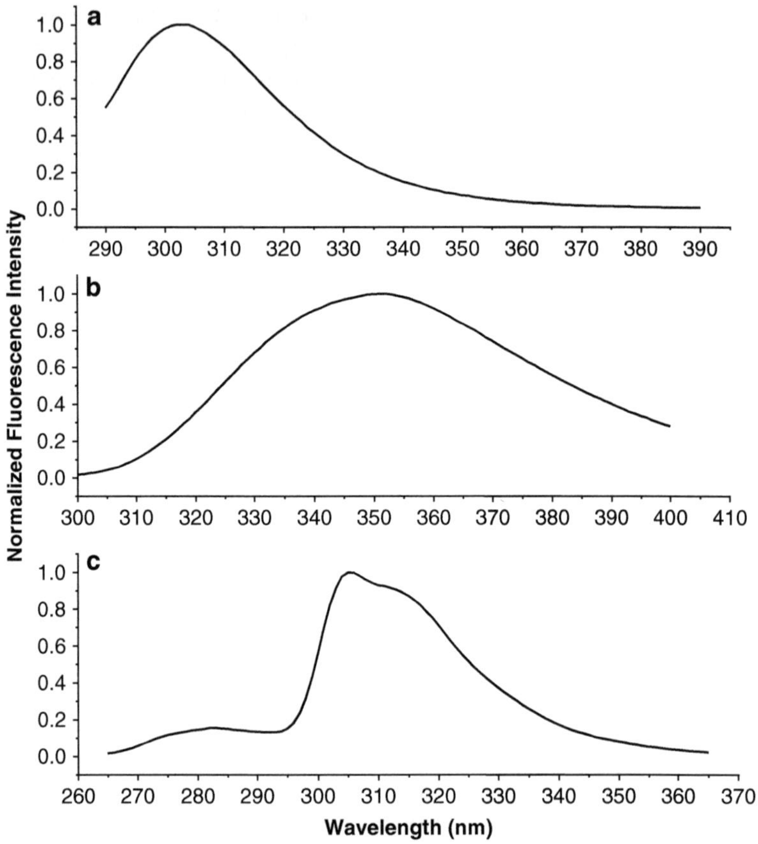

Fig. 1 Intrinsic fluorescence emission spectra of three representative model peptides: (**a**) Angiotensin I contains single Phe and Tyr residues. Because there are no Trp residues present, Tyr emission at 303 nm (excitation at 280 nm) can now be seen, but no Phe emission can be resolved. The peak position of Tyr is not sensitive to its immediate environment, so no conclusion about solvent exposure can be made from these data. (**b**) α-Melanocyte-stimulating hormone (α-MSH) contains one each of Trp, Tyr, and Phe residues. As expected, fluorescence is dominated by Trp emission, which in this case has been enhanced by excitation of 295 nm. Nevertheless, excitation at lower wavelengths still produces a similar emission. The location of the peak ca. 350 nm indicated extensive exposure of the indole side chain to the solvent. (**c**) Substance P possesses two Phe residues (with no Trp or Tyr present). Phe emission (excitation at 256 nm) can be seen as a weak peak ca. 282 nm. The origin of the structured emission between 300 and 320 nm is unclear. Substance P contains a β-turn, which could place one or both Phe residues in unique environments which could shift the expected 282 nm emission. Alternatively, the peptide could contain contaminating amounts of other fluorophores such as Tyr, which are visible because of the very weak Phe emission

1.3.3 Data Analysis

1. When dealing with peptide samples possessing optical density values exceeding 0.05, fluorescence intensities should be corrected for inner filter effects using the following equation (*see* **Note 5**):

$$F_{corr} = F_{obs}\,antilog\left(\left(OD_{ex} + OD_{em}\right)/2\right) \quad (1)$$

where F_{corr} is the corrected fluorescence intensity, F_{obs} is the measured fluorescence intensity, and OD_{ex} and OD_{em} are the optical densities at the excitation and emission wavelength, respectively [2].

2. Data processing and background corrections can conveniently be performed using Microsoft Excel, GRAMS, or Origin software. Emission peak intensity and position data are obtained after background correction using either derivative polynomial fits or mean spectral center of mass methods as required by the nature of the data. Fluorescence intensity and peak positions as a function of stress conditions are frequently further analyzed by either method using a commercial software package such as Origin or GRAMS (*see* **Note 6**).

1.4 Notes

1. The fluorescence spectra of peptides are often less complicated than those of larger proteins due to the presence of fewer numbers of Trp, Tyr, and Phe residues. Tyr fluorescence in peptides can be of special interest because of its dramatic pH dependence.
2. Under conditions of either limited sample availability or samples at higher peptide concentration than those typically used in fluorescence experiments, reduced volume cuvettes can be employed to collect fluorescence spectra. It must be ensured, however, that the physical dimensions of the cuvettes do not interfere with the two light beams. Other cuvette-positioning geometries include front-face and off-center illumination. Off-center illumination functions by reducing the path length, which can alternatively be achieved by using shorter path length cuvettes. Front-face illumination is frequently achieved by positioning triangular cuvettes 30–60° with respect to the incident beam. Some instruments have this cuvette positioning ability built into their movable sample holders. The geometrical positioning angle of 45° should be avoided, if possible, to minimize interference from stray light due to reflection of light into the monochromator [1]. Triangular cuvettes can be used for the same purpose. Note, however, that emission data from front-face geometries will not, in general, be linear with peptide concentration.

3. If a T-format fluorometer with two detectors located 180° relative to one another is employed, the second detector can be used to simultaneously collect static light scattering intensities and optimized independently during emission spectra acquisition. This setup permits the collection of peptide association (aggregation) and/or dissociation data during the experiment. The emission slits for static light scattering analysis should be kept much narrower than that for fluorescence measurements, since strong, direct excitation light reaching the detector can be detrimental to the life of the detector due to its high intensity.
4. Increases in temperature cause a curvilinear decrease in fluorescence intensity due to the temperature dependence of non-radiative processes, which deactivate the excited state. Conformational alterations of peptides can be recognized as deviations from this behavior. This effect can be observed as a dramatic, abrupt change in plots of fluorescence intensity versus temperature or as more subtle changes in slope. Because of this intrinsic drop in fluorescence intensity with temperature, however, studies of shifts in peak position (wavelength) are often preferable in thermal analysis.
5. When samples have optical density values over 0.05 at both excitation and emission wavelengths, the incident and emitted light intensities are significantly attenuated. These attenuations, which result in absorption of incident and emitted light, are called primary and secondary inner filter effects, respectively. An in-depth description of primary and secondary inner filter effects can be found in the literature [2, 3].
6. A derivative polynomial methodology can be used to fit the data employing, for example, an 11-point Savitsky–Golay polynomial smoothing. The peak maximum can then be identified by obtaining the first derivative of the polynomial. This technique provides peak positions that are very close to the absolute (zero order) experimental value. It is not, however, efficient at locating emission peak maxima when the peak becomes noisy, especially when scattering occurs. The mean spectral center of mass methodology at least partially circumvents this problem by calculating a peak position based on curve areas using the following relationship (Eq. 2):

$$\text{C.M.} = \Sigma\lambda_{em} \times F_{\rm i} \,/\, \Sigma F_{\rm i} \qquad (2)$$

where $\lambda_{\rm em}$ is the emission wavelength and $F_{\rm i}$ is the fluorescence intensity at that wavelength. Although relatively precise, this method produces a peak that is typically shifted approximately 8–10 nm relative to the actual emission peak. Thus, its primary use is for comparison of data.

2 Time-Resolved Fluorescence

2.1 Introduction

Time-resolved fluorescence has been widely used in the biochemical and biophysical research arena for studying fast domain motions of macromolecules as well as in the investigation of interactions of fluorescence dyes with various molecules. There are two frequently used methods for time-resolved measurements: time domain [4] and frequency domain [5]. The time domain measurements, however, are more frequently used owing to their relative simplicity, more available instrumentation, and higher reliability. This discussion will therefore focus solely on time domain methods [1]. In time-resolved fluorescence anisotropy experiments, fluorophores having a $\cos^2\theta$ distribution of transition dipoles relative to the electric field vector of the polarized incident pulses are excited by using a pulsed laser source. Excited fluorophores will initially have an anisotropic distribution of transition dipoles, which will decay over time back to an isotropic distribution as a result of their rotational and translational motions. As the excited fluorophores undergo rotational diffusion, the polarization of the emitted radiation is measured at a time t after excitation. Time-correlated single-photon counting (TCSPC) has been one of the two major techniques employed to obtain time-resolved fluorescence data [6] and is discussed here.

Fluorescence lifetimes of native Trp, Tyr, and Phe residues in polypeptides are affected by numerous processes such as charge transfer quenching by proximate groups and rotamer interconversion [7]. These effects can provide insight into the dynamics of both the intra- and intermolecular behavior of peptides.

2.2 Materials

1. A high-repetition-rate mode-locked picosecond or femtosecond laser light source. For instance, dye lasers, Ti:sapphire lasers, pulsed-laser diodes, and pulsed LEDs can all be used as light sources in time-resolved fluorescence studies (*see* **Note 1**).
2. A frequency doubler/tripler, two polarizers, two monochromators, and neutral density filter wheel.
3. Two fast multichannel plate PMT modules.
4. SPC-630 PC board made by Becker and Hickl GmbH (Berlin, Germany).
5. A number of manufacturers offer completely assembled systems that can be used in a turnkey mode. We describe a home-built system here for illustration purposes only.

2.3 Methods

2.3.1 Instrument Preparation

1. Excitation pulses are typically obtained from a mode-locked, cavity-dumped, or diode laser. Two-MHz pulses can be obtained from a Ti:sapphire output by sending a radio-frequency signal for cavity dumping (*see* **Note 1**).

2. The output from the pulsed light source is frequency doubled/tripled to obtain light in the wavelength range of 285–300 nm using a frequency-doubling/tripling crystal. The excitation power is adjusted and maintained for the measurements using a neutral density filter wheel.
3. For anisotropy measurements, vertically and horizontally polarized emissions are simultaneously collected in a T-format configuration with two polarizers (*see* **Note 2**).
4. The emission from both polarizations is then spectrally resolved to select emission from desired emission wavelengths using monochromators. For Tyr and Trp lifetimes, emission wavelengths are set to values such as 303 and 335 nm, respectively.
5. Wavelength-resolved emissions in vertical and horizontal channels are then detected using fast MCP-PMT modules.
6. For lifetime measurements, only a single channel with a polarization angle of 56.7 Å is employed (*see* **Note 3**).
7. A reference pulse for the arrival time determination is obtained using the fundamental output of the pulsed light source using a photodiode.
8. PMT control, routing of the signal from different channels, and data acquisition are controlled by an SPC-630 PC board or a similar circuit.
9. Instrumental response functions are collected at a wavelength 5 nm blue-shifted (290 nm) from each excitation wavelength using a colloidal solution of SDS micelles (*see* **Note 4**).
10. The G factor is calculated by using a 0.1 mg/ml tryptophan amino acid solution and measuring the intensity in both vertical and horizontal channels (*see* **Note 3**).

2.3.2 Sample Preparation

1. Quartz cuvettes with the desired path lengths (typically 10 mm) are filled with an adequate volume of peptide solution, yielding an optical density value below 0.1 (preferably below 0.05) to avoid inner filter effects. The cuvette is placed in the sample cell holder in such a way that fluorescence can be observed in a right-angle geometry from a centrally illuminated cuvette. When analyzing samples showing higher optical densities (for example, at high peptide concentration), shorter path length cuvettes and/or other geometric positioning of cuvettes are employed (*see* **Note 5**).
2. Cuvettes are vigorously cleaned after each experiment especially after thermal perturbation studies, which may lead to peptide aggregation and adsorption to the glass surface. First, cells are rinsed with water and then soaked in a strong acid

solution for 1 h. Cuvettes are then rinsed with nano pure water, and soaked in a detergent solution followed by one or more additional rinsing steps with nano pure water.

2.3.3 Data Analysis

1. TCSPC data can be analyzed with a program written in Microsoft Excel (or other convenient software) based on iterative reconvolution, which involves a deconvolution of the instrumental response function with a model fitting function as follows (Eq. 3):

$$C(t) = \int_0^t I(t)\, IRF(t-\tau)\, \mathrm{d}\tau \qquad (3)$$

where $I(t)$ is the model fitting function, and IRF(t) is the instrument response function. Since the observed anisotropy decay does not reflect the actual anisotropy decay, especially at short time scales, the vertical and horizontal fluorescence intensity decays are fit simultaneously with this program. This iterative fitting routine relies on the minimized χ^2 to obtain a good fit. Similar programs are available with all commercial instrumentation.

2.4 Notes

1. The laser source must also have a high enough repetition rate to collect reasonable count levels (e.g., 20,000 counts/s) and yet still only collect a single photon per laser pulse. O'Connor and Phillips [6] have recommended that the count levels for TCSPC be kept below 1 % of the pulse repetition rate. Therefore, this system must have a repetition rate above 2 MHz. It is also crucial that the repetition rate be low enough to allow the fluorescence to decay completely before a second pulse is seen. For an exponential decay, the count level should not go below 1 % of its original amplitude until five lifetimes have passed. For a 100-ns lifetime, it is advisable to have a repetition rate below 2 MHz.
2. Monochromators have differential diffraction efficiencies for the horizontal and vertical components of the polarization. This differential response should be corrected to obtain an accurate anisotropy reading.
3. Since a polarizer is used for anisotropy decays and it may not be conveniently removed, and because the excitation laser is vertically polarized, one does not have a good way to collect the intensity decay without polarization effects. The emission is collected at a polarization angle of 54.7° with respect to the vertical excitation polarization. This experimentally averages the horizontal and vertical components of the decay so that the intensity decay can be collected.

4. Instrumental response functions are collected at a wavelength 5 nm blue-shifted to avoid any contribution from the fluorescence to the instrument response function.
5. The fluorescence lifetime of Trp residues has been primarily exploited to probe the effect of environment on the structure of peptides. Free Trp possesses a mean fluorescence lifetime of 3.1 ns in water. This value may vary in the range of 0.3–7 ns in Trp-containing peptides depending on the extent of solvent exposure. Both Tyr and Phe, however, are of limited use for probing environmental changes in peptides due to their reduced quantum efficiencies. They possess fluorescence lifetimes of 3.2 and 6.8 nm in water, respectively [1].

References

1. Lakowicz JR (2006) Principles of fluorescence spectroscopy, 3rd edn. Kluwer, New York
2. Wiechelman KJ (1986) Empirical correction equation for the fluorescence inner filter effect. Am Lab 18:49–53
3. Kubista M, Sjobakc R, Erksson S, Albinsson B (1994) Experimental correction for the inner-filter effect in fluorescence spectra. Analyst 119:417–419
4. Birch DJS, Imhof RE (1991) Time-domain fluorescence spectroscopy using time-correlated single-photon counting. In: Lakowicz JR (ed) Topics in fluorescence spectroscopy, vol 1, Techniques. Plenum, New York, pp 1–95
5. Lakowicz JR, Gryczynski I (1991) Frequency-domain fluorescence spectroscopy. In: Lakowicz JR (ed) Topics in fluorescence spectroscopy, vol 1, Techniques. Plenum, New York, pp 293–355
6. O'Connor DV, Phillips D (1984) Time-correlated single photon counting. Academic, London
7. Unruh JR, Liyanage MR, Johnson CK (2007) Tyrosyl rotamer interconversion rates and the fluorescence decays of *n*-acetyltyrosinamide and short tyrosyl peptides. J Phys Chem B 111:5494–5502

Chapter 17

Circular Dichroism of Peptides

Kunal Bakshi, Mangala R. Liyanage, David B. Volkin, and C. Russell Middaugh

Abstract

Circular dichroism measures the difference between the absorbance of left- and right-handed circularly polarized light, and can be used to monitor the secondary structure of peptides (far UV) and the tertiary structure of larger polypeptides (near UV). This technique is especially useful for helix–coil transitions and other aspects of structural alterations. Data from several low-resolution spectroscopic techniques, including CD, can be combined to generate an overall picture of peptide structure as a function of environmental conditions.

Key words Circular dichroism, Near UV CD, Far UV CD, Helix–coil transitions, Peptide, Conformational integrity, Stability

1 Introduction

Circular dichroism (CD) measures the difference between the absorbance of left and right circularly polarized light. The far-UV CD spectra (260–180 nm) resulting from light absorption by peptide bonds contain information pertaining to the secondary structure of peptides. The near-UV CD spectra (320–250 nm) originate from light absorption by optically active tryptophan, tyrosine, phenylalanine, and cystine residues. These spectra are sensitive to the tertiary and quaternary structure of larger polypeptides. Oligopeptides showing a propensity to form α-helices in solution have been extensively studied by far-UV CD techniques [1]. Peptides having a tendency to self-assemble by aligning β-sheet structures have also been characterized using far-UV CD [2]. Since many peptides are relatively small, however, they may possess far-UV CD spectra more characteristic of disordered structure. A major exception is small peptides that form a variety of β-turn

Kunal Bakshi and Mangala R. Liyanage have contributed equally to this chapter.

Andrew E. Nixon (ed.), *Therapeutic Peptides: Methods and Protocols*, Methods in Molecular Biology, vol. 1088,
DOI 10.1007/978-1-62703-673-3_17, © Springer Science+Business Media, LLC 2014

structures [5]. Helix–coil transitions and other aspects of structural alterations in peptides have been intensively studied using far-UV CD spectroscopy [5] and provide a major use for this technology.

2 Materials

1. A CD spectropolarimeter equipped with single/multi-position Peltier-controlled cell holder and a xenon lamp (Jasco, OLIS, Applied Photophysics, AVIV, etc.) and temperature control.
2. A set of 0.1, 0.05, 1, and 10 mm quartz cuvettes (Starna, Hellma).
3. Dedicated software for CD data acquisition and processing and Excel and/or Origin software for data analysis.

3 Methods

3.1 Instrument Preparation

1. The CD instrument is purged with pure nitrogen at the manufacturer's indicated flow rate for at least 5 min prior to starting the xenon lamp. Nitrogen flow must be continued until the end of the experiment. Water circulation and Peltier control are started to maintain a constant temperature while data are being collected (*see* **Note 1**).
2. Initial wavelength scans for far-UV (180–260 nm) and near-UV (250–340 nm) CD are usually performed at room temperature (or lower temperatures such as 10 °C) in quartz cuvettes with appropriate path lengths. The run parameters are optimized to obtain good signal-to-noise statistics by adjusting the following experimental parameters according to the desired signal levels: scan rate (10–30 nm/min), integration time (time constant) (1–3 s), bandwidth (<2 nm), and number of spectral acquisitions [2–5]. Examples of typical far-UV CD spectra from the three model peptides are shown in Fig. 1, and representative near-UV CD spectra of the same three model peptides are shown in Fig. 2. In these examples, the three different peptides differ in their content of aromatic amino acid residues (angiotensin I contains one Tyr and Phe, α-melanocyte-stimulating hormone peptide contains one Trp, Tyr, and Phe, while substance P contains two Phe residues) and display some minor differences in their far-UV and near-UV CD spectra (*see* **Note 2**).
3. The spectral bandwidth and the integration time are adjusted for near-UV CD spectra in such a way that the fine structural details typically available in such spectra can be obtained without compromising the signal-to-noise ratio (*see* **Note 3**). Derivative analysis can be used for this purpose. Taking the derivative of a spectrum is the simple process of calculating the

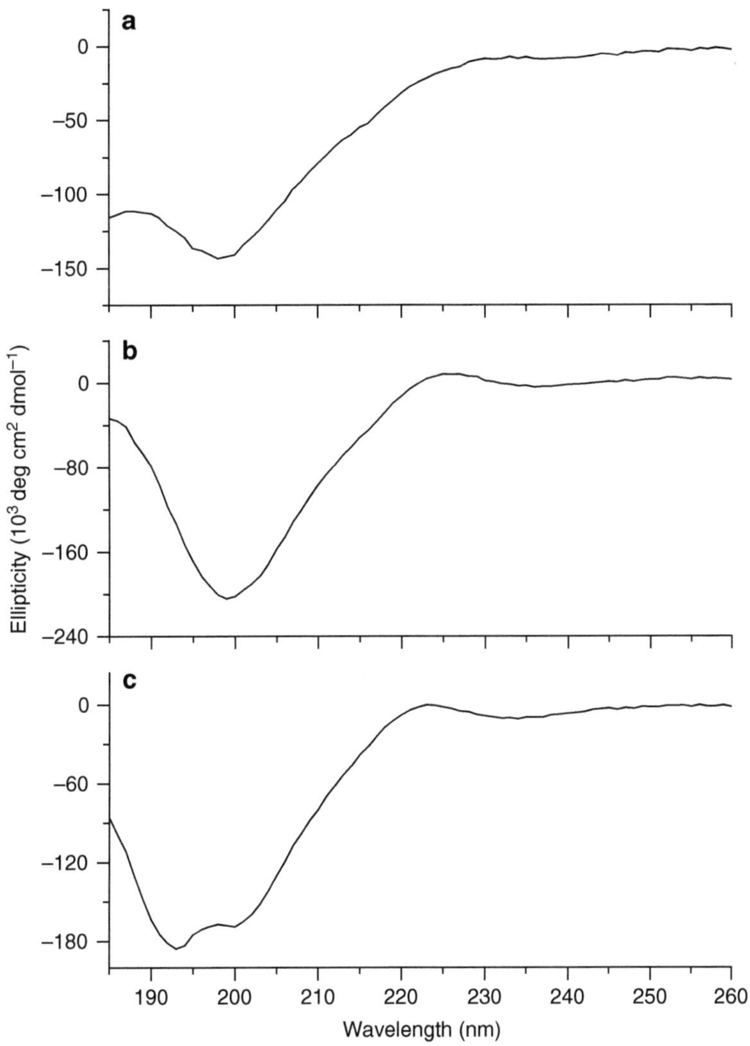

Fig. 1 Far-UV CD spectra of three model peptides: (**a**) Angiotensin I, (**b**) α-MSH, and (**c**) substance P. All three peptides display a far-UV CD spectral minimum near 198 nm characteristic of disordered structure. In addition, substance P has a second minimum near 193 nm presumably reflecting the presence of its β-turn. All three peptide spectra also show a weak positive peak near 220 nm that is usually thought to reflect the presence of some weak helical structure arising from polyproline II-like structure

slope or the gradient of an absorption band. This gradient ($dA/d\lambda$) is plotted against wavelength (λ) to produce a first-order derivative plot. An iteration of this process leads to higher order derivatives ($d^nA/d\lambda^n$). Temperature control parameters are set to the desired starting temperature with a thermal equilibration time of 2–5 min prior to each measurement. It is also possible to obtain thermal perturbation data by

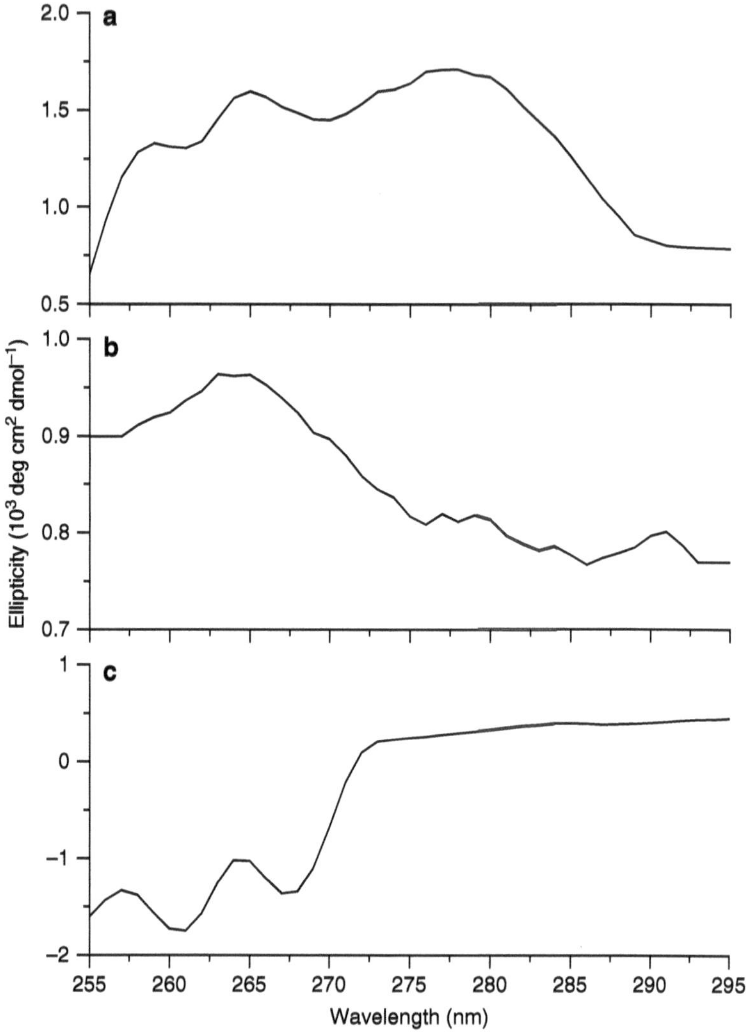

Fig. 2 Near-UV CD spectra of three model peptides: (**a**) Angiotensin I, (**b**) α-MSH, and (**c**) substance P. All three peptides display a near-UV CD spectra with weak transitions found primarily in the regions where the individual aromatic side chains most strongly absorb. Although not the situation here, local environments can sometimes result in the induction of enhanced optical activity in these transactions

slowly elevating the temperature in a continuous mode but temperature equilibrium may not be reached in such analyses. Data can be recorded at one or more fixed wavelengths or the entire spectra can be recorded at each temperature. The base line is recorded by employing a cuvette containing only the sample buffer (without peptide) and this spectrum is subtracted from each spectrum of the peptide-containing solution.

3.2 Sample Preparation

1. Peptide samples are filtered though suitable filters to remove any scattering particles. Peptide concentration must be accurately measured, preferably spectrophotometrically using UV absorbance and a previously determined molar extinction coefficient at 280 nm (using the Beer–Lambert law, which states that absorbance is directly proportional to path length, concentration, and extinction coefficient). Peptide concentrations must be maintained such that the total absorbance at measured wavelengths stays below unity to prevent saturation of the detector. For far-UV CD measurements, cells with path lengths from 0.05 to 0.2 cm and peptide concentrations of 0.1–1 mg/ml are typically employed. The sample volumes are dependent on the design of the cell and can range from 1 ml to 50 μl (*see* **Note 3**). For near-UV CD measurements, longer path lengths and higher peptide concentrations must be used due to the lower molar concentrations and optical activities of the chromophoric entities. Thus cuvettes with path lengths of 5–20 mm and peptide concentrations of 0.5–2 mg/ml are typically used for collecting near-UV CD spectra [4, 6] (*see* **Note 4**). Modern instruments often permit near and far UV spectra to be acquired in a single scan [7].
2. A cleaning step is necessary after every CD experiment using a rigorous protocol to ensure that the cuvettes are clean for the next experiment. Cuvettes are filled with a strong acid cleaning solution (e.g., Nochromix) and allowed to stand for at least an hour. The cuvettes are then emptied and rinsed with water followed by a detergent (10 % Neutrad) cleaning step. Finally, the cuvettes are rinsed thoroughly with nano pure water (*see* **Notes 5** and **6**).

3.3 Data Analysis and Interpretation

1. The CD signals are conveniently converted to the concentration-independent parameter of mean residue molar ellipticity $[\theta]$ using the following formula (Eq. (1)):

$$[\theta] = \frac{[(\theta/10)\times m]}{l \times C} \quad (1)$$

where θ is the measured value from the instrument in millidegrees after subtracting the value of ellipticity associated with the buffer; m is the mean residue molecular mass determined for the peptide, obtained by dividing molecular weight of the peptide by the number of amino acids in the peptide chain; l is the cell path length in cm; and C is the concentration of the sample in g/ml. Alternatively, molar ellipticity based on the molecular weight of the peptide can be used.

2. Results from CD thermal experiments are usually presented by plotting the molar or the mean residue ellipticities at corresponding wavelengths versus the temperature.

4 Notes

1. Purging with nitrogen is required to create an oxygen-free environment to prevent conversion of oxygen to ozone by far-UV radiation, which in turn can damage the optical surfaces. This purging with nitrogen also permits CD measurements below 200 nm. Ozone absorbs UV radiation from 200 to 300 nm, obstructing observation at these wavelengths. A noble gas, such as He and Ar, is not suitable for purging since it will leak into the interior of the PMT and reduce its vacuum, causing impaired performance. The system will stabilize in about 30 min and then the nitrogen flow rate can be reduced to 5–7 L/min. It also takes 5–10 min for the system to stabilize each time the sample compartment is open and closed.
2. A description of the signature absorbance bands typically belonging to different secondary structures of peptides is described below: (a) α-helix possesses three major absorbance bands with a positive maximum between 190 and 195 nm and two negative minima at approximately 208 and 222 nm; (b) the features of β-sheet are mainly limited to two major absorbance bands including a negative band in the spectral range of 217–218 nm and a positive one near 195–197 nm; (c) random coil structures of peptides feature a strong negative band below 200 nm. Some peptides also manifest a positive/negative weak shoulder at 220 nm, and (d) several different types of β-turns may be present in the secondary structure of peptides. Frequently observed types include type I and type II displaying multiple bands in far-UV CD spectra as discussed in detail elsewhere [8].
3. The spectral bandwidth should be set under 2 nm to minimize spectral distortion. This threshold is especially critical in near-UV CD measurements for resolving fine structural information in this region of the spectrum. The product of scan speed and integration time should be kept below 0.5 nm to minimize possible distortion of the spectrum [3].
4. Light absorption due to buffers, solvents, and other supporting electrolytes should be minimized by choosing appropriate buffers and other components. Absorption due to a wide variety of aqueous components arises in the far-UV region. A thorough description of absorption profiles of commonly used buffers and other reagents can be found in the literature [3]. It should be noted that one of the most interfering buffer components is chloride ion due to strong absorption below 195 nm. Therefore, chloride ions are highly undesirable as a buffer counterion in CD experiments. Salts with fluoride and sulfate ions are better choices to maintain the ionic strength since they possess better

UV transparency. Due to buffer and other components, it is often not possible to obtain data below 200 nm.

5. Artifacts in CD spectra: Similar issues to those seen with UV absorbance spectra are relevant with absorption flattening especially problematic for particulate samples (*see* **Note 6**). In addition, differential scattering of left- and right-handed circularly polarized light can produce artifactual signals. Simply moving the sample cell closer to the PMT to gather more of the transmitted or the scattered light can sometimes minimize such artifacts, although they are relatively rare [9].

6. As the size of particles in a sample becomes larger, light scattering corrections may not be sufficient. Absorption flattening is a phenomenon in which high local concentrations of peptides (due to their aggregation) create an inhomogeneous distribution of sample chromophores in the light path. Absorption flattening typically leads to reduced signal intensities and red shifts in absorption bands and may be erroneously interpreted as peptide structural changes [9]. The use of short-path-length cuvettes or disruption of aggregates (reduction in size) by a technique such as sonication can often be used to reduce or eliminate this artifact.

References

1. Dyson HJ, Wright PE (1991) Peptide conformation and protein folding. Annu Rev Biophys Biophys Chem 20:519–538
2. Aggeli A, Bell M, Boden N, Keen JN, Knowles PF, McLeish TCB, Pitkeathly M, Radford SE (1997) Responsive gels formed by the spontaneous self-assembly of peptides into polymeric β-sheet tapes. Nature 386:259–262
3. Martin SR (1996) Circular dichroism in proteins labfax (Price N.C., Ed.) BIOS scientific publishers Ltd., Oxford, p 195–204
4. Kelly SM, Price NC (2000) The use of circular dichroism in the investigation of protein structure and function. Curr Protein Pept Sci 1:349–384
5. Kallenbach NR, Lyu P, Zhou H, Tlstra L, Mattice WL, Perczel A, Hollósi M (1996) CD spectroscopy and the helix-coil transition in peptides and polypeptides. In: Fasman GD (ed) Circular Dichroism and the conformational analysis of biomolecules. Plenum Press, New York, NY, pp 201–380
6. Martin SR, Bayley PM (2002) Absorption and circular dichroism spectroscopy. Methods Mol Biol 173(2):43–55
7. Hu L, Olsen C, Maddux N, Joshi SB, Volkin DB, Middaugh CR (2011) Investigation of protein conformational stability employing a multimodal spectrometer. Anal Chem 83(24): 9399–9405
8. Sreerama N, Woody RW (2000) Circular dichroism of peptides and proteins in circular dichroism: principles and applications, 2nd ed. (Berova N, Nakanish k, & Woody R.W.,Ed.), John Wiley & Sons, Inc., New York. pp 601–620
9. Schneider AS (1973) Analysis of optical activity spectra in turbid biological suspensions. Methods Enzymol 27:751–767

Chapter 18

Fourier Transform Infrared Spectroscopy of Peptides

Kunal Bakshi, Mangala R. Liyanage, David B. Volkin, and C. Russell Middaugh

Abstract

Fourier transform infrared (FTIR) spectroscopy provides data that are widely used for secondary structure characterization of peptides. A wide array of available sampling methods permits structural analysis of peptides in diverse environments such as aqueous solution (including optically turbid media), powders, detergent micelles, and lipid bilayers. In some cases, side chain vibrations can also be resolved and used for tertiary structure and chemical analysis. Data from several low-resolution spectroscopic techniques, including FTIR, can be combined to generate an empirical phase diagram, an overall picture of peptide structure as a function of environmental conditions that can aid in the global interpretation of large amounts of spectroscopic data.

Key words Fourier transform infrared spectroscopy, Peptide, Empirical phase diagram

1 Introduction

Fourier transform infrared (FTIR) spectroscopy is increasingly becoming a method of choice to characterize and study structural changes in peptides. The ready availability of a wide array of sampling methods based on transmission and reflection principles allows FTIR spectroscopy to circumvent sample-related constraints often encountered by structural methods such as NMR, X-ray crystallography, and circular dichroism. These sampling methods permit structural analysis of peptides in diverse environments such as aqueous solution, powders, detergent micelles, and lipid bilayers. Due to the use of long-wavelength radiation that minimizes scattering problems, FTIR measurements can also be performed in optically turbid media, a common problem often encountered during the study of peptides. Additionally, the high speed of data acquisition, high signal-to-noise ratio, and low cost

Kunal Bakshi and Mangala R. Liyanage have contributed equally to this chapter.

Andrew E. Nixon (ed.), *Therapeutic Peptides: Methods and Protocols*, Methods in Molecular Biology, vol. 1088,
DOI 10.1007/978-1-62703-673-3_18, © Springer Science+Business Media, LLC 2014

of instrumentation make FTIR spectroscopy a highly desirable method for secondary structure characterization of peptides as described below.

2 Materials

1. *A research-grade FTIR instrument*: Some of the principal manufacturers are Perkin-Elmer (Waltham, MA), Jasco (Easton, MD), Bruker Optics (Billerica, MA), and Varian (Palo Alto, CA). The typical FTIR spectrometer requirements for peptide analysis include moderate resolution (0.5–4 cm^{-1}) combined with a high-sensitivity nitrogen-cooled mercury–cadmium–telluride (Hg–Cd–Te) detector. A diverse variety of sampling geometries and assemblies are available and the reader should refer to individual manufacturer's manuals for further details. Three of the most commonly used sampling techniques for peptide samples are transmittance, attenuated total reflectance, and diffuse reflectance and are summarized below.
2. *Transmittance*: This approach is the most conventional sampling method. Cells are constructed from two IR transparent windows with a spacer in between to define the path length. A number of cell options are available that include sealed cells, demountable cells with a choice of different IR window materials (CaF_2, KBr) depending on sample compatibility, and different spacers for different path lengths. A good choice of window material is calcium fluoride, which is transparent in the region 1,000–5,000 cm^{-1} and insoluble in water. Both tin and Teflon spacers are available. Solid peptides can also be pressed in the presence of KBr into IR transparent discs for spectral analysis.
3. *Attenuated total reflectance (ATR)*: One form of ATR sampling consists of a horizontal plate commonly made of zinc selenide (ZnSe), germanium (Ge), or diamond crystals. The sample is loaded on top of the horizontal plate and the IR light is passed through the bottom of this high-refractive-index material such that an evanescent wave is produced that penetrates the sample. This sampling method allows samples in a variety of physical states (solution, suspension, films, and gels) to be probed with ease. A major drawback of this technique is that peptide samples may be adsorbed on the surface of the ATR unit and thus the spectral features may be altered. Furthermore, the observed spectrum consists primarily of material near the surface itself.
4. *Diffuse reflectance infrared Fourier transform spectroscopy (DRIFTS)*: Spectra of powdered samples can be recorded with no or minimal sample preparation using this geometry. Powder can be directly placed into a sample cup or on the top of an

infrared inert material such as KBr, which is placed in the sample cup. The DRIFTS cell contains mirrors that direct the beam onto the surface of the powder sample. The mirrors may need to be aligned by following the manufacturer's instructions. Spectra must be corrected for the dependence of the spectra on wavelength by a procedure known as the Kubelka–Munk correction, which is typically available with the instrument's software.

5. In addition to these more common techniques, a broad range of accessories such as photoacoustic modalities are also available to overcome sample-related limitations.
6. Standard supplies to be used with FTIR experiments should always be of highest quality available to avoid contamination problems.

3 Methods

3.1 Preparation of the Instrument

1. The instrument should be turned on and allowed to stabilize for at least 60 min before collecting spectra. The instrument and the sample cell should also be allowed to equilibrate for 15–20 min at the set temperature.
2. The instrument should always be purged with dry nitrogen or dry air at a pressure between 20 and 40 psi. The detector should be cooled by filling the detector-cooling system with liquid nitrogen. Deuterated triglycine sulfate (DGTS) or other detectors can also be used but they are less sensitive.
3. Any purge shutters should be closed whenever the sample compartment lid is open to minimize water vapor and CO_2 in the optical system. Modern instruments are often designed in such a way to avoid exposure of the sample compartment to the air.
4. Prior to data collection, the interferogram or single-beam spectrum should be checked for energy throughput and spectral quality with an empty cell.
5. The transmittance cell must be thoroughly cleaned immediately after use, as instructed by the supplier. Generally, this involves removing the sample with a tissue, followed by several solvent washes and a final ethanol wash. The use of a polishing kit to further polish the window surface may also be necessary. Wear gloves to prevent fogging. The cleaned surface should be clear and free from scratches.
6. ATR crystals are cleaned by gently clearing the sample surface using a lint-free tissue soaked in solvent. Peptide samples can be removed by using 10 % SDS solution, followed by wash steps with water (repeated four to five times), and a final step with isopropanol.

3.2 Sample Preparation

1. FTIR sample preparation is dependent upon the sampling technique to be employed during spectral acquisition. For transmission spectroscopy, peptide solutions in H_2O or D_2O can be directly loaded into liquid transmittance or ATR cells. In dried form, peptides can be prepared as KBr pellets or placed directly in DRIFTS cells.
2. Liquid samples are spread as thin, even films to maintain a constant path length. For demountable cells, the sample is placed on one plate and the second plate is placed on top, with circular motions to form an even film. Modern liquid transmittance cells possess one or two loading/unloading ports with appropriate parts.
3. For aqueous samples, due to strong interfering absorbance bands in the amide I region (1,600–1,700 cm^{-1}), shorter path length (2–10 μm) cells and higher peptide concentrations (~2–10 mg/ml) have traditionally been employed. Recently, the increased sensitivity of spectrometers combined with ATR cells has permitted much lower aqueous peptide concentrations (<0.1 mg/ml) to be used to obtain high-quality spectra. Use of D_2O also permits lower peptide concentrations to be examined. Thus, longer path lengths (up to 100 μm) and lower peptide concentration (~1 mg/ml) are sufficient to obtain quality spectra with transmittance geometry. Similarly, for most organic solvents that do not have strong absorbance in the amide I region, lower peptide concentrations and longer path lengths can be used.
4. Solid sample preparation involves grinding the sample into a fine powder and mixing it with IR transparent matrix material (e.g., KBr) and pressing a pellet. The concentration of sample in the KBr should be between 0.3 and 1 %. Typically, 1–3 mg of sample is mixed with 350 mg of KBr. Since a pellet is generally much thicker (0.5–1 mm) than a liquid film, a lower concentration is typically sufficient. Accessories are available for homogenous grinding and mixing of solid samples. The pellet can be pressed using a hydraulic or a hand press. The procedure should lead to a clear glassy pellet or disc. DRIFTS provides a simpler method for more rapid measurement of powder samples. Powdered samples can be simply placed in the sample cup. If the sample is absorbent, it can be mixed with a non-absorbent matrix (e.g., KBr, KCl) (*see* **Note 1**).
5. Solid samples can also be grounded and mixed with Nujol (mineral oil) into a paste and spread over transmission cell windows. Typically, 5–10 mg of finely ground sample is placed on one cell window and a drop of mineral oil is added. A uniform paste is made by using circular motions with the second cell window.

6. Since the transmission method involves passing the radiation through sample and measuring absorption, sample homogeneity is crucial. Thus, sample preparation can be tedious, particularly for solid samples. Comparatively, minimal to no sample preparation is required in ATR spectroscopy and this technique is compatible with a wide variety of solid and liquid samples.
7. Liquids, pastes, or semisolid samples can be directly poured and spread onto the ATR crystal. The entire crystal must be covered. Powdered sample is spread in a thin layer placing just enough sample to cover the ATR crystal surface. To maintain good optical contact, a pressure arm can be used to clamp the sample onto the crystal surface to create a more intimate contact (*see* **Note 2**).
8. Other FTIR methods based on photoacoustic detection (PAD) can also be used to study peptide samples, although PAD has not been commonly employed with peptides despite its ability to be applied to a wide variety of sample states, presumably due to its lower sensitivity.

3.3 Data Acquisition

1. Depending on the information sought, a variety of different types of experiments may be performed. For example, peptide stability analysis can be performed under different structurally perturbing conditions like temperature, pH, and solute additions. In addition, peptide conformational changes and dynamics can be examined using hydrogen–deuterium exchange. Small peptides with flexible conformations will manifest rapid exchange rates while more conformationally restrained peptides may show reduced rates of exchange at some amide positions. This technique usually involves following the disappearance of the amide II band (N–H bending) as the band is shifted to a higher wave number as N–D moieties are produced. As another example, evaluating changes in peptide structure upon interaction with lipids and lipid bilayers has been a common application of this technique.

   A two-dimensional infrared (2D-IR) correlation spectrum can also be acquired to increase the resolution of complex/overlapped peptide spectra. 2D-IR spectroscopy can provide information on peptide local structure (for example, aggregation, solvation) and peptide dynamics in addition to secondary structure content [4].
2. The typical parameters to be adjusted for experiments include the following: (a) the number of scans (for good signal/noise ratio). This is partially dependent on the peptide concentration used. An increased number of scans can considerably improve signal/noise ratio (S/N) levels. With this approach, since the S/N falls as the square root of the number of scans, a few

hundred scans is generally sufficient; (b) a resolution at 2–4 cm^{-1} is usually quite adequate since the bands of interest are quite broad; and (c) although the complete spectral range of 4,000–650 cm^{-1} is usually measured due to the Fourier transform nature of the measurements, only the amide I and II (1,500–1,700 cm^{-1}) regions are usually examined in detail for most peptides.

3. The temperature of FTIR cells is usually controlled through either external water circulation or a Peltier device. The temperature of the sample is best reported by using a thermocouple inserted in the sample or the wall of the cell.
4. A few scans should be run to see whether all of the parameters are appropriate with no sample in the sample chamber. These scans serve as a background spectrum against which the sample spectrum is measured and also serves to confirm that the transmission cell or ATR crystal is clean.
5. For accurate data, subtraction of a buffer spectrum acquired under identical conditions (ideally the same number of scans, temperature, path length, resolution, etc.) is vital, especially in aqueous conditions because of the considerable liquid and vapor H_2O contributions. The buffer scans are ideally performed before sample analysis (*see* **Notes 3** and **4**).
6. To obtain a high-quality IR spectrum, an optimal intensity and good resolution are needed. The most prominent peak in the sample should have an intensity of ~ 5 % T (transmission of 5 % is equivalent to an absorption of 1.3). For higher intensity peaks, the detector will be saturated and the more intense peaks will be flattened or cut off while the small peaks or impurities will become more prominent.

3.4 Data Analysis

A typical FTIR spectrum is a plot of absorption intensity or percent transmittance on the *Y*-axis versus frequency in wave numbers (cm^{-1}) on the *X*-axis. The infrared spectra of peptides display multiple bands arising from peptide bonds in addition to the other expected signals. The two most commonly examined regions of the spectrum are the amide I and amide II peaks in aqueous buffers. In D_2O solutions, the corresponding bands are termed amide I′ and amide II′ and are shifted by approximately 2–12 cm^{-1} and 100 cm^{-1}, respectively [3]. The amide band shifts observed upon deuteration can be used to resolve ambiguities in secondary structure assignments (*see* **Note 5**). The more intense of the two, the amide I band (1,600–1,700 cm^{-1}), originates primarily from C=O stretching vibrations with minor contributions from C–N stretching. The less intense amide II band (1,480–1,580 cm^{-1}) originates from bending motions of the N–H groups and stretching of the C–N bonds of the peptide bonds. Because the C=O and N–H

Table 1
Summary of key IR frequency assignments in amide region of peptides

Secondary structural elements and amino acid side chains	Amide I frequency (cm^{-1}) range	Amide II frequency (cm^{-1}) range
Random coil	1,645–1,655	1,520–1,545
β-sheet	1,620–1,640, 1,670–1,695	1,525–1,530
α-helix	1,650–1,658	1,540–1,545
Left-handed α-helix	1,660–1,670	
3_{10}-helix	1,660–1,662	1,533–1,531
β-turns	1,660–1,690, 1,635–1,645	
Aggregation (intermolecular β-sheet)	1,610–1,620, 1,690–1,695	
Asn/Gln/Arg (ionized)	1,660–1,685	
Lys (ionized)/α-NH3+	1,620–1,640	
Tyr (ionized)/His (very weak)	1,590–1,605	

bonds of peptides are directly involved in the backbone conformation and hydrogen bond patterns of peptides, the amide I and II bands are very sensitive to secondary structure content and changes therein (*see* **Note 6**). The remaining amide bands (amide A, B, and III–VII) are employed less frequently for peptide analysis, but may be useful in certain situations.

The accepted characteristic frequency of the amide I and II bands originating from different secondary structure types and some side chains are summarized in Table 1. In addition, examples of typical FTIR spectra for the three model peptides used in Chapters 15–18 are shown in Fig. 1. In these examples, the three different peptides differ in their secondary structure as well as their aromatic amino acid side chain content. Thus, they display marked the differences in their FTIR spectra.

The fundamental problem encountered in the spectral analysis and band assignments of peptides is that the widths of the various contributing bands are usually much greater than the peak separations. As shown in Fig. 1 with the three model peptides, typical FTIR peptide spectra in the amide I region consist of a single, broad peak with small shoulders and inflection points. The individual components and their relative contributions are thus difficult to resolve and identify from the zero-order spectrum. The most important step during data analysis of the FTIR spectra of peptides involves the resolution of the broad amide bands into their individual components. The two most common resolution enhancement methods are (a) calculation of higher order

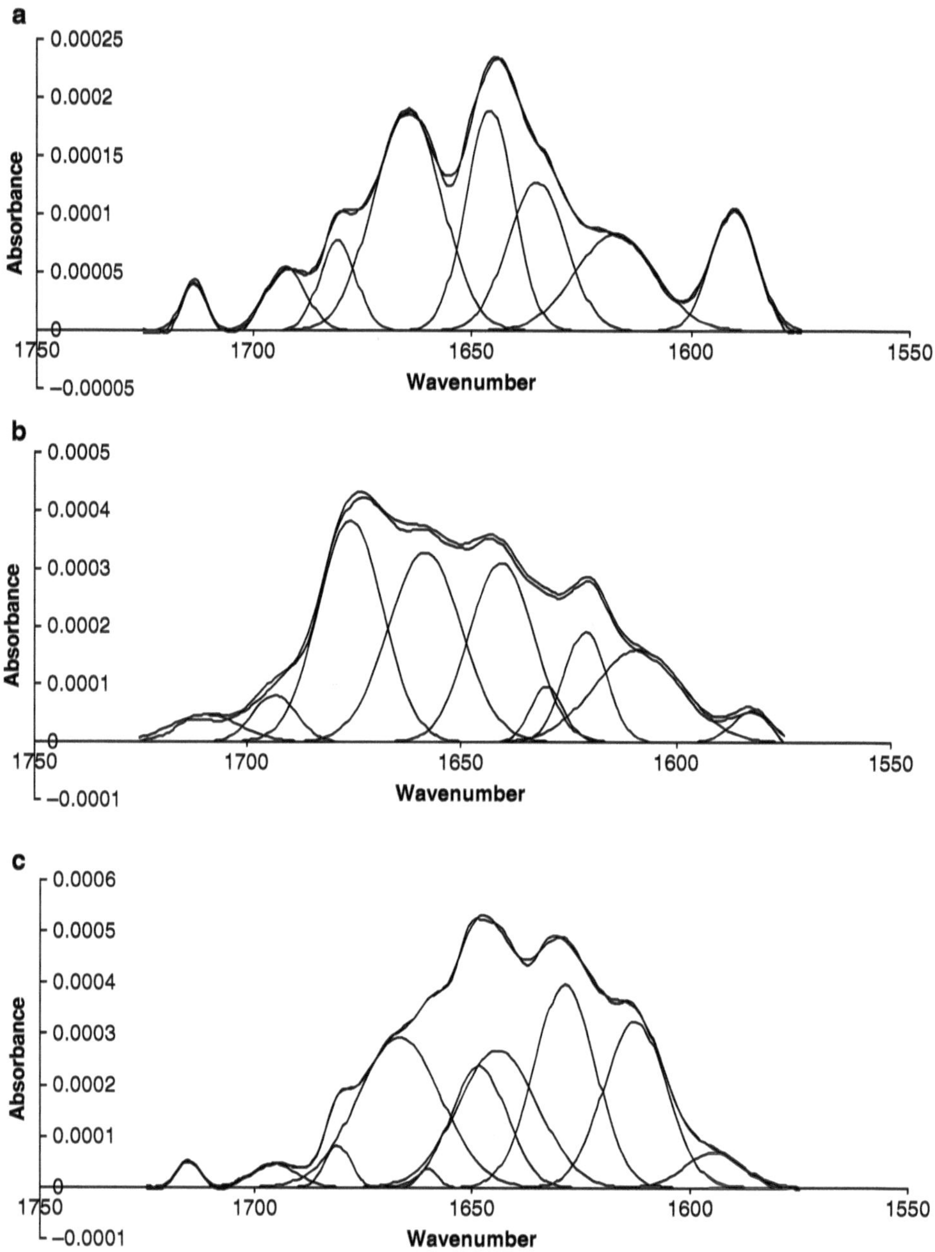

Fig. 1 Fourier transform infrared (FTIR) spectra in the amide I′ region of three model peptides acquired using an ATR cell: (**a**) Angiotensin I at ~4 mM, (**b**) α-melanocyte-stimulating hormone (α-MSH) at~5 mM, and (**c**) substance P at ~5 mM in solution (D_2O) at room temperature. The peaks obtained from curve fitting of the amide I′ bands show contributions from amino acid side chains of Asp (1,713 cm^{-1}), Glu (1,709 cm^{-1}), Arg (1,635 cm^{-1}), Lys (1,629 cm^{-1}), Tyr (1,617 cm^{-1}), Gln (1,612 cm^{-1}), Phe (1,608 cm^{-1}), and His (1,590 cm^{-1}) and structural elements such as β-turns (1,660–1,690 cm^{-1}), intermolecular β-sheets (1,620, 1,690–1,695 cm^{-1}), and random coil (1,640–1,645 cm^{-1})

(first to fourth) derivatives of the absorption spectrum (taking the derivative of a spectrum is the simple process of calculating the slope or the gradient of an absorption band). This gradient ($dA/d\lambda$) is plotted against wavelength (λ) to produce a first-order derivative plot. An iteration of this process leads to higher order derivatives ($d^nA/d\lambda^n$)), and (b) Fourier self-deconvolution (FSD), another band narrowing procedure. For typical FTIR data analysis, these procedures involve the following steps:

1. Spectral subtraction of buffer/solvent scans from sample scans, both acquired under identical conditions is always necessary. Most commercially available data analysis software has the option of variable subtraction factors that are especially useful in the case of aqueous samples. An accurate subtraction can often be achieved by targeting the H_2O association band near 2,150 cm^{-1} to achieve a flat baseline. This approach is best done by increasing the subtraction factor until over-subtraction is observed. The subtraction factor is then slowly decreased until a flat baseline is obtained in the 1,800–2,400 cm^{-1} region.
2. The subtracted spectrum is then usually truncated to the region of interest. For peptides, this generally encompasses the amide I and II regions (1,800–1,500 cm^{-1}).
3. A further baseline correction step must be used that fits a straight baseline to the non-peak areas of the truncated spectrum. This procedure is done to correct for the sloping, curving, shifted, or undesirable characteristics of the spectral baseline. After baseline correction, this open region should ideally show flat and near-zero absorbance (100 % transmittance). Both manual and automatic baseline correction options are available for this purpose with most instruments.
4. Smoothing the spectrum (typically using the Savitsky–Golay algorithm) is recommended to decrease the noise in the spectra if an S/N ratio of less than 1,000:1 is obtained. The degree of smoothing depends on the number of points employed (generally 5–25) with the greater number of points resulting in a greater degree of smoothing. Excess smoothing, however, may lead to loss of resolution. A change in resolution of the sharper bands in the spectrum before and after smoothing is a good indicator of a loss in resolution. After subtraction and smoothing, the broad amide bands then need to be deconvoluted (resolution enhanced) before the spectra can be further analyzed. It is sometimes argued that the following steps should be performed on unsmoothed spectra, but we have generally found that not to be a problem.
5. Deconvolution is usually performed by using the FSD method. This procedure is a mathematical enhancement of spectral resolution by band narrowing and can effectively partially resolve

the overlapping bands that make up the broad amide peaks. FSD parameters to be optimized interactively are the bandwidth (σ) at half height and an enhancement factor (K). Typical values of σ are between 10 and 20 cm^{-1} and 1.0 and 2.5 for the K value. Generally, the two parameters are varied until an over-deconvolution is seen. This effect is manifested as an oscillating baseline outside the region of absorption. The parameters are then readjusted to achieve the best resolution without over-deconvolution.

6. A second common method of spectral resolution involves computing derivative spectra from zero-order spectra. We generally find that the second derivative is the most useful spectral form. Drawbacks associated with the deconvolution and derivative procedures include potential amplification of random noise and appearance of side lobes (*see* **Notes** 7 and **8**) leading to distortions in spectra prior to applying curve fitting.
7. Curve fitting of the resolution-enhanced spectra (from deconvolution/derivative analysis) uses linear combinations of individual bands and an iterative adjustment in height width and position of contributing bands. Generally 100–500 iterations are sufficient. χ^2 (close to 0) and r^2 values (close to 1) are indicative of the goodness of fit. The fractional area of each contributing band gives the contribution of the individual secondary structure elements. Either Gaussian or Lorentzian functions can be used, with a two-component Gaussian/Lorentzian function providing the most general approach. Changes in peptide structure as a result of variations in pH, temperature, solvent composition, concentration, exposure to membranes, etc. are commonly investigated by monitoring changes in the frequency and intensity of the constituent bands. Quantitative analysis can also be performed using the area under each of the component bands. Using the assignments in Table 1, the relative content of an individual component (i.e., individual peak area) is divided by the summed total band areas to obtain a percent contribution to a peptide's secondary structure.

3.5 Data Interpretation

Quantitative estimation of a peptide's secondary structure is based on a number of assumptions of which the user must be aware to avoid common pitfalls encountered during FTIR data interpretation.

1. The number and position of IR bands obtained by deconvolution or derivation are indicative of composite absorptions arising from amide groups in similar environments, but will not generally resolve all of the groups in their unique environments even in small peptides. Thus, estimates of secondary structure are approximate with an overall average error of ~2–5 %.
2. Infrared band shapes are largely assumed to be Lorentzian, as is the case for simple molecules. For more complex peptides,

however, this assumption may not hold true and line shapes may be significantly dependent on local environment.

3. For peptide IR band assignments, it is often assumed that only the peptide amide groups contribute to absorption in this region. In fact, significant contributions to the amide region may also come from the side chains of tyrosine, glutamine, arginine, phenylalanine, and lysine [1]. In peptides where a significant number of these amino acid residues are present, side chain contributions must be taken into account to avoid over- or underestimation of secondary structure elements (*see* **Note 9**). One approach is to subtract their contribution based on the IR spectra of the individual side chains [1].
4. Assignments of amide I bands to specific secondary structures should be done with significant caution, and this is especially the case for absorption arising from turn segments which are difficult to assign due to the broad range of hydrogen bonding characteristics of the different types of turns. For example, absorption from turns has been reported at 1,630, 1,645, 1,660, 1,680 cm^{-1} and a series of weak features between 1,670 and 1,700 cm^{-1} [3]. This makes it difficult to correctly assign and quantitatively analyze turn data, although this may be feasible in shorter peptides (*see* **Note 9**).
5. The appearance of bands at 1,615 and 1,695 cm^{-1} in dried peptides and/or aggregated samples is often indicative of intermolecular β-sheet formation [2]. One recent study has, however, reported no aggregation in certain samples even though the FTIR spectra show these characteristic peaks. It is suggested that these changes in FTIR spectra may not be entirely due to secondary structural alteration but rather involve contributions from water removal during lyophilization [5] (*see* **Notes 10** and **11**).

In summary, FTIR has become an established tool for structural characterization of peptides. For achieving accurate results, the strengths and limitations of the technique must be well understood. Only high-quality spectra with high signal-to-noise ratios should be analyzed and interpreted. Ideally, FTIR should be combined with complementary methods of analysis (e.g., CD, NMR) to increase confidence in any conclusions reached. For further details, a review concerning the critical interpretation of FTIR spectra is suggested reading [2].

3.6 Empirical Phase Diagrams

The FTIR spectroscopic approach described in this chapter, along with the other spectroscopic techniques described in the three previous chapters in this volume (UV absorption and fluorescence spectroscopies and circular dichroism in Chapters 15, 16 and 17, respectively), is capable of examining various aspects of peptide conformation and structure. When used in combination, they can provide comprehensive insight into peptide structure and solution behavior under a range of conditions (e.g., temperature, pH, solvent composition, concentration). The large amount of data generated by such studies, however, may be difficult to interpret. An empirical phase diagram (EPD) approach provides one convenient method to integrate all of these diverse data sets into a visual picture that can aid in the global interpretation of such data. The detailed theory and calculation process for the EPD method has been described previously [6, 7]. Briefly, the collected data sets are represented as *n*-dimensional vectors (where n is the number of different data sets such as UV absorption peak positions, intrinsic fluorescence peak positions, CD intensities) in an experimental measurement parameter space. This space can be defined under variable conditions such as temperature, solution pH, and peptide concentration. Each individual point in the space thus has *n* corresponding data values associated with it which can be expressed in matrix form. This matrix is then converted into a density matrix from which eigenvalues and eigenvectors are calculated. The eigenvectors corresponding to the three largest eigenvalues are then used to define a new three-dimensional coordinate system and each of the eigenvectors is assigned red, green, and blue (RGB) colors. The linear combination of the three eigenvectors (RGB) at each point in the measurement parameter space produces a color plot based on the magnitude of each eigenvector at each point in the experimental parameter space. Regions of abrupt color changes thus identify changes in peptide physical state while the continuous color spaces suggest uniform behavior. Note that equilibrium is not implied in this analysis and that the colors themselves have no particular structural significance. Typical EPDs for a peptide are shown in Fig. 2 [8]. In our experience, EPDs provide an effective approach to display complex and diverse data sets and have been extensively employed in our laboratory as a pharmaceutical formulation tool for a wide variety of macromolecular systems including peptides. Recent developments in using this approach are summarized in references [9–12].

4 Notes

Sample Preparation

1. The KBr pellet may be cloudy if the potassium bromide was not thoroughly ground. Excessive grinding may also lead to absorption of moisture and result in high background noise. To avoid this situation, the sample can be ground before adding KBr to allow less exposure.
2. For solid powder samples in an ATR unit, it is ideal to use just enough sample to cover the crystal. A sample height of few millimeters is more than enough.
3. Extreme care must be taken to avoid water vapor during sample preparation and data acquisition. Nevertheless, although small

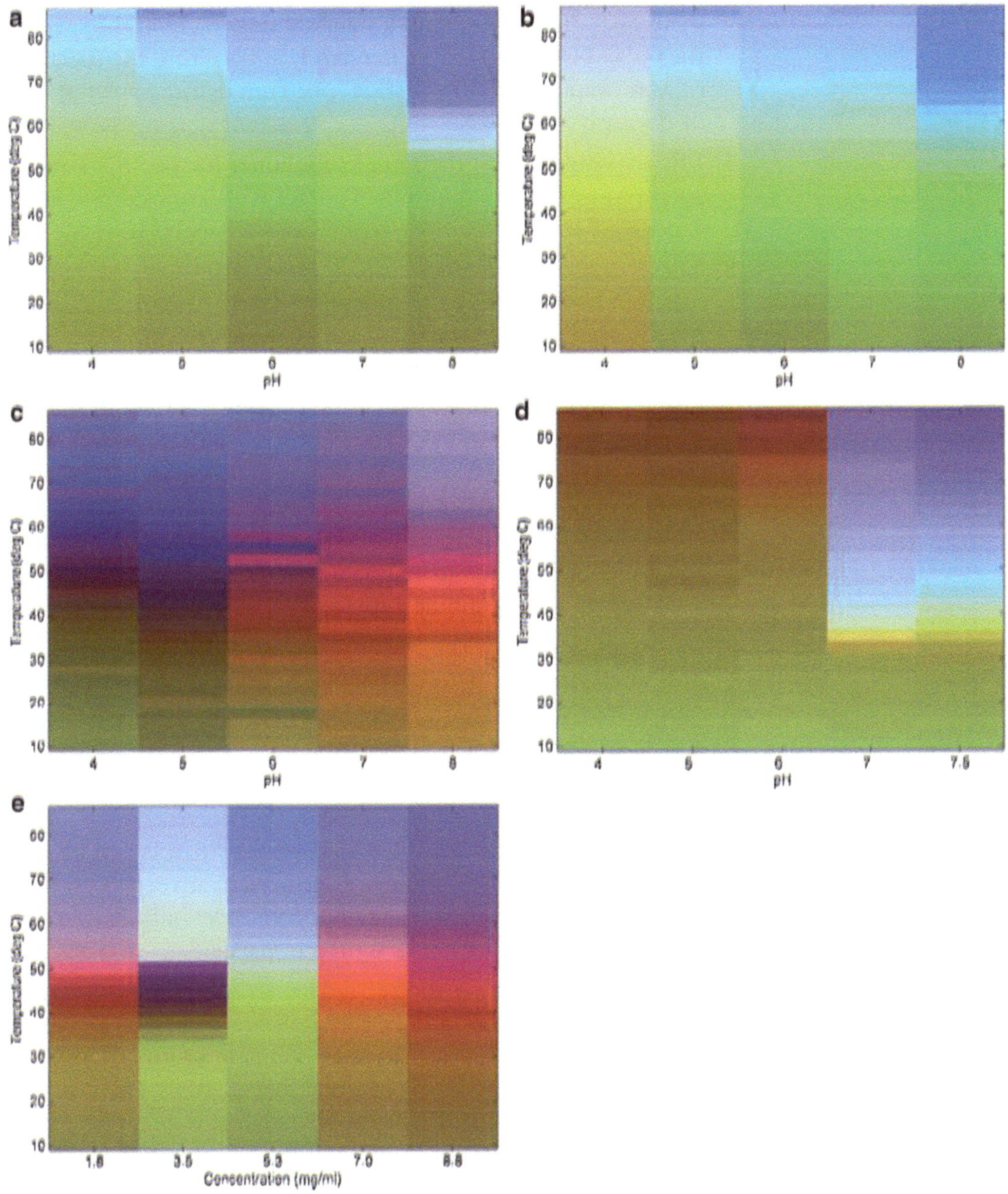

Fig. 2 Empirical phase diagrams (EPDs) of the peptide pramlintide at low and high concentration and its concentration dependence at pH 4. Low concentration studies reflect EPD generated from (**a**) second-derivative UV absorbance peak shift and OD at 350 nm; (**b**) UV absorbance peak shift, OD at 350 nm, fluorescence intensity and peak shift; and (**c**) UV absorbance peak shift, OD at 350 nm, fluorescence intensity and peak shift, and CD change at 204 nm. EPD generated from (**d**) UV absorbance peak shift, OD at 350 nm and fluorescence intensity and peak shift data for the peptide at high concentration (8.8 mg/mL), and (**e**) UV absorbance peak shift, OD at 350 nm, and fluorescence peak shift and intensity data at pH 4 as a function of peptide concentration [8]

sharp lines due to water vapor may still be present in the spectra, these can be eliminated by iteratively subtracting the spectrum of water vapor from the peptide spectrum before applying FSD or derivative methods.

4. It is advisable to seal the sample container for peptide samples prepared in D_2O to minimize contact with atmospheric water vapor and avoid conversion to species such as HOD.
5. There is significant ambiguity in assigning amide bands to α-helix (1,650–1,658 cm^{-1}) or disordered structures (1,645–1,655 cm^{-1}). One way to clarify these assignments is to exchange the peptide N–H groups with deuterium. The N–H groups in disordered structure are more susceptible to deutration and will show larger shifts than the N–H groups in more stable helical regions.
6. Sharp peaks from the absorption bands of water vapor (e.g., 1,644 cm^{-1}) in the amide I region and trifluoroacetic acid (TFA) (1,673 cm^{-1}) strongly absorb and may erroneously be assigned to peptide amide I modes.
7. It is important that the deconvolution and derivative procedures be applied only to spectra with very high signal-to-noise ratios. It is a common practice to apply both deconvolution and derivative procedures to the raw spectra and assign only the peaks that are observed using both methods to avoid problematic assignments.
8. Over-deconvolution can introduce artifacts that could be mistaken as signals. The deconvolution/derivative process may lead to disproportionate amplification (over-subtraction and under-subtraction) of water signals in particular.
9. The determination of the secondary structural conformation of peptides using the amide I and amide II bands is complicated by the absorption bands of the side chains of tyrosine (1,498, 1,518, 1,602 cm^{-1}), glutamine (1,670, 1,610 cm^{-1}), arginine, (1,673, 1,633 cm^{-1}), phenylalanine (1,515, 1,560, 1,598, 1,631, 1,740 cm^{-1}), and lysine (1,629, 1,526 cm^{-1}). In peptides where a significant number of these residues are present, side chain contributions should either be subtracted (based on the spectra of the individual amino acid side chains) [1] or at least recognized during deconvolution analyses.
10. The relative intensities of bands arising from β-turns ($I_{1,660-1,690}$:$I_{1,635-1,645}$ ~ 2:1–3:2) differ considerably from β-sheet signals ($I_{1,660-1,690}$:$I_{1,620-1,640}$ ~ 1:10–1:8) and can be used to clarify such assignments.
11. In small peptides containing proline, β-turns may also produce a band near 1,628–1,632 cm^{-1}.

References

1. Venyaminov SY, Kanin NN (1990) Quantitative IR spectrophotometery of peptide compounds in water (H_2O) solutions. Spectral parameters of amino acid residue absorption bands. Biopolymers 30:1243–1257
2. Surewicz WK, Mantsch HH, Chapman D (1993) Determination of protein secondary structure by Fourier transform infrared. Biochemistry 32:389–394
3. Vass E, Hollósi M, Besson F, Buchet R (2003) Vibrational spectroscopic detection of beta and gamma turns in synthetic and natural peptides and proteins. Chem Rev 103:1917–1954
4. Ganim Z, Chung HS, Smith AW, Deflores LP, Jones KC, Tokmakoff A (2008) Amide I two-dimensional infrared spectroscopy of proteins. Acc Chem Res 41(3):432–441
5. Van de Weert M, Haris PI, Hennink WE, Crommelin DJA (2001) Fourier transform infrared spectrometric analysis of protein conformation: effect of sampling method and stress factors. Anal Biochem 297(2):160–169
6. Fan H, Ralston J, Dibiase M, Faulkner E, Middaugh CR (2005) Solution behavior of IFN-beta-1a: an empirical phase diagram based approach. J Pharm Sci 94:1893–1911
7. Kueltzo LA, Ersoy B, Ralston JP, Middaugh CR (2003) Derivative absorbance spectroscopy and protein phase diagrams as tools for comprehensive protein characterization: a bGCSF case study. J Pharm Sci 92:1805–1820
8. Nonoyama A, Laurence JS, Garriques L, Qi H, Le T, Middaugh CR (2008) A biophysical characterization of the peptide drug pramlintide (AC137) using empirical phase diagrams. J Pharm Sci 97(7):2552–2567
9. Maddux NR, Joshi SB, Volkin DB, Ralston JP, Middaugh CR (2011) Multidimensional methods for the formulation of biopharmaceuticals and vaccines. J Pharm Sci 100(10): 4171–4197
10. Maddux NR, Rosen IT, Hu L, Olsen CM, Volkin DB, Middaugh CR (2012) An improved methodology for multidimensional high-throughput preformulation characterization of protein conformational stability. J Pharm Sci 101(6):2017–2024
11. Kim JH, Iyer V, Joshi SB, Volkin DB, Middaugh CR (2012) Improved data visualization techniques for analyzing macromolecule structural changes. Protein Sci 21(10): 1540–1553
12. Iyer V, Maddux N, Hu L, Cheng W, Youssef AK, Winter G, Joshi SB, Volkin DB, Middaugh CR (2013) Comparative signature diagrams to evaluate biophysical data for differences in protein structure across various formulations. J Pharm Sci 102(1):43–51

Index

A

B

C

Andrew E. Nixon (ed.), *Therapeutic Peptides: Methods and Protocols*, Methods in Molecular Biology, vol. 1088,
DOI 10.1007/978-1-62703-673-3, © Springer Science+Business Media, LLC 2014

D

E

F

G

H

T

U

V

W

X

Z

MIX
Papier aus verantwortungsvollen Quellen
Paper from responsible sources
FSC® C105338

If you have any concerns about our products,
you can contact us on
ProductSafety@springernature.com

In case Publisher is established outside the EU,
the EU authorized representative is:
Springer Nature Customer Service Center GmbH
Europaplatz 3, 69115 Heidelberg, Germany

Printed by Libri Plureos GmbH
in Hamburg, Germany